Foreword

The U.S. Army is the most highly sophisticated professional military force in the world. Capable of conducting worldwide operations, deeply experienced from decades of deployments in harm's way, battle hardened from over eight years of war, strengthened by sacrifice, and resolved to defeat the enemies of our nation, the U.S. military today stands as the most combat experienced in this nation's history.

The world today is governed by geopolitical and economic interests, and fraught with natural disasters and humanitarian challenges. Because of this, America will face an era of drawn-out confrontations between state and non-state actors, and a wide range of adversaries, all using diffuse technologies in a complex distributed battlespace. Described by the Chief of Staff of the Army, General George Casey Jr., as "Persistent Conflict," these confrontations will be fueled by the affects of globalization, competition for natural resources, demographic trends, climate change, proliferation of weapons of mass destruction, and failed or failing states that provide safe havens for terrorists.[1]

Although the Army remains ready to face a myriad of global challenges—including support to civil authorities at home and abroad to mitigate the effects of natural and manmade disasters, like Haiti—its emphasis has rightfully remained on the ongoing wars in Afghanistan and Iraq. Major engagements in Afghanistan and continued pressure in Iraq have significantly degraded al-Qaeda, with fledgling democracies emerging in both countries. While popular support for the group in the Muslim world remains relatively low, the resurgence of the Taliban in Afghanistan, the recalcitrance of al-Qaeda cells in Pakistan, the return of foreign fighters to al-Qaeda franchises worldwide, and the rise of Hizballah, Jaysh al-Mahdi, the Special Groups, and other Iranian proxies, all suggest that there is still much work to be done. Many of America's adversaries and enemies have no national standing, no government, no uniformed military, no allegiance to international law, no interest in humanity, and no restrictions on the weapons they are willing to use to achieve economic, political, and theocratic domination of the world.

Recognizing this, Admiral Michael Mullen, Chairman of the Joint Chiefs of Staff, outlined three strategic priorities for the U. S. military: continue to improve stability and defend our vital national interests in the broader Middle East and South Central Asia; continue efforts to reset, reconstitute, and revitalize our Armed Forces; and, continue to balance global strategic risks in a manner that enables America to deter conflict and be prepared for future conflicts.[2]

In support of these priorities, the Army's mission is to "fight and win our Nation's wars by providing prompt, sustained land dominance across the full range of military operations and spectrum of conflict in support of combatant commanders."[3] Army forces provide the capability—by threat, force, or occupation—to promptly gain, sustain, and exploit comprehensive control over land, resources, and people. One of the ways the Army accomplishes its mission is through the organizing, equipping, and training of its forces for the conduct of prompt and sustained combat operations on land. The role of the U.S. Army must continuously evolve to meet new threats and challenges. Today, the U.S. Army is in the midst of executing the most profound transformation since World War II, all while continuing to shoulder a heavy burden worldwide, particularly in the Middle East and South Central Asia. While the fight against the global extremist network that attacked America on September 11, 2001, is ongoing and will continue to be a top concern for the U.S. military and policymakers, the United States cannot afford to lose its conventional qualitative edge or accept technological inferiority.

U.S. Arm [illegible] compendium of the most innovative and important U.S. Army weapon systems and initiatives. This intuitive, readable, and attractive handbook focuses not only on specific U.S. Army weapon systems, but also promotes a greater understanding of the Army's major acquisition programs, describes system interdependencies by outlining which other weapon systems or components the main system works in concert with or relies upon for its operation, summarizes program schedules, and offers information regarding contractors, teaming arrangements, and technical maturity. Finally, because the U.S. military cannot face these global challenges alone—it benefits greatly from networks of partners and allies throughout the world—this handbook provides information on foreign military sales.

In the wake of September 11, 2001, and after eight years at war, the American public yearns to understand not only the nation's strategic purpose, but also its military. As you page through this book, you will see the combat systems, weapons, and equipment used by America's Army. What you will not see, however, is the most important part of the equation—the men and women of the U.S. Army. Despite being the most technologically advanced Army on the globe, it is the soldier that makes America's Army the most respected force in the world—it is the American soldier, not the weapons, that make the U.S. Army the most lethal army in history.

—Lieutenant Colonel William D. Wunderle,
U.S. Army (Retired)

1 General George W. Casey Jr., speech at the Brookings Institution, January 28, 2010.
2 Posture Statement of Admiral Michael G. Mullen, Chairman of the Joint Chiefs of Staff, Before the 111th Congress, Senate Appropriations Subcommittee on Defense, June 9, 2009.
3 Headquarters, Department of the Army (HQDA), FM 1, The Army, Washington, DC: Department of the Army, June 2005, p 2–7.

Table of Contents

How to Use this Book....VI
About the 2010 Edition VII

Introduction 1

Weapon Systems 13
2.75" Family of Rockets....14
Abrams Upgrade....16
AcqBusiness 18
Advanced Field Artillery Tactical Data System (AFATDS) 20
Aerial Common Sensor (ACS)....22
Air Warrior (AW) 24
Air/Missile Defense Planning and Control System (AMDPCS) 26
Airborne Reconnaissance Low (ARL) 28
All Terrain Lifter Army System (ATLAS) 30
Armored Knight 32
Armored Security Vehicle (ASV) 34
Army Key Management System (AKMS) 36
Artillery Ammunition....38
Aviation Combined Arms Tactical Trainer (AVCATT)....40
Battle Command Sustainment Support System (BCS3) 42
Biometric Enterprise Core Capability (BECC) 44
Biometric Family of Capabilities for Full Spectrum Operations (BFCFSO)....46
Black Hawk/UH-60....48
Bradley Upgrade 50
Calibration Sets Equipment (CALSETS)....52
Chemical Biological Medical Systems–Diagnostics 54
Chemical Biological Medical Systems–Prophylaxis 56
Chemical Biological Medical Systems–Therapeutics 58
Chemical Biological Protective Shelter (CBPS) 60
Chemical Demilitarization 62
Chinook/CH-47 Improved Cargo Helicopter (ICH) 64
Close Combat Tactical Trainer (CCTT) 66
Combat Service Support Communications (CSS Comms) 68
Command Post Systems and Integration (CPS&I)....70
Common Hardware Systems (CHS) 72
Common Remotely Operated Weapon Station (CROWS) 74
Counter-Rocket, Artillery and Mortar (C-RAM)....76
Countermine....78
Defense Enterprise Wideband SATCOM Systems (DEWSS)....80
Distributed Common Ground System (DCGS–Army) 82
Distributed Learning System (DLS) 84
Dry Support Bridge (DSB) 86
Early Infantry Brigade Combat Team (E-IBCT) Capabilities 88
Excalibur (XM982)....92
Extended Range Multipurpose (ERMP) Unmanned Aircraft System (UAS)....94
Family of Medium Tactical Vehicles (FMTV) 96
Fixed Wing 98
Force Protection Systems 100
Force Provider (FP)....102
Force XXI Battle Command Brigade-and-Below (FBCB2)....104
Forward Area Air Defense Command and Control (FAAD C2)....106
Future Tank Main Gun Ammunition....108
General Fund Enterprise Business Systems (GFEBS)....110
Global Combat Support System–Army (GCSS–Army) 112
Global Command and Control System–Army (GCCS–A)....114
Ground Soldier System (GSS)....116
Guardrail Common Sensor (GR/CS) 118

Guided Multiple Launch Rocket System (GMLRS) ... 120
Heavy Expanded Mobility Tactical Truck (HEMTT)/ HEMTT Extended Service Program (ESP) ... 122
Heavy Loader ... 124
HELLFIRE Family of Missiles ... 126
Helmet Mounted Enhanced Vision Devices ... 128
High Mobility Artillery Rocket System (HIMARS) ... 130
High Mobility Engineer Excavator (HMEE) ... 132
High Mobility Multipurpose Wheeled Vehicle (HMMWV) ... 134
Improved Ribbon Bridge (IRB) ... 136
Improved Target Acquisition System (ITAS) ... 138
Installation Protection Program (IPP) Family of Systems ... 140
Instrumentable–Multiple Integrated Laser Engagement System (I–MILES) ... 142
Integrated Air and Missile Defense (IAMD) ... 144
Integrated Family of Test Equipment (IFTE) ... 146
Interceptor Body Armor ... 148
Javelin ... 150
Joint Air-to-Ground Missile (JAGM) ... 152
Joint Biological Point Detection System (JBPDS) ... 154
Joint Biological Standoff Detection System (JBSDS) ... 156
Joint Cargo Aircraft (JCA) ... 158
Joint Chem/Bio Coverall for Combat Vehicle Crewman (JC3) ... 160
Joint Chemical Agent Detector (JCAD) ... 162
Joint Chemical Biological Radiological Agent Water Monitor (JCBRAWM) ... 164
Joint Effects Model (JEM) ... 166
Joint High Speed Vessel (JHSV) ... 168
Joint Land Attack Cruise Missile Defense Elevated Netted Sensor System (JLENS) ... 170
Joint Land Component Constructive Training Capability (JLCCTC) ... 172
Joint Light Tactical Vehicle (JLTV) ... 174
Joint Nuclear Biological Chemical Reconnaissance System (JNBCRS) ... 176
Joint Precision Airdrop System (JPADS) ... 178
Joint Service General Purpose Mask (JSGPM) ... 180
Joint Service Personnel/Skin Decontamination System (JSPDS) ... 182
Joint Service Transportable Decontamination System (JSTDS)–Small Scale (SS) 184
Joint Tactical Ground Stations (JTAGS) ... 186
Joint Tactical Radio System Airborne, Maritime/Fixed Station (JTRS AMF) ... 188
Joint Tactical Radio System Ground Mobile Radios (JTRS GMR) ... 190
Joint Tactical Radio System Handheld, Manpack, and Small Form Fit (JTRS HMS) ... 192
Joint Tactical Radio System Network Enterprise Domain (JTRS NED) ... 194
Joint Warning and Reporting Network (JWARN) ... 196
Kiowa Warrior ... 198
Light Tactical Trailer (LTT) ... 200
Light Utility Helicopter (LUH)/UH-72A Lakota ... 202
Lightweight 155mm Howitzer (LW155) ... 204
Lightweight .50 cal Machine Gun ... 206
Lightweight Laser Designator Range Finder (LLDR) ... 208
Line Haul Tractor ... 210
Load Handling System Compatible Water Tank Rack (Hippo) ... 212
Longbow Apache ... 214
Maneuver Control System (MCS) ... 216
Medical Communications for Combat Casualty Care (MC4) ... 218
Medical Simulation Training Center (MSTC) ... 220
Medium Caliber Ammunition ... 222
Medium Extended Air Defense System (MEADS) ... 224
Meteorological Measuring Set–Profiler (MMS–P) ... 226
Mine Protection Vehicle Family (MPVF) ... 228

Table of Contents

Mine Resistant Ambush Protected Vehicles (MRAP) 230
Mobile Maintenance Equipment Systems (MMES) 232
Modular Fuel System (MFS) 234
Mortar Systems 236
Mounted Soldier 238
Movement Tracking System (MTS) 240
Multifunctional Information Distribution System (MIDS)–Joint Tactical Radio System (JTRS) 242
Multiple Launch Rocket System (MLRS) M270A1 244
NAVSTAR Global Positioning System (GPS) 246
Non-Intrusive Inspection (NII) Systems 248
Non Line of Sight–Launch System (NLOS–LS) 250
Nuclear Biological Chemical Reconnaissance Vehicle (NBCRV)–Stryker 252
One Semi-Automated Forces (OneSAF) 254
Paladin/Field Artillery Ammunition Supply Vehicle (FAASV) 256
Palletized Load System (PLS) and PLS Extended Service Program (ESP) 258
PATRIOT (PAC-3) 260
Precision Guidance Kit 262
Prophet 264
Raven Small Unmanned Aircraft System (SUAS) 266
Rough Terrain Container Handler (RTCH) 268
Screening Obscuration Device (SOD)–Visual Restricted (Vr) 270
Secure Mobile Anti-Jam Reliable Tactical–Terminal (SMART–T) 272
Sentinel 274
Single Channel Ground and Airborne Radio System (SINCGARS) 276
Small Arms–Crew Served Weapons 278
Small Arms–Individual Weapons 280
Small Caliber Ammunition 282
Sniper Systems 284
Spider 286
Stryker 288
Surface Launched Advanced Medium Range Air-to-Air Missile (SLAMRAAM) 290
Shadow Tactical Unmanned Aerial Vehicle (TUAV) 292
Tactical Electric Power (TEP) 294
Tank Ammunition 296
Test Equipment Modernization (TEMOD) 298
Thermal Weapon Sight 300
Transportation Coordinators' Automated Information for Movement System II (TC-AIMS II) 302
Tube-Launched, Optically-Tracked, Wire-Guided (TOW) Missiles 304
Unit Water Pod System (Camel) 306
Warfighter Information Network–Tactical (WIN–T) Increment 1 308
Warfighter Information Network–Tactical (WIN–T) Increment 2 310
Warfighter Information Network–Tactical (WIN–T) Increment 3 312
Weapons of Mass Destruction Elimination 314

Science & Technology (S&T) 316
S&T Investment—Future Force Technology Areas 317
Force Protection 318
- Kinetic Energy Active Protection System 318
- Tactical Wheeled Vehicle Survivability 318
- Threat and Minefield Detection Payload for Shadow Tactical Unmanned Aerial Vehicle 319
- Detection for In-Road Threats 319
- Extended Area Protection & Survivability (EAPS) Integrated Demo 319

Intelligence, Surveillance, Reconnaissance 320
- All-Terrain Radar for Tactical Exploitation of Moving Target Indicator and Imaging Surveillance (ARTEMIS) 320
- Battlespace Terrain Reasoning Awareness—Battle Command 320

Target Location Designation System 321
Flexible Display Technology for Soldiers and Vehicles 321
Multi-Spectral Threat Warning 322
Command, Control, Communications, and Computers (C4) 322
Network-Enabled Command and Control 322
Tactical Mobile Networks 322
Collaborative Battlespace Reasoning and Awareness 323
RF Adaptive Technologies Integrated with Communications and Location (RADICAL) 323
Lethality 324
Non Line of Sight-Launch System Technology 324
Advanced Lasers and Unmanned Aerial System Payloads 324
Applied Smaller, Lighter, Cheaper Munitions Components 324
Scalable Technology for Adaptive Response 325
Medical 325
Psychological Resetting after Combat Deployment: Advanced Battlemind 325
Damage Control Resuscitation 325
Drug for the Treatment of Traumatic Brain Injury (TBI) 326
Prophylactic Drugs to Prevent Drug Resistant Malaria 326
Unmanned Systems 327
Robotic Vehicle Technologies Control Architecture for BCT Modernization 327
Safe Operations of Unmanned Systems for Reconnaissance in Complex Environments 327
Soldier Systems 328
Soldier Planning Interfaces & Networked Electronics 328
Soldier Blast and Ballistic Protective System Assessment and Analysis Tools 328
Enhanced Performance Personnel Armor Technology 328
High-Definition Cognition (HD-COG) In Operational Environments 328
Logistics 329
Power for the Dismounted Soldier 329
Wheeled Vehicle Power and Mobility 329
High Performance Lightweight Track 329
Prognostics and Diagnostics for Operational Readiness and Condition-Based Maintenance 329
JP-8 Reformation for Alternate Power Sources 330
Advanced Simulation 330
Research for Scalable Embedded Training and Mission Rehearsal 330
Simulated Severe Trauma for Medical Simulation 331
Basic Research 331
S&T Role in Formal Acquisition Milestones 332
Summary 333

Appendices 334
Army Combat Organizations 335
Glossary of Terms 336
Systems by Contractors 340
Contractors by State 350
Points of Contact 354

How to Use this Book

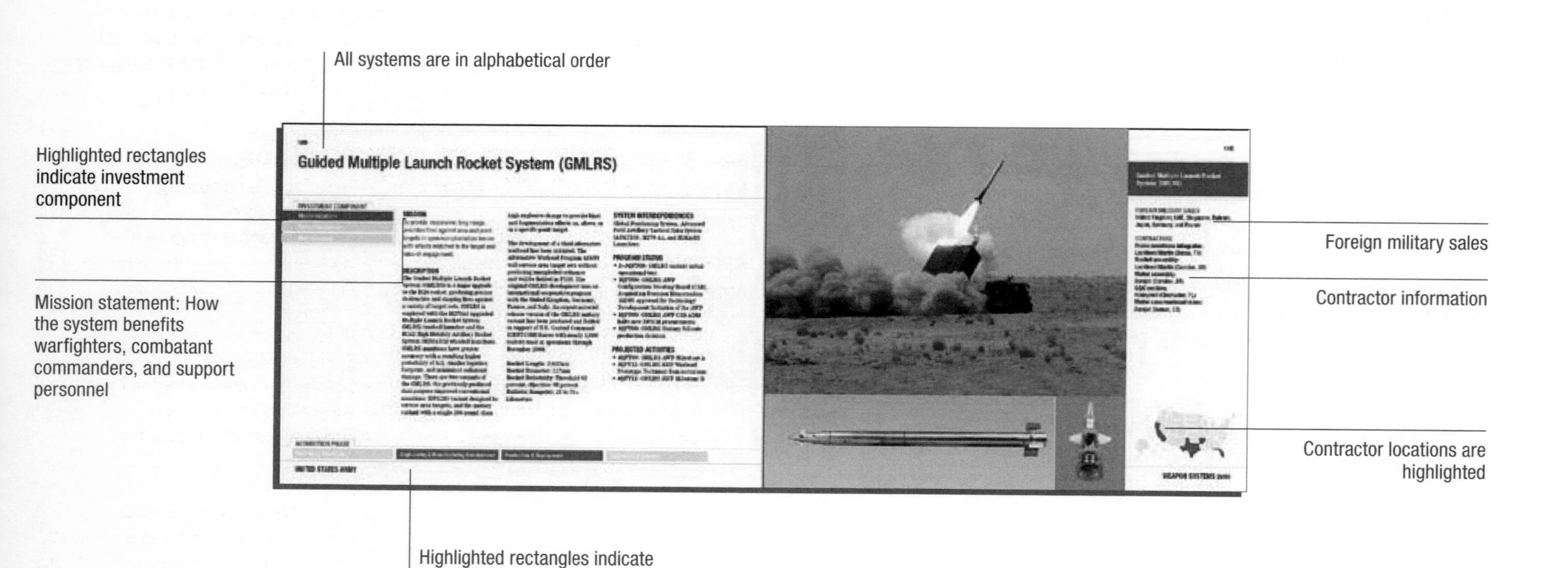

About the 2010 Edition

THE CHANGES

Readers familiar with this publication will notice a few changes this year, maintaining the intuitiveness, readability, and attractiveness of the *U.S. Army Weapon Systems 2010* handbook.

In this year's edition:

- We have added an additional heading to the narrative spreads, "System Interdependencies." The goal of this addition is to outline which other weapon systems or components (if any)the main system works in concert with or relies upon for its operation.
- Also, the names of two of the acquisition phases have changed, from "Concept & Technology Development" to "Technology Development," and from "Systems Development & Demonstration" to "Engineering & Manufacturing Development," reflecting upgrades in the systems development on the acquisition end.

For explanations of each of the elements on a typical system spread, see the example on the left.

WHAT ARE INVESTMENT COMPONENTS?

Modernization programs develop and/or procure new systems with improved warfighting capabilities.

Recapitalization programs rebuild or provide selected upgrades to currently fielded systems to ensure operational readiness and a zero-time, zero-mile system.

Maintenance programs include the repair or replacement of end items, parts, assemblies, and subassemblies that wear out or break.

For additional information and definitions of these categories, please see the Glossary.

WHAT ARE ACQUISITION PHASES?

Technology Development refers to the development of a materiel solution to an identified, validated need. During this phase, the Mission Needs Statement (MNS) is approved, technology issues are considered, and possible alternatives are identified. This phase includes:

- Concept exploration
- Decision review
- Component advanced development

Engineering & Manufacturing Development is the phase in which a system is developed, program risk is reduced, operational supportability and design feasibility are ensured, and feasibility and affordability are demonstrated. This is also the phase in which system integration, interoperability, and utility are demonstrated. It includes:

- System integration
- System demonstration
- Interim progress review

Production & Deployment achieves an operational capability that satisfies mission needs. Components of this phase are:

- Low-rate initial production (LRIP)
- Full-rate production decision review
- Full-rate production and deployment

Operations & Support ensures that operational support performance requirements and sustainment of systems are met in the most cost-effective manner. Support varies but generally includes:

- Supply
- Maintenance
- Transportation
- Sustaining engineering
- Data management
- Configuration management
- Manpower
- Personnel
- Training
- Habitability
- Survivability
- Safety, Information technology supportability
- Environmental management functions

Because the Army is spiraling technology to the troops as soon as it is feasible, some programs and systems may be in all four phases at the same time. Mature programs are often only in one phase, such as operations and support, while newer systems are only in concept and technology development.

PROVIDING WARFIGHTERS WITH THE DECISIVE EDGE

UNITED STATES ARMY

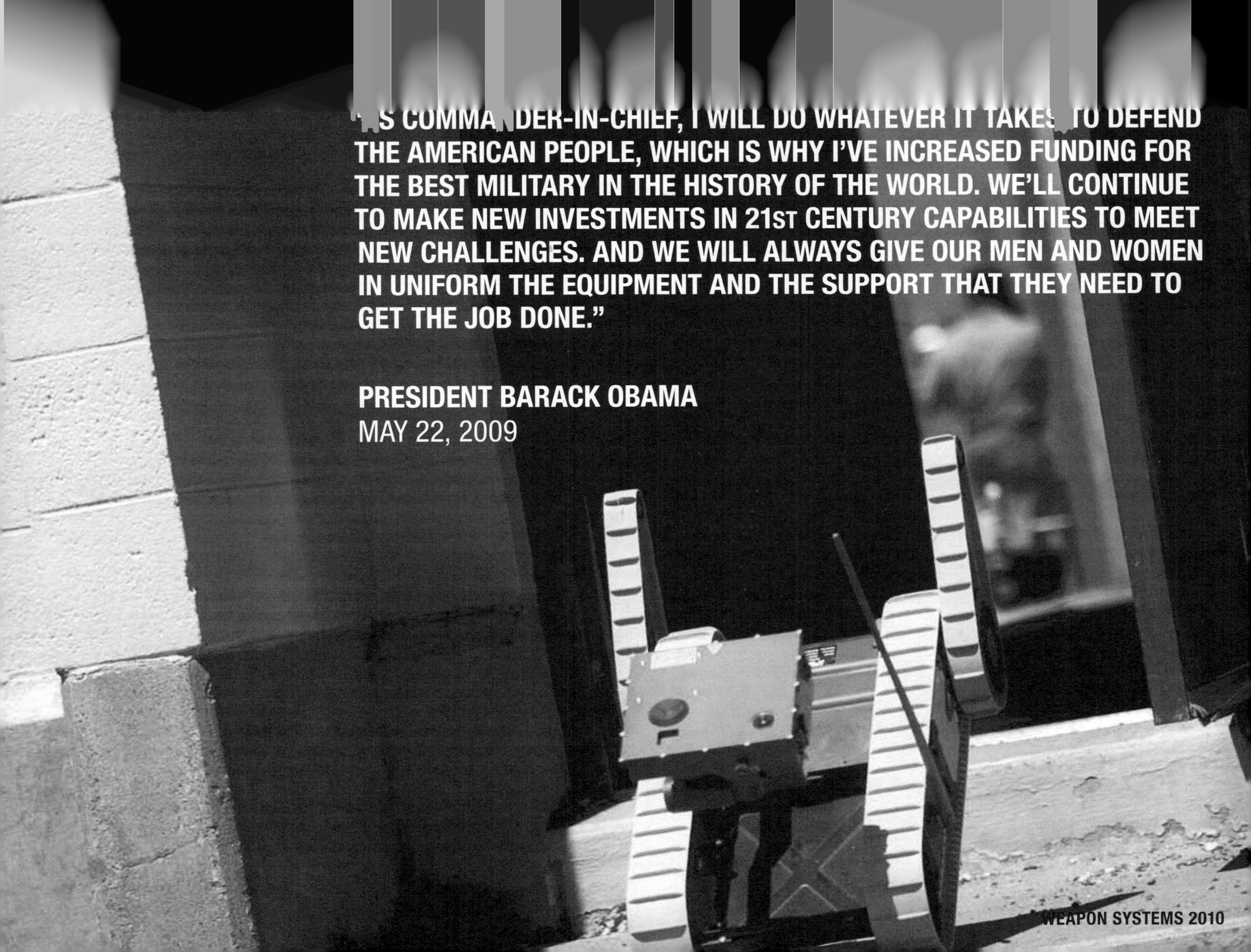

"AS COMMANDER-IN-CHIEF, I WILL DO WHATEVER IT TAKES TO DEFEND THE AMERICAN PEOPLE, WHICH IS WHY I'VE INCREASED FUNDING FOR THE BEST MILITARY IN THE HISTORY OF THE WORLD. WE'LL CONTINUE TO MAKE NEW INVESTMENTS IN 21ST CENTURY CAPABILITIES TO MEET NEW CHALLENGES. AND WE WILL ALWAYS GIVE OUR MEN AND WOMEN IN UNIFORM THE EQUIPMENT AND THE SUPPORT THAT THEY NEED TO GET THE JOB DONE."

PRESIDENT BARACK OBAMA
MAY 22, 2009

PROVIDING WARFIGHTERS WITH THE DECISIVE EDGE

Our mission is to effectively and efficiently develop, acquire, field, and sustain materiel by leveraging domestic and international, organic, and commercial technologies and capabilities to meet the Army's current and future mission requirements. Our vision is clear: **To equip and sustain the world's most capable, powerful, and respected Army**.

The Army's ability to achieve this vision rests on the Army Acquisition Workforce, fully employed and deployed worldwide in support of our Soldiers. The men and women who make up this workforce serve under the direction of 11 Program Executive Offices, two Joint Program Executive Offices, seven Deputy Assistant Secretaries, one Deputy for Acquisition and Systems Management, three Direct Reporting Units, and several major subordinate commands of the U.S. Army Materiel Command.

These professionals perform a wide range of responsibilities which include: research and development; program management; contracting; and systems engineering. They also develop and oversee Army-wide policy for procurement, logistics, chemical weapons destruction and demilitarization, science and technology, defense exports and cooperation, and many other areas. To expedite delivery of vital warfighting systems and services, we are continuing to improve our acquisition processes. At the same time, we are developing and institutionalizing new processes to improve our effectiveness, efficiency, transparency, collaboration, and our overall ability to rapidly procure the equipment and technologies that our Warfighters require.

WARFIGHTERS: OUR FOCUS

Soldiers...Warfighters...are the heart of everything we do. They are over a million strong—men and women, Active and Reserve—steeled by eight years of war. We now have a generation of Soldiers not seen in over 30 years: hardened by battle, strengthened by sacrifice, and resolved to defeat the enemies of our Nation. Embodying the strength of the Nation, they will face a dangerous, uncertain operational environment for the foreseeable future.

As we transition from major operations in Iraq to Afghanistan, while facing complex global challenges elsewhere, our responsibility to prepare our Warfighters grows in importance and magnitude. The systems and platforms described in the Army's *Weapon Systems 2010* handbook are vital to our Warfighters. 150 of the 650 programs we currently manage are described in this handbook. These programs enable the Army to equip, reset, and modernize the force. For this reason, they represent our highest priority systems and platforms. In addition, you will find descriptions of critical joint programs, as well as business information technology systems (which are improving transparency; sharing of reliable, authoritative data; efficiency; and the overall effectiveness of the Department of the Army).

STRATEGIC CONTEXT

Several factors combine to create the context in which we are accomplishing our mission:

OBJECTIVE REALITY OF WAR

America's Army is the Strength of the Nation. Deployed on a global scale, our Warfighters are engaged in protracted combat in two theaters and in other operations in many regions. Our operational demands and high personnel tempo outpace our ability to fully restore readiness across the Army.

STRATEGIC UNCERTAINTY

In the years ahead, the United States will continue to face unanticipated strategic challenges to our national security and the collective security of our international partners. These challenges will occur in many forms and will be waged across the spectrum of conflict—ranging from peaceful competition to challenges posed by hybrid threats to wartime contingency scenarios of varying scale and complexity. In addition, the Nation may be engaged in simultaneous military operations in all operational domains: land, sea, air, space, and cyberspace.

FISCAL CONSTRAINT AND ACQUISITION EXCELLENCE

We will continue to execute our acquisition programs in an increasingly constrained fiscal environment. Our efforts will remain highly visible and a subject of national attention. The Army, and our Acquisition Workforce, must fully institutionalize its continuous process improvement initiatives to obtain greater effectiveness and efficiencies—while embracing the tenets of Acquisition Reform and enhancing the overall capacity and capability of our acquisition professionals.

ENABLING ARMY MODERNIZATION GOALS

We are working to build a versatile mix of tailorable and networked organizations, operating on a rotational cycle, to provide a sustained flow of trained and ready forces for full spectrum operations and to hedge against unexpected contingencies at a sustainable tempo for our All-Volunteer Force. We seek to speed the fielding of successes from our research and development base to improve our current capabilities, while leveraging what we have learned during eight years of war to develop future capabilities. We foresee three broad goals:

UPGRADE AND MODERNIZE SELECTED SYSTEMS TO BEST PREPARE SOLDIERS FOR COMBAT

Our objective is to ensure that every Soldier, in every theater, receives the proper type and amount of equipment needed to accomplish their full spectrum of missions. We have replaced our old tiered readiness approach—which resulted in some units always well equipped, others less equipped—all based on a static Master Priorities List. The goal of ensuring that every Soldier and every unit have all of their equipment all of the time is neither achievable nor required. Instead, we are "equipping to mission," as we have been doing for some years now. To provide trained, ready forces to the combatant commanders, we work to ensure that our Soldiers have the equipment they need, in the right amount and at the right level of modernization, to accomplish their missions—whether in combat...training for combat...preparing units for combat via our Generating Force...supporting civilian authorities...or securing the homeland.

INCORPORATE NEW TECHNOLOGIES INTO OUR BRIGADE COMBAT TEAMS

We are working to deliver the most immediately relevant technologies developed through Future Combat Systems research and development to all our Brigade Combat Teams (and other priority combat formations), rather than focusing primarily on producing unique capabilities for a small set of Brigade Combat Teams. We are accelerating our efforts to field these key technologies (described in this handbook) to selected Brigade Combat Teams in 2011. These technologies link manned systems, unmanned systems, sensors, and munitions through the use of the integrated communications Network we are building.

KEY TECHNOLOGIES INCLUDE:
Interceptor Body Armor (See page 148)
Non Line of Sight-Launch System (NLOS-LS) (See pages 90, 250, 324)
Tactical Unattended Ground Sensors (T-UGS) and (U-UGS) (See page 90)
Ground Soldier System (See page 116)
Small Unmanned Ground Vehicle (SUGV) Block 1 (See page 90)
Class I Unmanned Aerial Vehicle (CL I UAV) (See pages 90, 324)

BETTER ENABLE ***ALL*** OF OUR FORMATIONS THROUGH CONTINUOUS UPGRADES AND MODERNIZATION

We are continuing our work to modernize all our formations – consisting of over 300 brigades (both Brigade Combat Teams and Support Brigades)—to increase the depth and breadth of our overall capacity. We are applying the lessons of war to build a more versatile, more readily deployable mix of networked formations to better leverage mobility, protection, information, and precision fires to improve our operational effectiveness. Across the force, we are also continuing modular conversion (to complete our conversion from a division-based to a brigade-based Army), rebalancing the size and capabilities of our active and reserve components, and stabilizing people in units for longer periods of time. We will to improve our capability for irregular warfare and the full spectrum of challenges our Soldiers will face while conducting offensive, defensive, and stability operations simultaneously.

STRATEGIC DIRECTION

We have established a set of key strategic initiatives to guide the efforts of the acquisition community to achieve our mission, realize our vision, and enable the Army's broad modernization goals. These initiatives provide the enduring, unifying focus for our collective effort.

DELIVER MATERIEL AND SERVICES NEEDED TO PROVIDE WARFIGHTERS WITH THE DECISIVE EDGE

To underwrite our ability to accomplish National Security, National Defense, and National Military strategic objectives, we provide our Warfighters with the best equipment and support the Nation can deliver. We fulfill this purpose through the effort and innovation of our military and civilian workforce and our collective ability to plan, program, and execute our acquisition programs accordingly. We are continuing our work to respond rapidly and flexibly to time-sensitive requirements. At the same time, we are complying fully with ethical standards of conduct and the laws that create the context for our responsibilities, relationships, and fiscal and environmental stewardship requirements.

To enable the accomplishment of our vital mission, we must sustain an independent acquisition function. We must fully leverage the skills and capabilities of our professional workforce and strengthen collaboration with our key partners and stakeholders to perform effective, efficient life cycle functions for design, development, deployment, sustainment, and other areas.

This initiative is overarching. It supports and is enabled by the following initiatives.

LEVERAGE THE FULL POTENTIAL OF TECHNOLOGY TO EMPOWER SOLDIERS

The American Soldier—the most potent of our Nation's weapons—is enabled by technology. We must sustain the technological superiority of our Soldiers by creating unprecedented capabilities for them. Underpinning this imperative is a robust, dynamic Army Science and Technology community—of people and laboratories—that seeks to achieve radical scientific and technological breakthroughs to ensure our Soldiers maintain a decisive edge over our enemies.

The Army's scientists, engineers, and integrated product teams of acquisition professionals have been at the forefront in adapting technology for urgent operational needs. They are enhancing our Warfighters' capabilities, as exemplified by the newly fielded First Strike Ration, which reduces by 40–50 percent the weight of the daily combat food ration carried by Soldiers during initial periods of high intensity conflict.

Our scientists and engineers continuously harvest materiel solutions from past investments, such as the development of mine detection ground penetrating radar technology. They also provide extraordinary technical expertise which has resulted in the development and integration of technologies such as new lightweight armor. This armor has dramatically enhanced the survivability of Mine Resistant Ambush Protected and other combat vehicles in the face of constantly evolving threats. Sufficient, sustained, and predictable investment in research and development and science and technology is needed to provide our Soldiers with the decisive edge.

CONTINUALLY IMPROVE AND ACHIEVE EXCELLENCE IN OUR ACQUISITION PROCESSES

Supporting an Army at war is critical, both tactically and strategically. From a tactical standpoint, we work with our joint, international, and industry partners to provide the weapon systems, software, and equipment our Soldiers need to accomplish their missions decisively. Strategically, as we meet ongoing requirements, we work to collapse the timelines required to get weapon systems and equipment to our Soldiers. Our goal is to compress the concept-to-combat cycle to best meet Soldiers' needs.

To enhance the value and relevance of our products and services, we are continually reviewing our internal processes and procedures and strengthening our internal and external interfaces. We strive to achieve acquisition excellence by reinforcing our history as good stewards of taxpayer dollars and remaining accountable to Congress, the President, the American Public—and our Soldiers who depend on us. We are committed to making progress in two key areas—human capital enrichment and portfolio integration—to keep our Army the world's preeminent landpower.

We cannot have a 21st Century operational force generated and supported by 20th Century processes. To meet future challenges, we must achieve a high level of continuous, measurable improvement in our core acquisition and logistics business processes. By "taking work out" of our processes—reducing waste in all its forms—we will accelerate our transformation. In addition, in the face of downward fiscal pressure, we will continue to enable our Army to best direct resources to our most compelling wartime needs.

CONTINUALLY IMPROVE OUR CAPACITY TO DESIGN, DEVELOP, DELIVER, DOMINATE—AND SUSTAIN

We must further embrace the interdependencies of systems and platforms—both under development and in sustainment—to best manage the resource, scheduling, and operational impacts of program adjustments. We are improving our coordination across programs, over time, formation by formation. We are also improving linkages to both our force generation and planning, programming, and budgeting processes.

We are working to improve our systems engineering capacity, to rebuild and revitalize our Governmental workforce of systems engineers, and to integrate these improvements across our entire acquisition and program management framework. To provide the skill sets needed to manage our complex acquisition portfolio as a collaborative team, we are working to attract and retain the finest scientists, engineers, program managers, logisticians, business, and contracting professionals.

We are continuing to improve how we manage systems of systems across their entire life cycle. We are also improving how we work with the Training and Doctrine Command, other Army entities, and combatant commanders—to better understand, anticipate, and respond to emerging requirements for warfighting capabilities.

The platforms and systems in the *U.S. Army Weapon Systems 2010* handbook are not stand-alone systems. Each depends on other systems to produce capabilities for Soldiers. We are strengthening and investing in our system of systems portfolio approach to best synchronize, integrate, and deliver the capabilities our deploying formations need to accomplish their missions. To realize our broader objectives for improving systems engineering, we are examining each of our core processes. These processes include: engineering; acquisition program management; configuration management; testing and validation; force integration; and planning, programming, and budgeting.

The Apache Block III attack helicopter program exemplifies the complexities of the interdependencies we manage. The attack helicopter, a system in itself, is actually a "system of systems." For this reason, Program Executive Office (PEO) Aviation does not act *independently* to field an aircraft. In fact, its efforts are wholly *interdependent* with other PEO organizations. To fully field and employ this system, this single PEO must synchronize its efforts with many other PEO portfolios—each of which has different delivery dates for the numerous products or services it provides to the Army or the Joint Force.

As the Apache Interoperability chart to the right depicts, the "system of systems" known as Apache Block III requires interaction with at least seven different categories of programs and platforms. These include (beginning at the left of the diagram and working clockwise): (1) Ground; (2) Air; (3) Supporting; (4) Hosting; (5) Weapons Systems and Munitions; (6) Communications (to employ Net Centric doctrine); and (7) Intelligence, Surveillance, and Reconnaissance.

In practical terms, this means that, among others, PEO Aviation must work closely with all of the 11 ASA(ALT) PEOs and two Joint PEOs, each of whom are responsible for the timing of a range of programs, some of which are depicted here. In sum, fielding an attack helicopter requires a "systems of systems" approach to ensure that the helicopter is able to: interact with ground, air, transporting, and hosting platforms; employ its onboard weapons systems; receive its supply and resupply of ammunition; communicate (through voice, digital, satellite, and other means); and receive and transmit imagery, position locating, and intelligence information.

Achieving excellence in acquisition also involves demonstrating continuous stewardship and superb management of highly sensitive and visible programs for which we have executive agent authority, such as the Nation's chemical weapons disposal program.

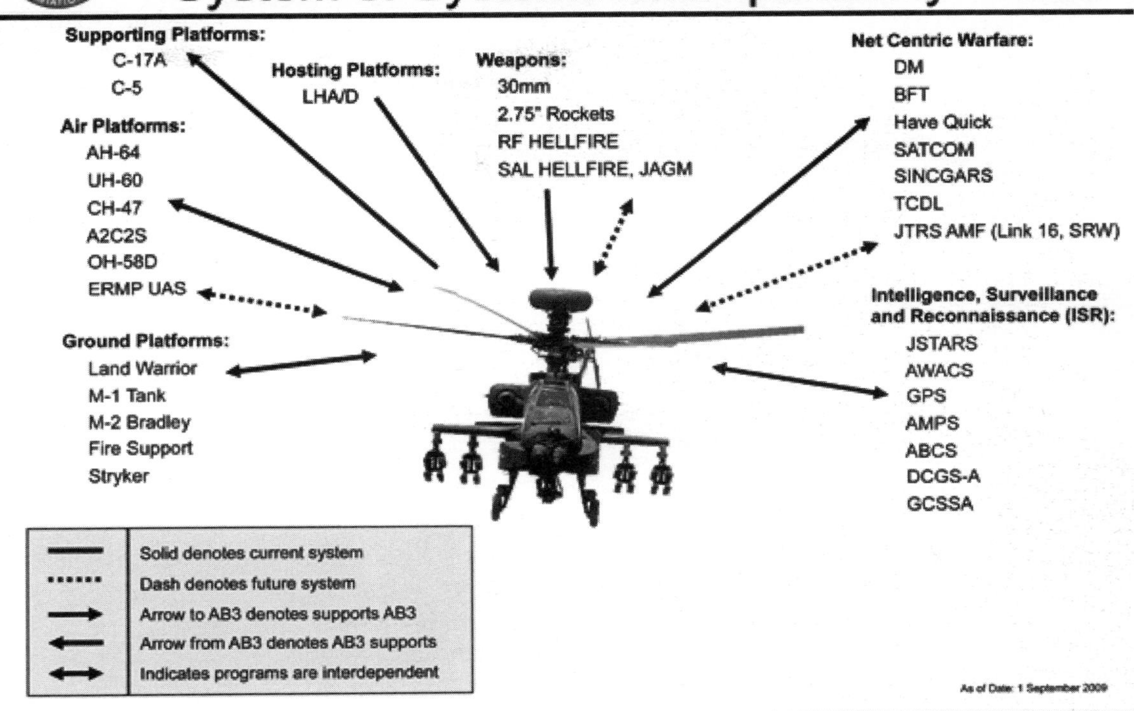

KEY TO ACRONYMS

Program/Platform

- A2C2S: Army Airborne Command and Control System
- ABCS: Army Battle Command System
- AH-64: Apache Helicopter
- AMPS: Aviation Mission Planning System
- AWACS: Airborne Warning and Control System
- BFT: Blue Force Tracker
- C-17A: Globemaster Cargo Aircraft
- C-5: Galaxy Cargo Aircraft
- CH-47: Chinook Helicopter
- DCGS-A: Distributed Common Ground System-Army
- DM: Distribution Management
- ERMP UAS: Extended Range Multi-Purpose Unmanned Aircraft System
- Fire Support: Artillery Systems
- GCSSA: Global Combat Support System Army
- GPS: Global Positioning System
- Have Quick: Frequency-hopping Radio
- JAGM: Joint Air Ground Munitions
- JSTARS: Joint Surveillance and Target Attack Radar System–Air
- JTRS AMF: Joint Tactical Radio System Airborne Maritime Fixed
- LHA/D: Landing Helicopter Assault/Dock
- Land Warrior: Ground Soldier Ensemble
- LHA-D: Amphibious Assault Ship–Dock
- Link 16: Radio Type
- M-1: Abrams Tank
- M-2: Bradley Fighting Vehicle
- OH-58D: Kiowa Warrior
- SATCOM: Satellite Communications
- SINCGARS: Single Channel Ground and Airborne Radio System
- SRW: Soldier Radio Waveform
- Stryker: Armored Combat Vehicle
- TCDL: Tactical Common Data Link
- UH-60: Black Hawk Helicopter

The U.S. Army Chemical Materials Agency (CMA), using acquisition processes as its baseline, works with private industry, academia, and other interested policy and environmental stakeholders to eliminate America's obsolete chemical weapons. CMA also responds to discoveries of non-stockpile chemical weapons and safely stores those weapons until their disposal. Moreover, CMA partners with the Federal Emergency Management Agency to prepare local communities to deal with potential emergencies involving those weapons.

LEVERAGE LESSONS LEARNED TO SUPPORT THE FULL RANGE OF ARMY MODERNIZATION AND EQUIPPING INITIATIVES

The Army's enduring mission is to protect and defend our vital security interests and to provide support to civil authorities in response to domestic emergencies. This requires an expeditionary, campaign capable Army able to dominate across the full spectrum of conflict, at any time, in any environment, and against any adversary—for extended periods of time. To support this requirement, we are continually reviewing and adapting our structure, organization, and capabilities.

As an example, we are applying the lessons learned from Future Combat Systems—the value of spin-outs and increments, systems of systems engineering, networked operations, and others—to continuously improve "how we do business" to support Soldiers. Just as the Army applies the DOTMLPF construct (Doctrine, Organization, Training, Materiel, Leadership and Education, Personnel, and Facilities) to develop and adapt its operational capabilities, we apply this same construct to our acquisition processes to enable us to evolve on pace with the Warfighters we support.

To enhance our contributions, we are continuing our efforts to bring the Army's acquisition and sustainment communities closer together to focus seamlessly on the entire life cycle of our weapon systems and equipment. By strengthening collaboration among all partners and stakeholders, and implementing numerous improvements to our life cycle management process, we are furnishing products to Soldiers faster, making good products better, and reducing costs.

REBUILD AND REBALANCE THE CAPABILITY OF THE ACQUISITION WORKFORCE

In the Army, our people are our most important asset. During the last decade, we witnessed a steady decline in the size of the Army Acquisition Workforce—in the face of a wartime workload increasing in both size and complexity. The civilian and military members of our Acquisition Workforce now total approximately 41,000, a significant reduction from the Cold War era. These acquisition professionals are located in our PEOs, in various commands, and in other organizations across the Army. During 2008, this workforce managed over one-quarter of every Federal dollar spent on contracts. Every day, they make a direct impact on the products and services we procure for Soldiers.

To better support the Army, enable our combatant commanders, and alleviate the stress of doing more with less, we are rebuilding (growing) and rebalancing (aligning the right skills to the work) the Army Acquisition Workforce. On April 6, 2009, in discussing the proposed Fiscal Year 2010 Defense Budget, Secretary of Defense Robert M. Gates said, "this budget will… increas[e] the size of the Defense Acquisition Workforce, converting 11,000 contractors to full-time government employees, and hiring 9,000 more government acquisition professionals by 2015, beginning with 4,100" in Fiscal Year 2010. We are working aggressively to implement Defense Acquisition Workforce growth. The purpose is clear: to ensure the Department of Defense is well positioned to produce best value for the American taxpayer and for the Soldiers, Sailors, Airmen, and Marines who depend on the weapons, products, and services we buy.

The objectives of the growth strategy are to: rebalance the acquisition total force; grow the Government Acquisition Workforce 15 percent by 2015; improve acquisition capabilities and capacities; improve defense acquisition oversight; close workforce gaps; strategically reshape acquisition training; and target incentives appropriately. We are well underway in our work to properly resource this growth. By the end of Fiscal Year 2010, we plan to have hired and insourced a total of 2,600 civilian acquisition employees.

We are also continuing our work to achieve the intent of Section 852 of the National Defense Authorization Act (NDAA) of 2008, Public Law No. 110-181. Section 852 directed the establishment of the Defense Acquisition Workforce Development Fund. This fund enables the Defense Department to better recruit, hire, develop, recognize, and retain its acquisition workforce. The Army is building and executing a program of nearly $1 billion focused on: hiring acquisition interns, journeymen, and highly qualified experts; offering new education, training, and developmental programs; and funding recognition and retention incentives. These initiatives are helping us to enhance the overall stature, development, and professionalism of those who fill our ranks.

As we work to rebuild and rebalance the force, we are also strengthening the unique identity of our Army Acquisition Workforce. In addition, we are accelerating our work to institutionalize Contingency Contracting as a core competency—to better provide the Army-wide program management and logistics skills needed in expeditionary operations.

IMPROVE OUR CAPABILITY AND CAPACITY TO ARTICULATE OUR STRATEGIC INITIATIVES AND COMPELLING NEEDS

We are continuing our work to more fully develop the ability to communicate more effectively with both our internal and external stakeholders. We serve both the Soldier and the American Public—and must remain connected to both. We are working aggressively to:

- **Build Awareness** of ASA(ALT)'s strategic direction and priorities to advance understanding of our organizational mission and the execution of Army acquisition programs;
- **Build Cooperative Relationships** with ASA(ALT) stakeholders to ensure effective, efficient execution of organization priorities and programs; and,
- **Build Advocacy** for Army and ASA(ALT) priorities and initiatives through carefully focused activities intended to educate and inform key stakeholders. Our efforts in this realm are intended to increase the likelihood of achieving our strategic goals. We seek to create "champions" and obtain sufficient, sustained, and predictable resourcing needed to ensure program stability and enable better program management.

Ultimately, to accomplish our mission for Warfighters, we are working—as part of an overarching Department of the Army effort—to better communicate with our stakeholders in clear, unambiguous terms.

PATH FORWARD

The likelihood of continuing conflict and the resilience of ruthless, determined, and adaptive enemies form the basis of our requirement to modernize. Continuous modernization is the key to transforming Army capabilities and maintaining a technological advantage over our adversaries across the full spectrum of conflict. We have received extraordinary funding support through wartime Overseas Contingency Operations funds, but they have only enabled us to sustain the current fight. We look forward to continued Congressional support to achieve our broad modernization goals.

The systems listed in this book are not isolated, individual products. Rather, they are part of an integrated investment approach to make the Army of the future able to deal successfully with the challenges it will face. Each system and each capability is important. These systems represent today's investment in tomorrow's security—to ensure our Army can continue to successfully defend our Nation.

WEAPON SYSTEMS

LISTED IN ALPHABETICAL ORDER

2.75" Family of Rockets

INVESTMENT COMPONENT
Modernization
Recapitalization
Maintenance

MISSION

To provide air-to-ground suppression, illumination, and direct/indirect fires to defeat area, materiel, and personnel targets at close and extended ranges.

DESCRIPTION

The Hydra 70 Rocket System of 2.75 inch air-launched rockets is employed by tri-service and special operating forces on both fixed wing and rotary wing aircraft. This highly modular rocket family incorporates several different mission-oriented warheads for the Hydra 70 variant, including high-explosive, multipurpose submunition, red phosphorus smoke, flechette, visible light illumination flare, and infrared illumination flare.

Diameter: 2.75 inches
Weight: 23–27 pounds (depending on warhead)
Length: 55–70 inches (depending on warhead)
Range: 300–8,000 meters

SYSTEM INTERDEPENDENCIES

None

PROGRAM STATUS

Hydra 70

- **Current:** Producing annual replenishment requirements for training and war reserve

PROJECTED ACTIVITIES

Hydra 70

- **Continue:** Hydra 70 production and Safety Reliability and Producibility (SRAP) Program activities.

ACQUISITION PHASE
Technology Development | Engineering & Manufacturing Development | Production & Deployment | Operations & Support

2.75" Family of Rockets

FOREIGN MILITARY SALES

Hydra 70: Kuwait, the Netherlands, Colombia, Singapore, Thailand, United Arab Emirates, and Japan

CONTRACTORS

Hydra 70:

General Dynamics (Burlington, VT)

General Dynamics Armament and Technical Products (GDATP) (Camden, AR)

Grain:

Alliant Techsystems (Radford, VA)

Fuzes:

Action Manufacturing (Philadelphia, PA)

Fin and nozzle:

General Dynamics Ordnance and Tactical Systems (Anniston, AL)

Abrams Upgrade

INVESTMENT COMPONENT

Modernization

Recapitalization

Maintenance

MISSION

To provide mobile, protected firepower for battlefield superiority

DESCRIPTION

The Abrams tank provides the lethality, survivability, and fightability to defeat advanced threats on the integrated battlefield using mobility, firepower, and shock effect. The 120mm main gun on the M1A1 SA (Situational Awareness) and M1A2 SEP (System Enhancement Program) v2 and the 1,500-horsepower AGT turbine engine and special armor make the Abrams tank particularly lethal against heavy armor forces. The Abrams Modular Tank fleet includes two variants, the M1A1 SA and the M1A2 SEP v2.

M1A1 SA: Improvements include Block I forward-looking infrared (FLIR) and far-target locator. Lethality improvements include Stabilized Commander's Weapon Station (SCWS) and ballistic solution upgrades for the M829A3 kinetic and the M1028 canister rounds. Common Abrams modifications include Blue Force Tracking (BFT), which is a digital command and control system that gives Army commanders across the battlefield current information about their location relative to friendly forces; and the Power Train Improvement and Integration Optimization Program (TIGER engine and improved transmission), which provides more reliability, durability, and a single standard for the vehicle's power train. Survivability improvements include frontal armor and turret side armor upgrades.

M1A2SEP v2: Upgrades include survivability, automotive power pack, computer systems, and night vision capabilities. Lethality improvements include Common Remotely Operated Weapon Station (CROWS) and ballistic solution upgrades for the M829A3 kinetic and the M1028 canister rounds. The M1A2 SEP v2 has improved microprocessors, color flat panel displays, improved memory capacity, better soldier-machine interface, and a new open operating system designed to run the Common Operating Environment (COE) software. Both the Gunner's Primary Sight (GPS) and the Commander's Independent Thermal Viewer (CITV) on the M1A2SEP tank include the improved thermal imaging capabilities of the new Block I second-generation FLIR technology. The M1A2 SEP has improved frontal and side armor for enhanced crew survivability. The M1A2 SEP is also equipped with a battery-based auxiliary power unit, the total integrated revitalization (TIGER) engine, and upgraded transmission for improved automotive reliability and durability.

SYSTEM INTERDEPENDENCIES

None

PROGRAM STATUS

- **Current:** The 1st Cavalry Division and 1st Brigade, 1st Armored Division are equipped with the Abrams M1A2 SEP v2.
- **Current:** Abrams production of M1A1 SA and M1A2SEP v2 tanks continue for both the Active Army and the Army National Guard (ARNG) to meet the Army's modularity goals by 2013.

PROJECTED ACTIVITIES

- **FY10–11:** M1A1 SA fielding continues to the ARNG, 1st Infantry Division, 2nd Infantry Division, Training and Doctrine Command/ Combined Arms Support Command, ARNG Regional Training Site-Maintenance (RTSM) units, and Army Prepositioned Stock 5 (Kuwait) and Army Prepositioned Stock 4 (Korea).
- **FY10–12:** M1A2 SEP v2 multiyear contract production continues.
- **4QFY09–2QFY10:** 1st, 2nd, and 3rd Brigades, 4th Infantry Division will be fielded with the Abrams M1A2 SEP v2 tank.
- **1QFY10–2QFY10:** 4th Brigade 1st Cavalry Division and the 1st Armored Division will be fielded with the Abrams M1A2 SEP v2 tank.
- **3QFY10:** Army Prepositioned Stock 5 (Southwest Asia) will be fielded with the Abrams M1A2 SEP v2 tank.
- **2QFY11–4QFY12:** 3rd Infantry Division and the 116th Army National Guard will be fielded with the Abrams M1A2 SEP v2 tank.
- **FY10–12:** TIGER production continues

ACQUISITION PHASE

Technology Development | Engineering & Manufacturing Development | Production & Deployment | Operations & Support

Abrams Upgrade

FOREIGN MILITARY SALES
M1A1: Australia (59), Egypt (1,005), Iraq (140)
M1A2: Kuwait (218), Saudi Arabia (315)

CONTRACTORS
General Dynamics (Sterling Heights, MI; Warren, MI; Muskegon, MI; Scranton, PA; Lima, OH; Tallahassee, FL)
Honeywell (Phoenix, AZ)
Simulation, Training, and Instrumentation Command (STRICOM) (Orlando, FL)
Anniston Army Depot (ANAD) (Anniston, AL)

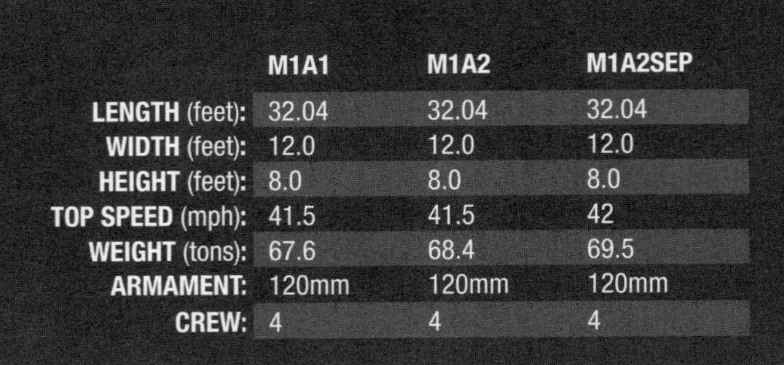

	M1A1	M1A2	M1A2SEP
LENGTH (feet):	32.04	32.04	32.04
WIDTH (feet):	12.0	12.0	12.0
HEIGHT (feet):	8.0	8.0	8.0
TOP SPEED (mph):	41.5	41.5	42
WEIGHT (tons):	67.6	68.4	69.5
ARMAMENT:	120mm	120mm	120mm
CREW:	4	4	4

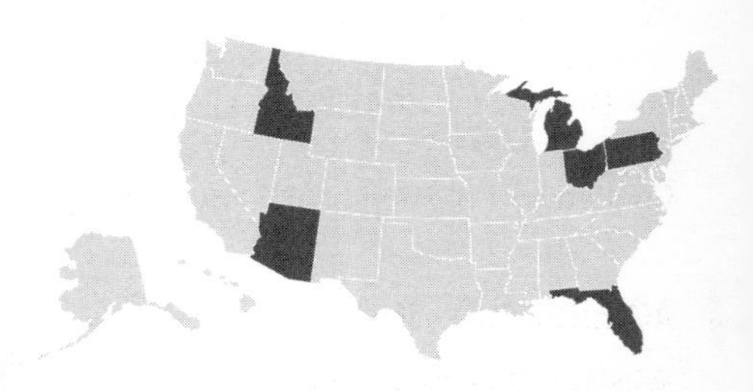

AcqBusiness

INVESTMENT COMPONENT

Modernization

Recapitalization

Maintenance

MISSION

Provide an enterprise, service-oriented, business environment populated with the Information Management (IM) systems and services that bring the right information to the right people at the right time.

DESCRIPTION

The AcqBusiness Program provides Information Management capabilities that support Acquisition community needs for Acquisition data, data management services and Enterprise Business applications. These capabilities enable the consistent, effective and efficient conduct of the acquisition business. Planning and development of additional capabilities are ongoing with rapid prototyping, user involvement and rapid capability distribution as core elements of the program strategy. The Acquisition Business Program is not a traditional program; it consists of a continuing series of independent software projects managed to cost, schedule and user requirements. This program strategy is modeled after best industry practices for rapid development and distribution of enterprise software solutions. The initial Service Oriented Environment has been deployed and provides an initial set of capabilities.

Initial Capabilities include:

- **PM Toolkit:** A collection of project management, risk management, collaboration, market Research and analysis tools focused on enabling project managers to better accomplish their mission.
- **Collaboration Tools:** A collection of tools (e.g., Defense Collaboration On-Line, Green Force Tracker, Oracle Collaboration Tool , milWiki, etc.) that enable acquisition users to find each other and collaborate in a variety of ways to better accomplish their mission.
- **Acquisition Personnel Management:** A collection of Acquisition personnel and career management solutions that support the management of acquisition career professionals.
- **AcqReadiness:** A suite of tools that provides total asset visibility, real time reporting and analysis of financial planning, execution, supply and readiness data.
- **Acquisition Information Management Tools:** The suite of legacy acquisition reporting and support tools that enable PMs to support Acquisition oversight and reporting responsibilities.
- **AcqTech:** An enterprise solution for the Science and Technology community that supports the management of Army Technology Objectives, traceability to Warfighter outcome requirements, alignment of Small Business Innovative Research Programs with Army Technology Objectives, and access to International Cooperative Agreements.

Future Capabilities

AcqBusiness is developing, procuring and planning to augment, enrich and extend these existing capabilities to enable effective and efficient conduct of the Acquisition Business. Some of the near-term capabilities that are in the planning and prototyping stages include:

- **Expeditionary Contracting:** A suite of procurement capabilities designed to operate in the tactical environment (low band-width, disconnected) to support the procurement process from development of procurement requests through contract fulfillment.
- **Additions and enhancements to the existing PM Toolkit:** Planned enhancements include Earned Value Management, Integrated Scheduling, Requirements Management, Contract Data Requirements List (CDRL) Management, and Contractor Task Management tools that are focused on enhancing the PM's ability to manage his/her program.
- **Interfaces with other Army Enterprise Solutions:** (e.g. General Funds Enterprise Business Systems (GFEBS), Logistics Modernization Program (LMP), Defense Integrated Military Human Resource System (DIMHRS), Global Combat Support System–Army (GCSS–Army)

ACQUISITION PHASE

Technology Development | Engineering & Manufacturing Development | Production & Deployment | Operations & Support

- **Common Operating Picture:** A collection of dashboards populated by authoritative Acquisition and Army data that are focused on enhancing decision making and providing acquisition leaders with accurate and relevant data, on time and when needed.
- **Tools to support to the Army Force Generation (ARFORGEN) /RESET process:** Tools that enable PMs to better manage their support to the ARFORGEN/RESET process by providing visibility into the Warfighter pre-deployment and redeployment requirements, and by synchronizing PM fielding schedules with Warfighter training schedules.

SYSTEM INTERDEPENDENCIES

None

PROGRAM STATUS

- **1QFY09:** Increments I and II deployed
- **1QFY10:** Increment III in development w/IOC scheduled
- **2–3QFY10:** Increments IV and V in prototyping w/IOC tentatively scheduled
- **3–4QFY10:** Increments VI and VII in Requirements Development w/ IOC tentatively scheduled
- **1QFY11:** Increment VIII in Concept Development w/IOC tentatively scheduled

PROJECTED ACTIVITIES

Continued pursuit of enterprise business tools and services that enrich the acquisition business environment and provide for enhanced decision making, consistency in business process, and access to authoritative acquisition data.

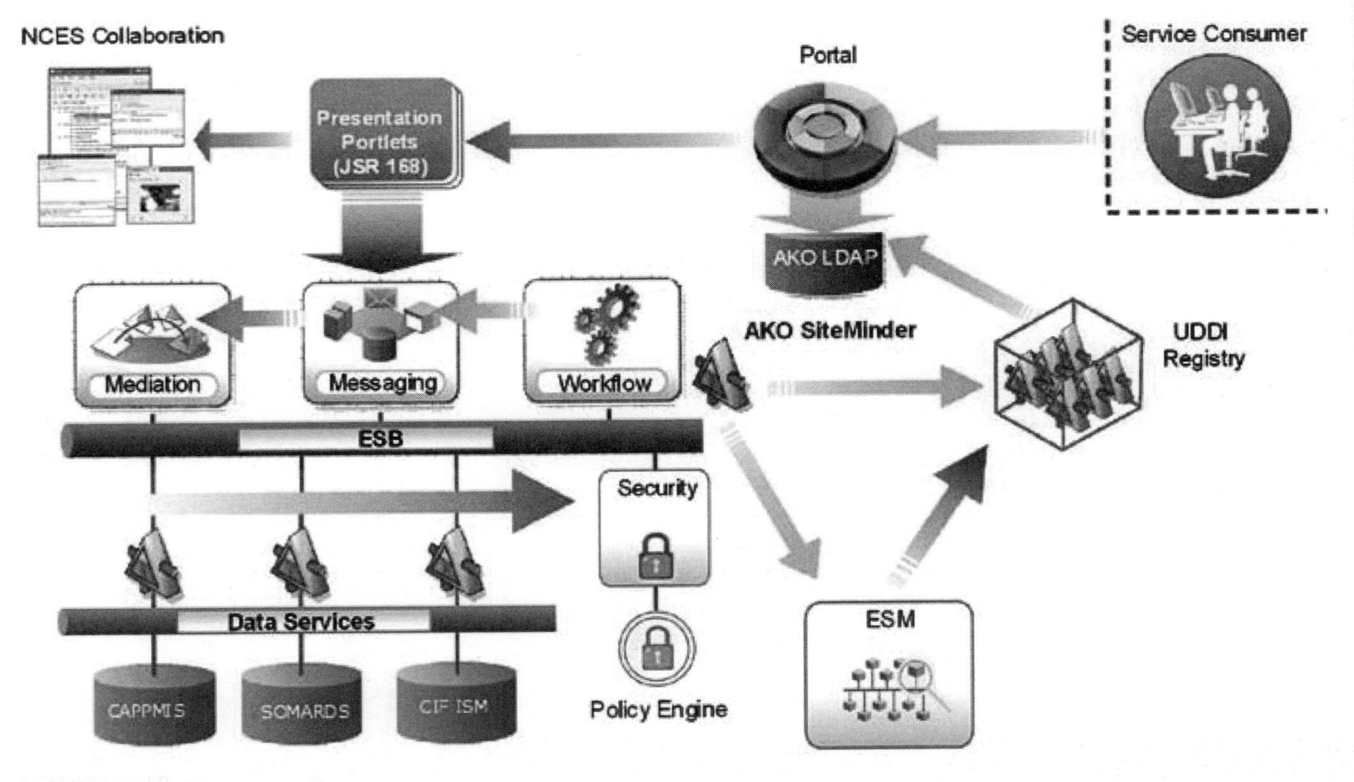

Ref: https://soa.army.mil

AcqBusiness

Foreign Military Sales:
None

Contractors
Booz Allen Hamilton (McLean, VA)
Computer Sciences Corp. (CSC) (Falls Church, VA)
Deloitte LLP (McLean, VA)

Advanced Field Artillery Tactical Data System (AFATDS)

INVESTMENT COMPONENT

Modernization

Recapitalization

Maintenance

MISSION

To provide the Army, Navy, and Marine Corps automated fire support command, control and communications.

DESCRIPTION

The Advanced Field Artillery Tactical Data System (AFATDS) performs the attack analysis necessary to determine optimal weapon-target pairing to provide automated planning, coordination, and control for maximum use of fire support assets (field artillery, mortars, close air support, naval gunfire, attack helicopters, and offensive electronic warfare).

AFATDS performs the fire support command, control, and coordination requirements of field artillery and maneuver from echelons above corps to battery or platoon in support of all levels of conflict. The system is composed of common hardware and software employed in varying configurations at different operational facilities (or nodes) and unique system software interconnected by tactical communications in the form of a software-driven, automated network.

AFATDS will automatically implement detailed commander's guidance in the automation of operational planning, movement control, targeting, target value analysis, and fire support planning. This project is a replacement system for the Initial Fire Support Automated System, Battery Computer System, and Fire Direction System. AFATDS is designed to interoperate with the other Army battle command systems; current and future Navy and Air Force command and control weapon systems; and the German, French, British, and Italian fire support systems.

SYSTEM INTERDEPENDENCIES

Forward Entry Device (FED), Pocket-Sized Forward Entry Device (PFED), Joint Automated Deep Operations Coordination System (JADOCS), Paladin, Multiple Launch Rocket System (MLRS), Theater Battle Management Core System (TBMCS), Gun Display Unit (GDU)/Gun Display Unit–Replacement (GDU–R), Force XXI Battle Command, Brigade-and-Below (FBCB2), Non Line of Sight–Cannon (NLOS–C)/Non Line of Sight–Launch System (NLOS–LS), Excalibur, All Source Analysis System (ASAS)/ Distributed Common Ground System–Army (DCGS–A)

PROGRAM STATUS

- **3QFY07:** Conditional materiel release (CMR) of AFATDS 6.4.0.1
- **4QFY07:** CMR of AFATDS 6.4.0.2
- **1QFY09:** Full materiel release (FMR) of AFATDS 6.5.0

PROJECTED ACTIVITIES

- **3QFY09:** FMR of AFATDS 6.5.1 (Windows)
- **1QFY10:** FMR of AFATDS 6.6.0 (Marshall Build)
- **1QFY11:** FMR of AFATDS 6.7.0 (MacArthur Build)
- **1QFY12:** FMR of AFATDS 6.8.0 (Eisenhower Build)

ACQUISITION PHASE

Technology Development | Engineering & Manufacturing Development | Production & Deployment | Operations & Support

Advanced Field Artillery Tactical Data System (AFATDS)

FOREIGN MILITARY SALES
Bahrain, Egypt, Portugal, Turkey, Taiwan

CONTRACTORS

Software:
Raytheon (Fort Wayne, IN)

Hardware:
General Dynamics (Taunton, MA)

Technical support:
Computer Sciences Corp. (CSC) (Eatontown, NJ)

New equipment training:
Engineering Professional Services (Lawton, OK)
Titan Corp. (Lawton, OK)

Aerial Common Sensor (ACS)

INVESTMENT COMPONENT

Modernization

Recapitalization

Maintenance

MISSION

To provide global, real-time, multi-intelligence precision targeting information to joint land, maritime, and air combat commanders across the full spectrum of military operations.

DESCRIPTION

The Aerial Common Sensor (ACS) is essential to the tactical warfighter. It fills a critical capability gap by providing actionable intelligence directly to ground commanders with the timeliness and accuracy they require. ACS is a multi-intelligence, manned, fixed-wing, Reconnaissance Surveillance and Target Acquisition (RSTA)/Intelligence, Surveillance, and Reconnaissance (ISR) system that carries multiple, highly accurate intelligence sensors, processing tools, air/ground/satellite communications, and onboard operators/analysts. Capable of worldwide deployment, ACS provides dedicated, persistent RSTA/ISR coverage over the depth and breadth of a tactical commander's battlespace. ACS is integrated in the greater Distributed Common Ground System–Army (DCGS–A). This unique combination of attributes (multi-intelligence sensing, persistence, wide-area coverage, reach, manned-unmanned teaming, Distributed common Ground System–Array (DCGS–A) connectivity, and battle command) provides the ground tactical commander a near-real-time operational view of unprecedented clarity, enabling tactical ground forces to operate at their highest potential in future joint operations.

ACS will replace the Airborne Reconnaissance Low (ARL) and Guardrail Common Sensor (GR/CS) airborne surveillance systems and will be fielded to the Army's Aerial Exploitation Battalions (AEB).

SYSTEM INTERDEPENDENCIES

Distributed Common Ground System–Army

PROGRAM STATUS

- **4QFY09:** Capabilities development document (CDD) approved by the Joint Capabilities Board (JCB)

PROJECTED ACTIVITIES

- **4QFY09:** Joint Requirements Oversight Council (JROC) consideration of the CDD
- **1QFY10:** Release of the Technology Development (TD) Request for Proposal (RFP)
- **2–4QFY10:** TD contract award
- **FY10–12:** TD phase

ACQUISITION PHASE

Technology Development | Engineering & Manufacturing Development | Production & Deployment | Operations & Support

Aerial Common Sensor (ACS)

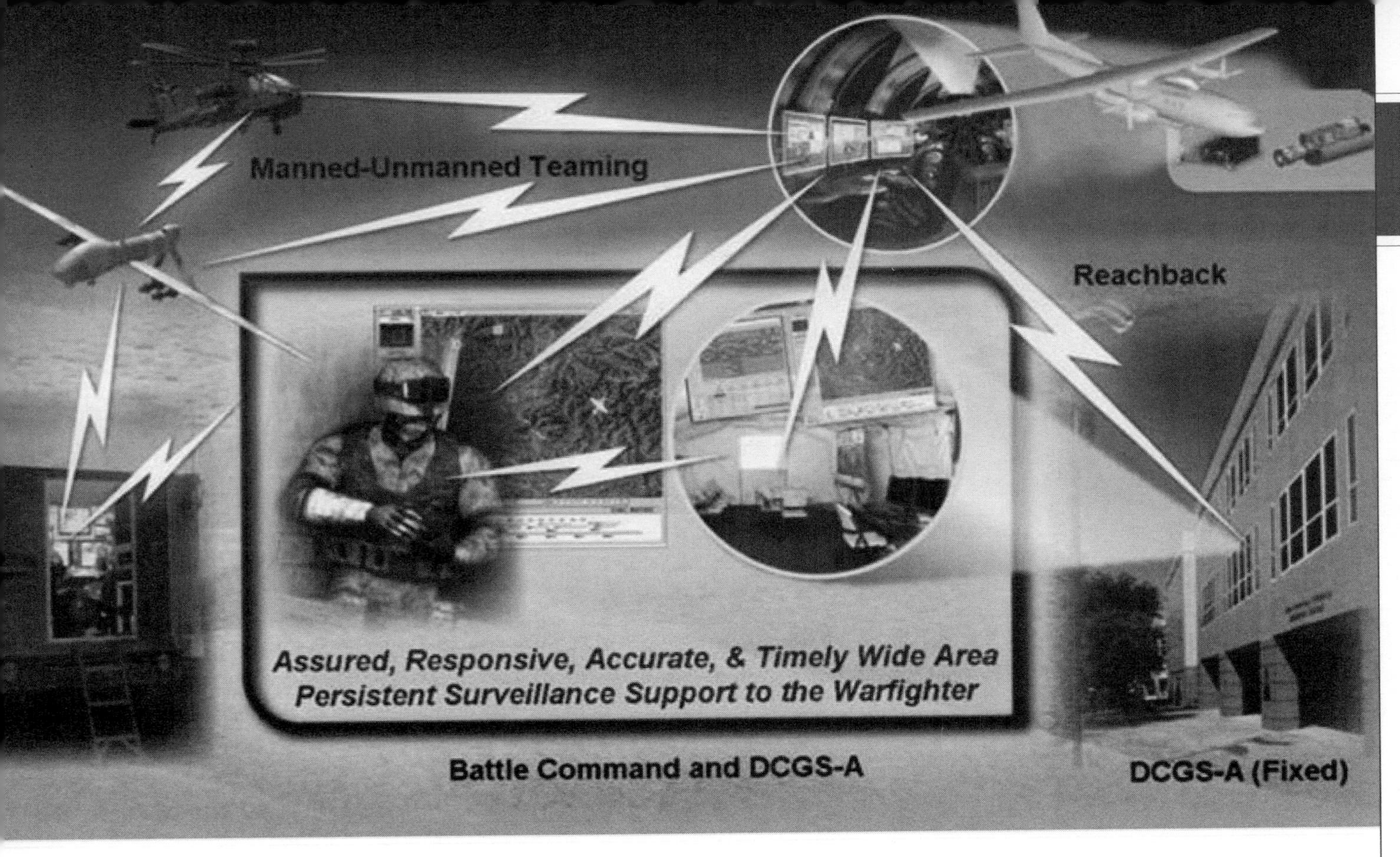

FOREIGN MILITARY SALES
None

CONTRACTORS
Radix (Mountain View, CA)
Institute for Defense Analysis (Alexandria, VA)
CACI (Eatontown, NJ)
MITRE (Eatontown, NJ)

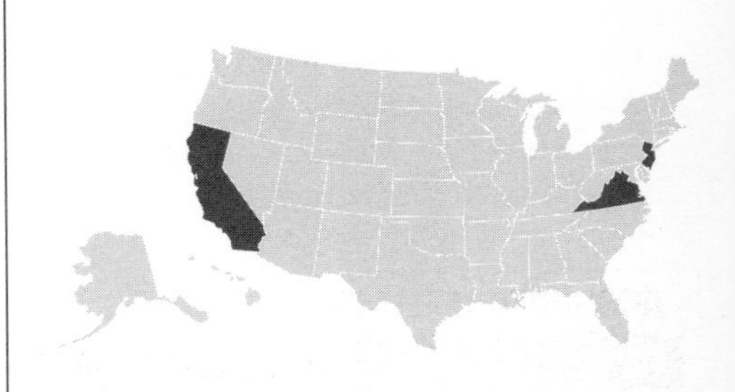

Air Warrior (AW)

INVESTMENT COMPONENT

Modernization

Recapitalization

Maintenance

MISSION

To improve the survivability, mobility, and sustainability of the Army aircrew member through a new generation, modular, integrated Aviation Life Support Equipment ensemble. Enhance Soldier safety, comfort, crew synergy, and capability.

DESCRIPTION

Air Warrior (AW) integrates advanced situational awareness and environmental, ballistic, chemical and biological (CB) protection into a single system comprising rapidly tailorable, mission-configurable modules. Air Warrior addresses interoperability and leverages several joint-service technology efforts to effectively integrate survival, evasion, and escape capabilities. AW maximizes human performance and safety without encumbering the Soldier. Components include Microclimate Cooling System; digital connectivity and threat and friendly forces situational awareness; wireless crew communications; aircraft platform interface; chemical-biological protection; body armor; survival and escape and evasion tools; and overwater survival items.

Improvements to the Air Warrior system are incrementally provided through an evolutionary acquisition program to solve equipment shortcomings. Components include the following:

- Survival Equipment Subsystem, which integrates first aid, survival, signaling, and communications equipment with body armor and over-water survival subsystems
- Microclimate Cooling System, which increases effective mission duration in heat-stress environments by more than 350 percent
- Aircrew Integrated Helmet System, a lighter helmet with increased head and hearing protection
- Electronic Data Manager (EDM), a portable digital mission planning device for over-the-horizon messaging and enhanced situational awareness capabilities through connectivity to Ble Force Tracking, Aviation
- Aircraft Wireless Intercom System (AWIS) for secure cordless, hands-free aircrew intercommunications
- Go-Bag Assembly & Tie-Down Strap
- Hydration System
- Portable Helicopter Oxygen Delivery System

The Air Warrior system is the key ingredient to closing the performance gap between the aircrew and the aircraft. Air Warrior is answering the aviation warfighter challenges of today and tomorrow by developing affordable, responsive, deployable, versatile, lethal, survivable, and sustainable aircrew equipment.

SYSTEM INTERDEPENDENCIES

FBCB2's Blue Force Tracking–Aviation system.

PROGRAM STATUS

- **2QFY07:** Fielding continues to units deploying to Operation Iraqi Freedom and Operation Enduring Freedom
- **2QFY08:** U.S. Navy adopts and procures the AW Microclimate Cooling System for its H-53 helicopter fleet
- **3QFY08:** The AW Portable Helicopter Oxygen Delivery System enters production; the Army's project managers for Bradley and Abrams adopt and procure the AW Microclimate Cooling System for their deployed crewmembers; first Spiral 3 EDM fielded to the 28th Combat Aviation Brigade's 1-137 Aviation Company
- **4QFY08:** PM Stryker adopts and procures the AW Microclimate Cooling System for its deployed crewmembers

PROJECTED ACTIVITIES

- **1QFY11:** Begin fielding for the AW Go Bag and helmet external audio products
- **1QFY11:** Production decision for the encrypted Aircraft Wireless Intercom System

ACQUISITION PHASE

Technology Development | Engineering & Manufacturing Development | Production & Deployment | **Operations & Support**

Air Warrior (AW)

FOREIGN MILITARY SALES
Australia, Canada, UAE

CONTRACTORS
Carleton Technologies, Inc. (Orchard Park, NY)
BAE Systems (Phoenix, AZ)
Aerial Machine and Tool, Inc. (Vesta, VA)
Westwind Technologies, Inc. (Huntsville, AL)
Raytheon Technical Services, Inc. (Indianapolis, IN)
Secure Communications Systems, Inc. (Santa Ana, CA)
Telephonics Corp. (Farmingdale, NY)
General Dynamics C4 Systems, Inc. (Scottsdale, AZ)
Science and Engineering Services, Inc. (SESI) (Huntsville, AL)
Gibson and Barnes (Santa Clara, CA)
US Divers (Vista, CA)
Oxygen Generating Systems International (Buffalo, NY)
Gentex Corportation (Rancho Cucamonga, CA)
Mountain High Equipment and Supply Co. (Redmond, OR)
Taylor-Wharton (Huntsville, AL)

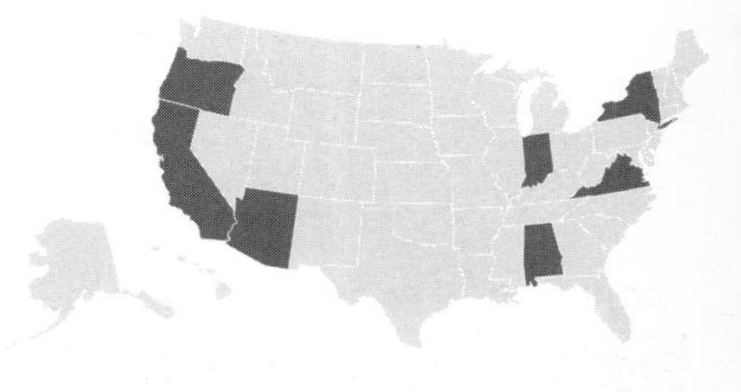

Air/Missile Defense Planning and Control System (AMDPCS)

INVESTMENT COMPONENT

Modernization

Recapitalization

Maintenance

MISSION

To provide an automated command and control system that integrates Air and Missile Defense planning and operations for Air Defense Airspace Management (ADAM) systems, Air Defense Artillery (ADA) Brigades, and Army Air and Missile Defense Commands (AAMDCs).

DESCRIPTION

The Air/Missile Defense Planning and Control System (AMDPCS) is an Army Objective Force system that provides integration of Air and Missile Defense (AMD) operations at all echelons. AMDPCS systems are deployed with ADAM Systems, ADA Brigades, and AAMDCs.

ADAM provides the commanders of Brigade Combat Teams (BCTs), fires brigades, combat aviation brigades, and division and corps tactical operations systems with situation awareness of the airspace, the third dimension of the battlefield. ADAM provides collaboration and staff planning capabilities through the Army Battle Command System and operational links for airspace coordination with joint, interagency, multi-national, and coalition forces.

AMDPCS in ADA Brigades and AAMDCs provide expanded staff planning and coordination capabilities for integrating defense of the air battlespace.

AMDPCS includes shelters, automated data processing equipment, tactical communications, standard vehicles, tactical power, and software systems for force operations and engagement operations including Air and Missile Defense Workstation (AMDWS) and Air Defense System Integrator (ADSI).

AMDWS is a staff planning and battlespace situational awareness tool that provides commanders with a common tactical and operational air picture. ADSI is a fire-control system that monitors and controls air battle engagement operations by subordinate or attached units.

SYSTEM INTERDEPENDENCIES

None

PROGRAM STATUS

- **4QFY08:** ADAM full materiel release and full-rate production decision
- **4QFY08:** Fielding one ADA brigade and 18 ADAMs procured in FY08
- **1QFY09:** Completed FY08 reset of 26 ADAMs

PROJECTED ACTIVITIES

- **4QFY09:** Fielding three ADA Brigades and 46 ADAMs procured in FY09
- **4QFY09:** Field ADSI Version 15.0 with 3D display
- **2QFY10:** Field AMDWS Versions 6.5 with software Block 3 capabilities

ACQUISITION PHASE

Technology Development | Engineering & Manufacturing Development | Production & Deployment | Operations & Support

Air/Missile Defense Planning and Control System (AMDPCS)

FOREIGN MILITARY SALES
None

CONTRACTORS
Northrop Grumman (Huntsville, AL)
Ultra, Inc. (Austin, TX)

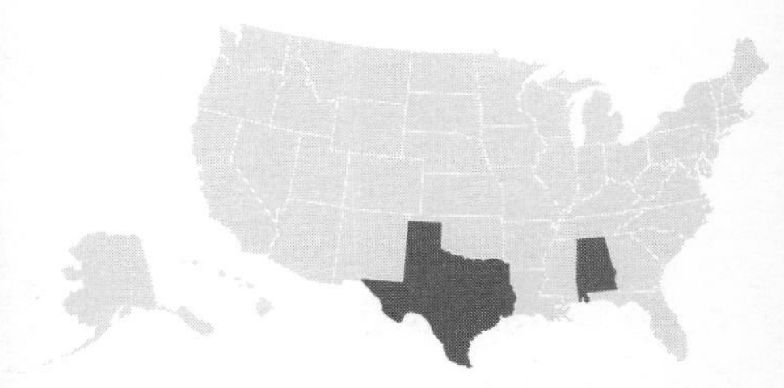

Airborne Reconnaissance Low (ARL)

INVESTMENT COMPONENT

Modernization

Recapitalization

Maintenance

MISSION

To detect, locate, and report threats using a variety of imagery, communications-intercept, and moving-target indicator sensor payloads.

DESCRIPTION

Airborne Reconnaissance Low (ARL) is a self-deploying, multi-sensor, day/night, all-weather reconnaissance, intelligence, system. It consists of a modified DeHavilland DHC-7 fixed-wing aircraft equipped with communications intelligence (COMINT), imagery intelligence (IMINT), and Synthetic Aperture Radar/Moving Target Indicator (SAR/MTI) mission payloads. The payloads are controlled and operated via on-board open-architecture, multi-function workstations.

Intelligence collected on the ARL can be analyzed, recorded, and disseminated on the aircraft workstations in real time and stored on board for post-mission processing. During multi-aircraft missions, data can be shared between cooperating aircraft via ultra high frequency air-to-air data links allowing multi-platform COMINT geolocation operations. The ARL system includes a variety of communications subsystems to support near-real-time dissemination of intelligence and dynamic retasking of the aircraft.

There are currently two configurations of the ARL system:

- Two aircraft are configured as ARL–COMINT (ARL–C), with a conventional communications intercept and direction finding (location) payload.
- Six aircraft are configured as ARL–Multifunction (ARL–M), equipped with a combination of IMINT, COMINT, and SAR/MTI payload and demonstrated hyperspectral imager applications and multi-intelligence (multi-INT) data fusion capabilities.

Southern Command (SOUTHCOM) operates one ARL–C and two ARL–M aircraft. United States Forces Korea (USFK) operates three ARL–M aircraft. Planned upgrades for ARL include baselining the fleet by providing a common architecture for sensor management and workstation man-machine interface. ARL–C systems will be converted from COMINT only to ARL–M multi-INT configuration. Planned sensor improvements include upgrading the radar to provide change detection and super-resolution SAR, upgrading the MX-20 electro-optical/infrared (EO/IR) subsystem to reflect current standards, including the addition of a laser illuminator, and the addition of digital pan cameras across the fleet for high-resolution imaging and change detection. A new and improved COMINT payload will be fielded, increasing frequency coverage and improving target intercept probability.

SYSTEM INTERDEPENDENCIES

None

PROGRAM STATUS

- **2QFY09** Phoenix Eye upgrade on ARL-M1
- **3QFY10** Convert ARL C1 into ARL M8

PROJECTED ACTIVITIES

- **FY09 and on:** Continued imagery, radar, COMINT, system interoperability, workstation architecture upgrades and C to M conversions

ACQUISITION PHASE

Technology Development | Engineering & Manufacturing Development | Production & Deployment | Operations & Support

Airborne Reconnaissance Low (ARL)

FOREIGN MILITARY SALES
None

CONTRACTORS
Sierra Nevada Corp. (Hagerstown, MD)
Aircraft survivability:
Litton Advanced Systems (Gaithersburg, MD)
COMINT subsystem:
BAE Systems (Manchester, NH)
EO/IR subsystem:
WESCAM (Hamilton, Ontario, Canada)
Engineering support:
CACI (Berryville, VA)
Radar subsystem:
Lockheed Martin (Phoenix, AZ)

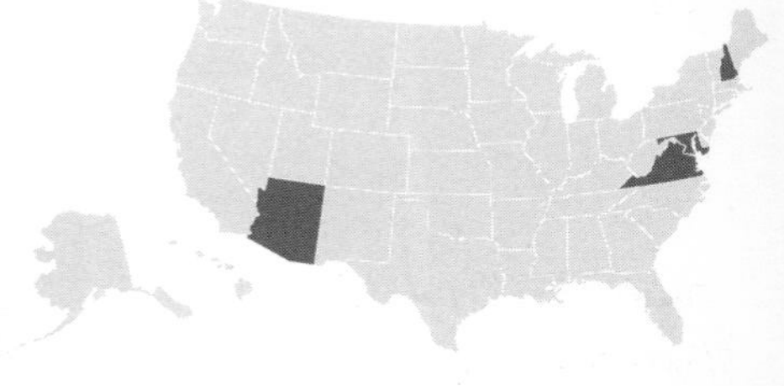

All Terrain Lifter Army System (ATLAS)

INVESTMENT COMPONENT

Modernization

Recapitalization

Maintenance

MISSION

To provide a mobile, variable-reach, rough-terrain forklift capable of handling all classes of supplies.

DESCRIPTION

The All Terrain Lifter Army System (ATLAS) is a C-130 air-transportable, 10,000-pound-capacity, variable-reach, rough-terrain forklift capable of stuffing and un-stuffing 20-foot International Standards Organization (ISO) containers and handling Air Force 463L pallets weighing up to 10,000 pounds. ATLAS supports units from the transportation, quartermaster, ordnance, missiles and munitions, engineer, aviation, and medical army branches. ATLAS's mobility allows it to support the Brigade Combat Teams, and it is a critical asset supporting an expeditionary Army.

ATLAS is a military-unique vehicle: commercial forklifts cannot meet military requirements. It is capable of lifting 4,000 pounds at a 21.5 feet reach, 6,000 pounds at 15 feet, and 10,000 pounds at four feet. It is equipped with two interchangeable fork carriages: a 6,000-pound carriage for stuffing and un-stuffing standard Army pallets with 24-inch load centers from 20-foot containers weighing up to 6,000 pounds; and a 10,000-pound carriage for handling loads weighing up to 10,000 pounds at 48-inch load center (Air Force 463L pallets)

ATLAS is a key component of the Army's Container Oriented Distribution System, which is essential to deployment of a continental U.S.-based Army and sustainment of a deployed force.

The ATLAS II is an EPA Tier III-compliant ATLAS with improved reliability, performance, survivability, and transportability.

ATLAS Features:
Length: 27.02 feet
Width: 8.35 feet (ATLAS II is four inches narrower)
Height: 8.92 feet
Weight: 33,500 pounds
Power Train: 165 horsepower Cummins diesel engine; Funk 1723 PowerShift (three-speed forward and reverse) mechanical transmission
Cruising range: 10 hours of operations before refueling
Road speed: 23 miles per hour
Force protection: Integrated armor

SYSTEM INTERDEPENDENCIES

None

PROGRAM STATUS

- **2QFY07:** ATLAS II contract award; ongoing production and fielding of ATLAS I

PROJECTED ACTIVITIES

- **3QFY09:** Full-rate production ATLAS II

ACQUISITION PHASE

Technology Development | Engineering & Manufacturing Development | Production & Deployment | Operations & Support

All Terrain Lifter Army System (ATLAS)

- TIER III Engine (ATLAS I has TIER I Engine)
- Significant Maintainability Improvements
- Improved Transportability
- 4 inches Narrower

- Integrated A/B Armor Kit
- Electronic Manuals
- Electronic Training Aid

FOREIGN MILITARY SALES
None

CONTRACTORS
JLG Industries, Inc. (McConnellsburg, PA)
An Oshkosh Corporation Company

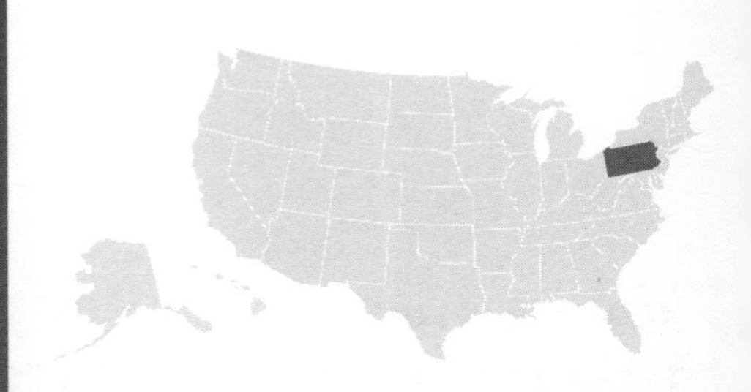

Armored Knight

INVESTMENT COMPONENT

- Modernization
- Recapitalization
- Maintenance

MISSION

To assist heavy and infantry Brigade Combat Teams to perform 24-hour terrain surveillance, target acquisition, target location, and fire support for combat observation lasing team missions.

DESCRIPTION

The M1200 Armored Knight provides precision strike capability by locating and designating targets for both ground- and air-delivered laser-guided ordnance and conventional munitions. It replaces the M707 Knight High Mobility Multipurpose Wheeled Vehicle (HMMWV) base and M981 fire support team vehicles used by combat observation lasing teams (COLTs) in both the heavy and infantry Brigade Combat Teams. It operates as an integral part of the brigade reconnaissance element, providing COLT and fire support mission planning and execution.

The M1200 Armored Knight is a M117 Armored Security Vehicle (ASV) chassis/hull with Add-on Armor fragmentation kits installed. This provides enhanced survivability and maneuverability over the unarmored M707 HMMWV based Armored Knight. The system includes a full 360-degree armored cupola and integrated Knight mission equipment package that is common with the M7 Bradley Fire Support Team (BFIST) vehicle/M707 Knight and the Stryker Fire Support Vehicle.

The mission equipment package includes: Fire Support Sensor System (FS3) mounted sensor, Targeting Station Control Panel, Mission Processor Unit, Inertial Navigation Unit, Defense Advanced Global Positioning System Receiver (DAGR), Power Distribution Unit, Rugged Hand-Held Computer Unit (RHC) Forward Observer Software (FOS).

Other Armored Knight specifications:
Crew: Three COLT members
Combat loaded weight: Approximately 15 tons
Maximum speed: 63 miles per hour
Cruising range: 440 miles
Target location accuracy: <20 meters circular error probable

SYSTEM INTERDEPENDENCIES

ASV Chassis, Lightweight Laser Designator Rangefinder (LLDR), FS3, FOS, Force XXI Battle Command–Brigade and Below (FBCB2), Advanced Field Artillery Tactical Data System (AFATAD3), Single Channel Ground to Air Radio Station (SINCGARS)

PROGRAM STATUS

- **1QFY10:** 294 Vehicle Systems produced. 278 Vehicle Systems to various deploying Heavy Brigade Combat Teams (HBCTs) and Infantry Brigade Combat Teams (IBCTs)

PROJECTED ACTIVITIES

- **2QFY10–FY12:** Procure additional 188 Vehicle Systems. 438 total systems fielded to next deployers, HBCT/IBCT in Active Component (AC) and Army National Guard (ARNG)
- **FY12:** Design/Integrate/Validate Targeting under Armor–On the Move/Remote Weapon Station (TUA-OTM/RWS) capability for M1200 Armored Knight

ACQUISITION PHASE

- Technology Development
- Engineering & Manufacturing Development
- Production & Deployment
- Operations & Support

Armored Knight

FOREIGN MILITARY SALES
None

CONTRACTORS
Precision targeting systems production/vehicle integration:
DRS Sustainment Systems, Inc. (DRS–SSI) (St. Louis, MO; West Plains, MO)
Common display unit:
DRS Tactical Systems (Melbourne, FL)
Slip ring:
Airflyte Electronics Co. (Bayonne, NJ)
Targeting station control panel:
Oppenheimer (Horsham, PA)
M1117 ASV Hull:
Textron Marine & Land Systems (New Orleans, LA)
FS3 Sensor:
Raytheon (McKinney, TX)
Inertial Navigation Unit:
Honeywell (Clearwater, FL)

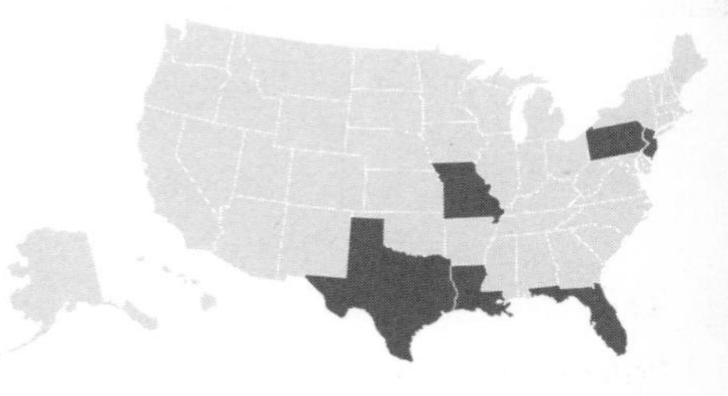

Armored Security Vehicle (ASV)

INVESTMENT COMPONENT

Modernization

Recapitalization

Maintenance

MISSION

To support the entire spectrum of military police missions and to protect convoys in hostile areas.

DESCRIPTION

The M1117 Armored Security Vehicle (ASV) is a turreted, lightly armored, all-wheel drive vehicle that supports military police and convoy missions, such as rear area security, law and order operations, convoy protection, battlefield circulation, and enemy prisoner of war operations, over the entire spectrum of war and operations other than war.

The ASV provides protection to the crew compartment, gunner's station, and the ammunition storage area. The turret is fully enclosed with an MK-19 40mm grenade launcher gun, a M48 .50-caliber machine gun, and a multi-salvo smoke grenade launcher. The ASV provides ballistic, blast, and overhead protection for its four-person crew. The ASV has a payload of 3,360 pounds and supports Army transformation with its 400-mile-plus range, top speed of nearly 70 miles per hour, and C-130 deployability.

SYSTEM INTERDEPENDENCIES

None

PROGRAM STATUS

- **Current:** Continued fielding to support military police companies and convoy protection units

PROJECTED ACTIVITIES

- **FY10:** Produce and field approximately 150 vehicles; upgrade turret capabilities

ACQUISITION PHASE

Technology Development | Engineering & Manufacturing Development | Production & Deployment | Operations & Support

Armored Security Vehicle (ASV)

FOREIGN MILITARY SALES
ASV variant delivered to Iraq, M1117 delivered to Iraq

CONTRACTORS
Textron Marine & Land Systems (New Orleans, LA)
BAE Systems (Phoenix, AZ)
Cummins Mid-South LLC (Memphis, TN)
Chenega (Panama City, FL)
Lapeer Industries Inc. (Lapeer, MI)

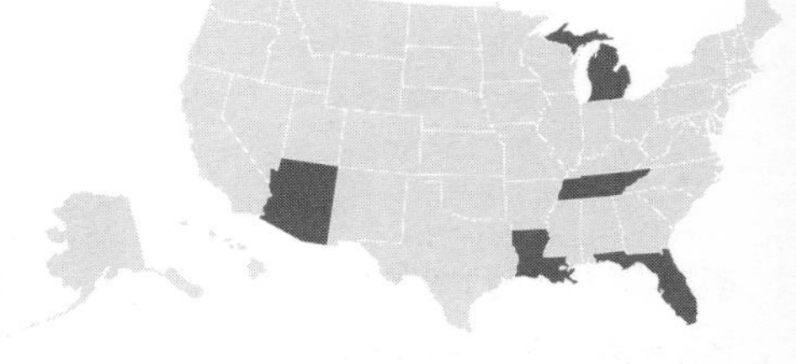

Army Key Management System (AKMS)

INVESTMENT COMPONENT

Modernization

Recapitalization

Maintenance

MISSION

To automate the functions of communication securities (COMSEC) key management, control, and distribution; electronic protection generation and distribution; signal operating instruction management; to provide planners and operators with automated, secure communications at theater/tactical and strategic/sustaining base levels.

DESCRIPTION

The Army Key Management System (AKMS) is a fielded system composed of three subsystems, Local COMSEC Management Software (LCMS), Automated Communications Engineering Software (ACES), and the Data Transfer Device (DTD)/ Simple Key Loader (SKL). Under the umbrella of the objective National Security Agency (NSA) Electronic Key Management System (EKMS), AKMS provides tactical units and sustaining bases with an organic key generation capability and an efficient secure electronic key distribution means. AKMS provides a system for distribution of communications security (COMSEC), electronic protection, and signal operating instructions (SOI) information from the planning level to the point of use in support of current, interim, and objective force at division and brigade levels.

The LCMS workstation provides automated key generation, distribution, and COMSEC accounting. The ACES, which is the frequency management portion of AKMS, has been designated by the Military Communications Electronics Board as the joint standard for use by all services in development of frequency management and cryptographic net planning and SOI generation. The SKL (AN/PYQ-10) is the associated support item of equipment that provides the interface between the ACES workstation, the LCMS workstation, the warfighter's End Crypto Unit (ECU), and the Soldier. It is a small, ruggedized hand-held key loading device.

Product Direct Network Operations (PD NetOps) will deploy and sustain the Coalition Joint Spectrum Management Planning Tool (CJSMPT), which began as a Joint Improvised Explosive Device Defeat (JIEDD) Task Force initiative. It will provide joint spectrum management deconfliction capabilities for both communications and electronic warfare spectrum users.

SYSTEM INTERDEPENDENCIES

AKMS systems are considered enabling systems for equipment/systems to receive key and frequency allotments.

PROGRAM STATUS

- **4QFY08:** Released version 6.0 of SKL software
- **4QFY08:** Released LCMS v5.0.3 to Army COMSEC custodians
- **4QFY08:** Procured over 40,000 SKLs for Army units
- **1QFY09:** Released version 1.9 of ACES software

PROJECTED ACTIVITIES

- **FY09–11:** Continue to procure and field SKLs for Army, Air Force, Navy and civilians
- **FY09:** SKL Software upgrade v6.0; ACES Software Upgrade 2.0; LCMS Software Upgrade v5.1
- **FY09:** Refresh LCMS hardware to all Army COMSEC custodians
- **FY09–10:** Train, deploy and sustain Coalition Joint Spectrum Management Planning Tool (CJSMPT) capability
- **FY10:** Refresh ACES hardware to current users

ACQUISITION PHASE

Technology Development | Engineering & Manufacturing Development | Production & Deployment | Operations & Support

Army Key Management System (AKMS)

FOREIGN MILITARY SALES
Canada, United Kingdom, Australia, New Zealand, Hungary, Germany, Turkey, Spain, Lithuania

CONTRACTORS
Sierra Nevada Corp. (Sparks, NV)
Science Applications International Corp. (SAIC) (San Diego, CA)
CACI (Eatontown, NJ)
Sypris (Tampa, FL)
CSS (Augusta, GA)

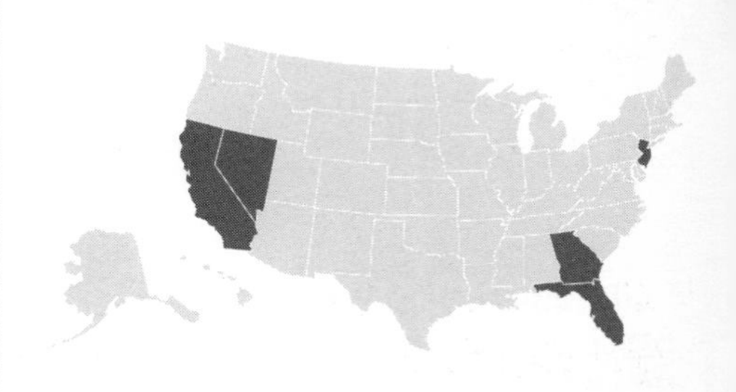

Artillery Ammunition

INVESTMENT COMPONENT

Modernization

Recapitalization

Maintenance

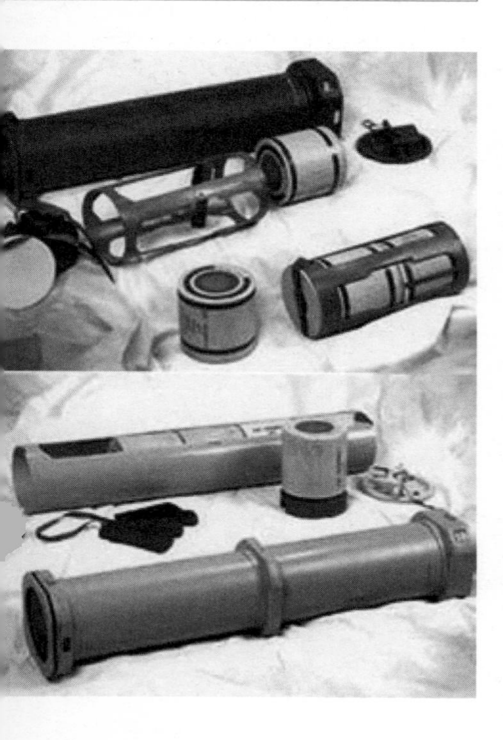

MISSION

To provide field artillery forces with modernized munitions to destroy, neutralize, or suppress the enemy by cannon fire.

DESCRIPTION

The Army's artillery ammunition program includes 75mm (used for ceremonies and simulated firing), 105mm, and 155mm projectiles and their associated fuzes and propelling charges.

Semi-fixed ammunition for short and intermediate ranges, used in 105mm howitzers, is characterized by adjusting the number of multiple propelling charges. Semi-fixed ammunition for long ranges contains a single bag of propellant optimized for obtaining high velocity and is not adjustable. The primer is an integral part of the cartridge case, and is located in the base. All 105mm cartridges are issued in a fuzed or unfuzed configuration. Both cartridge configurations are packaged with propellant.

Separate-loading ammunition, used in 155mm howitzers, has separately issued projectiles, fuzes, propellants, and primers, which are loaded into the cannon separately.

The artillery ammunition program includes fuzes for cargo-carrying projectiles, such as smoke, illumination, dual-purpose improved conventional munitions, and bursting projectiles, such as high explosive. This program also includes bag propellant for the 105mm semi-fixed cartridges and modular artillery charge system (MACS) for 155mm howitzers.

SYSTEM INTERDEPENDENCIES

None

PROGRAM STATUS

- **2QFY09:** Completed type classification of the 105mm M1064 Infrared (IR) Illumination cartridge

PROJECTED ACTIVITIES

- **4QFY09:** Complete type classification of the 155mm M1066 IR Illumination projectile.
- **2QFY10:** Complete full material release of the 105mm M1064 IR Illumination cartridge

ACQUISITION PHASE

Technology Development | **Engineering & Manufacturing Development** | Production & Deployment | Operations & Support

Artillery Ammunition

FOREIGN MILITARY SALES
Australia, Canada, Israel, and Lebanon

CONTRACTORS
General Dynamics Ordnance and Tactical Systems–Scranton Operations (Scranton, PA)
SNC Technologies (LeGardeur, Canada)
American Ordnance (Middletown, IA)
Alliant Techsystems (Janesville, WI)
Armtec Defense (Palm Springs, CA)

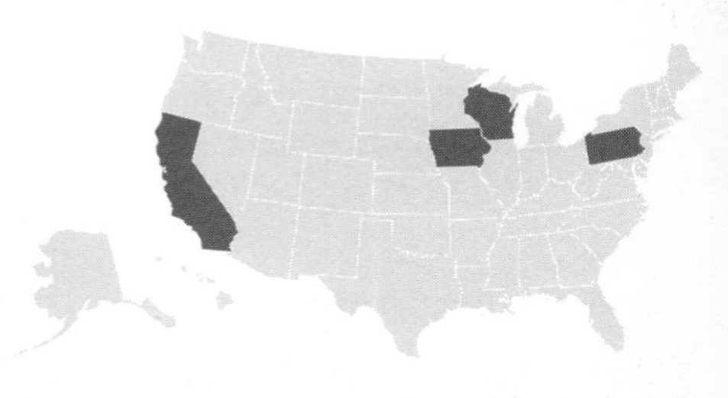

Aviation Combined Arms Tactical Trainer (AVCATT)

INVESTMENT COMPONENT

Modernization

Recapitalization

Maintenance

MISSION

To enable Army aviation units to rehearse and participate in a unit-collective and combined-arms simulated battlefield environment through networked simulation training.

DESCRIPTION

The Aviation Combined Arms Tactical Trainer (AVCATT) is a reconfigurable, transportable, combined-arms virtual training simulator that provides current and Future Force aviation commanders and units a dynamic, synthetic instructional environment. AVCATT enables realistic, high-intensity collective and combined arms training to aviation leadership, staff members and units, improving overall aviation task force readiness. It meets institutional, organizational, and sustainment aviation training requirements for Active and Reserve Army aviation units worldwide and enables geographic-specific mission rehearsals in both classified and unclassified modes before real-world mission execution. AVCATT is a critical element of the Combined Arms Training Strategy. It is distributive interactive simulation (DIS) and high-level architecture (HLA) compliant, and is compatible and interoperable with other synthetic environment systems. AVCATT supports role-player and semi-automated blue and opposing forces.

The AVCATT single suite of equipment consists of two mobile trailers that house six reconfigurable networked simulators to support the Apache, Apache Longbow, Kiowa Warrior, Chinook, and Black Hawk. An after-action review theater and battle master control station is also provided as part of each suite.

AVCATT builds and sustains training proficiency on mission-essential tasks through crew and individual training by supporting aviation collective tasks, including armed reconnaissance (area, zone, route); deliberate attack; covering force operations; downed aircrew recovery operations; joint air attack team; hasty attack; and air assault operations.

AVCATT is fully mobile, capable of using commercial and generator power, and is transportable worldwide.

SYSTEM INTERDEPENDENCIES

AVCATT requires Synthetic Environment Core (SE Core) to provide terrain databases and virtual models. The One Semi-Automated Forces (OneSAF) will provide a common SAF through SE Core in the future.

PROGRAM STATUS

- **1QFY09:** Fielding of 19 suites completed; includes support to Army National Guard as well as U.S. forces in Germany, Korea, and Hawaii

PROJECTED ACTIVITIES

- **2QFY09:** Initiate development of Digital Communications Baseline, Tactical Message Format
- **2QFY09:** Initiate development of Kiowa Warrior concurrency upgrade
- **3QFY09:** Field the first system with upgraded visual system: helmet-mounted display and image generator
- **3QFY09:** Field production suites 20, 21 and 22
- **4QFY09:** Field production suite 23
- **2QFY10:** Field the first system with Longbow Block I, Software 6.1 and Block II, Software 10.0 concurrency upgrade
- **2QFY10:** Complete fielding of Common Missile Warning System (CMWS) to all AVCATT systems

ACQUISITION PHASE

Technology Development | Engineering & Manufacturing Development | Production & Deployment | Operations & Support

Aviation Combined Arms Tactical Trainer (AVCATT)

AVCATT–A
2 Trailer Suite

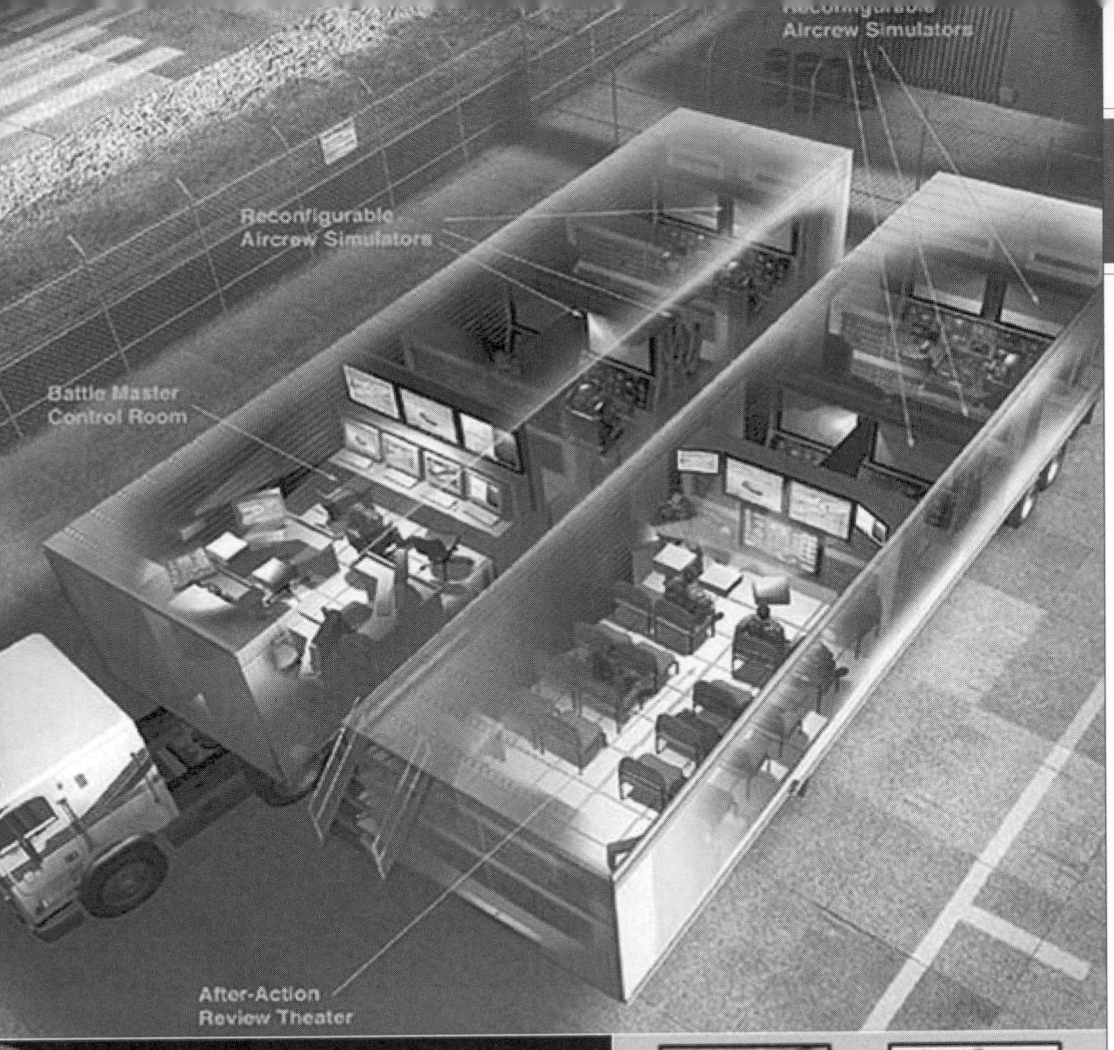

Manned Simulator

Apache
AH-64A

Kiowa Warrior
OH-58D

Chinook
Ch-47D

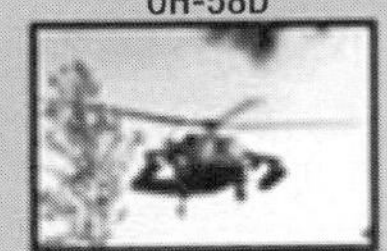

Black Hawk
UH-60A/L

Apache Longbow

FOREIGN MILITARY SALES
None

CONTRACTORS
L-3 Communications (Arlington, TX)

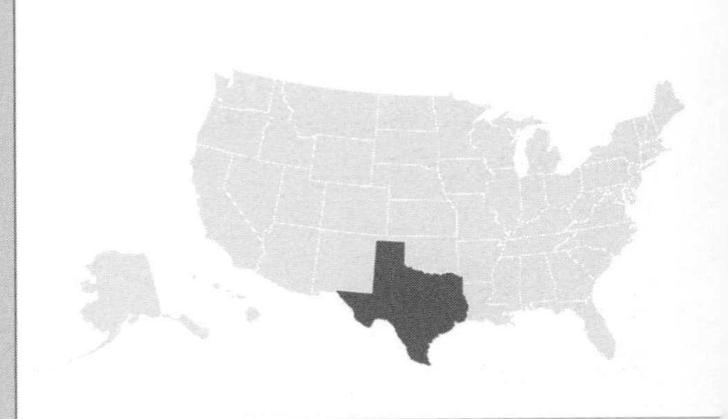

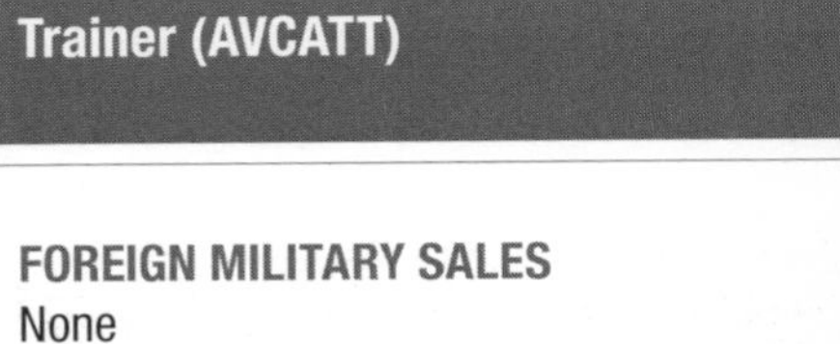

Battle Command Sustainment Support System (BCS3)

INVESTMENT COMPONENT

Modernization

Recapitalization

Maintenance

MISSION

To serve as the United States land forces' fusion center for Logistics command and control information, providing better situational awareness and decision-making capability to U.S. Land Component Forces at tactical, operational and strategic echelons.

DESCRIPTION

The Battle Command Logistics Command and Control (BCS3) system is employed at multiple echelons to fuse sustainment, in-transit, and force data to aid commanders in making critical tactical, operational, and strategic decisions. BCS3 is a force multiplier, a precision tool for logistics planning and execution that provides commanders with the situational awareness to make prudent decisions, rapidly and effectively, for today's fight and tomorrow's mission.

BCS3 is an integral part of Army Battle Command System (ABCS) automation, providing the logistics portion of the battle command common operational picture (COP).

BCS3 provides a Microsoft Windows-type COP for logistics that is modular, tailorable, and scalable to meet the full spectrum of battlefield logistics command and control requirements in near-real-time. It incorporates relevant technologies developed over the past 10 years with emerging logistics technologies and applications used today worldwide and supports U.S. land forces deployed in Iraq and Afghanistan. BCS3 will continue development while integrating into the Early Infantry Brigade Combat Team (E-IBCT) and Net Enabled Command Capability (NECC) architectures to provide commanders the capability to execute end-to-end distribution and deployment management for better situational awareness.

The system supports training, mission planning, rehearsal, and execution all in one tool and operates in both unclassified and classified environments. BCS3 interfaces with other Army, Joint Interagency and Multinational (JIM) command and control (C2) and logistics business systems. In effect, the system advances the goal to "increase Battle Command capability for U.S. land forces in joint full spectrum operations" as laid out in the Army's *Strategic Planning Guidance*.

BCS3's core competencies provide units, staffs, and commanders with the best warfighting capability now through the logistics COP; commodity visibility; convoy operations; reception, staging, onward movement; and logistics reporting.

SYSTEM INTERDEPENDENCIES

LIW/LOGSA, ILAP, SARSS, SAMS(E), SASS–MOD, PBUSE, EMILPO, MTS, RFID

PROGRAM STATUS

- **1QFY09:** Release of BCS3 BC08.10.02.03

PROJECTED ACTIVITIES

- **2QFY09:** Fielding to 92nd Brigade Combat Team
- **2QFY09:** Fielding to 6th Army
- **3QFY09:** Fielding to 5th Army
- **3QFY09:** Release of BCS3 BC08.10.02.04
- **4QFY09:** Release of BCS3 BC10.01.00
- **2QFY10:** Fielding to 167th Theater Sustainment Command
- **3QFY10:** Fielding to 135th Sustainment Command (Expeditionary)
- **3QFY10:** Fielding to 184th Sustainment Command (Expeditionary)

ACQUISITION PHASE

Technology Development | Engineering & Manufacturing Development | Production & Deployment | Operations & Support

Battle Command Sustainment Support System (BCS3)

FOREIGN MILITARY SALES
None

CONTRACTORS
Northrop Grumman (Carson, CA)
Tapestry Solutions (San Diego, CA)
L-3 Communications (Chantilly, VA)
Lockheed Martin (Tinton Falls, NJ)
Wexford Group International (Vienna, VA)

Biometric Enterprise Core Capability (BECC)

INVESTMENT COMPONENT

- **Modernization**
- Recapitalization
- Maintenance

MISSION

To serve as an enterprise biometric system acting as DoD's authoritative biometric repository enabling identity superiority.

DESCRIPTION

BECC will be developed with a system-of-systems architecture using multi-modal storage and matching using fingerprint, palm, iris, and face modalities.

SYSTEM INTERDEPENDENCIES

Joint Biometrics Identity Intelligence Program, Identity Dominance System, Biometric Family of Capabilities for Full Spectrum Operations

PROGRAM STATUS

- **4QFY08:** DoD Biometrics Acquisition Decision Memorandum directs Milestone B no later than FY10
- **1QFY09:** Biometrics in Support of Identity Management Initial Capabilities Document approved by Joint Requirements Oversight Council

PROJECTED ACTIVITIES

- **2–3QFY09:** Biometric analysis of alternatives
- **1QFY10:** Biometrics Capability Development Document(s) approved
- **3QFY10:** Milestone B, i.e. permission to enter system development and demonstration

ACQUISITION PHASE

- **Technology Development**
- Engineering & Manufacturing Development
- Production & Deployment
- Operations & Support

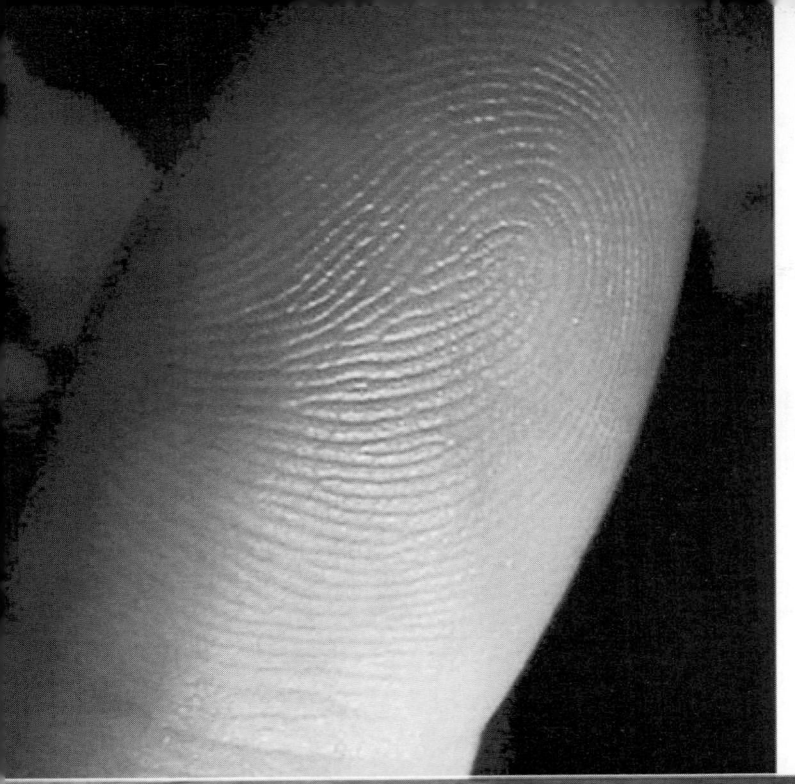

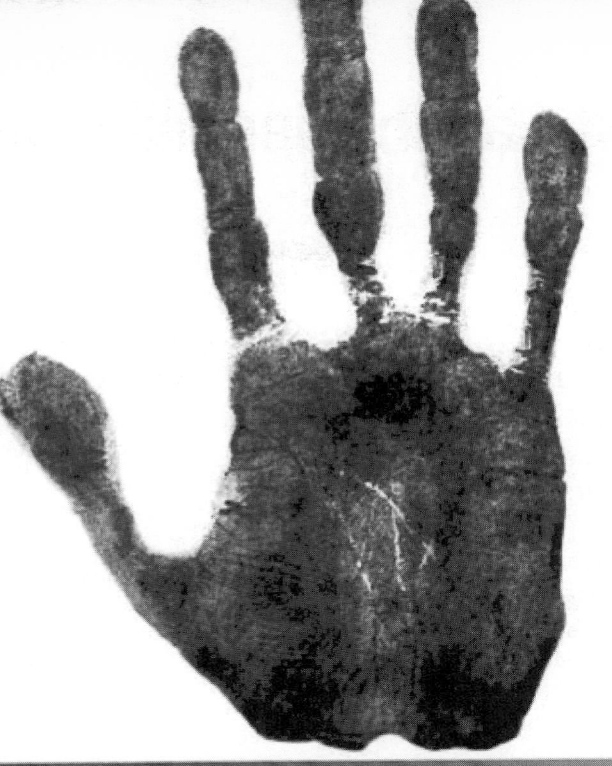

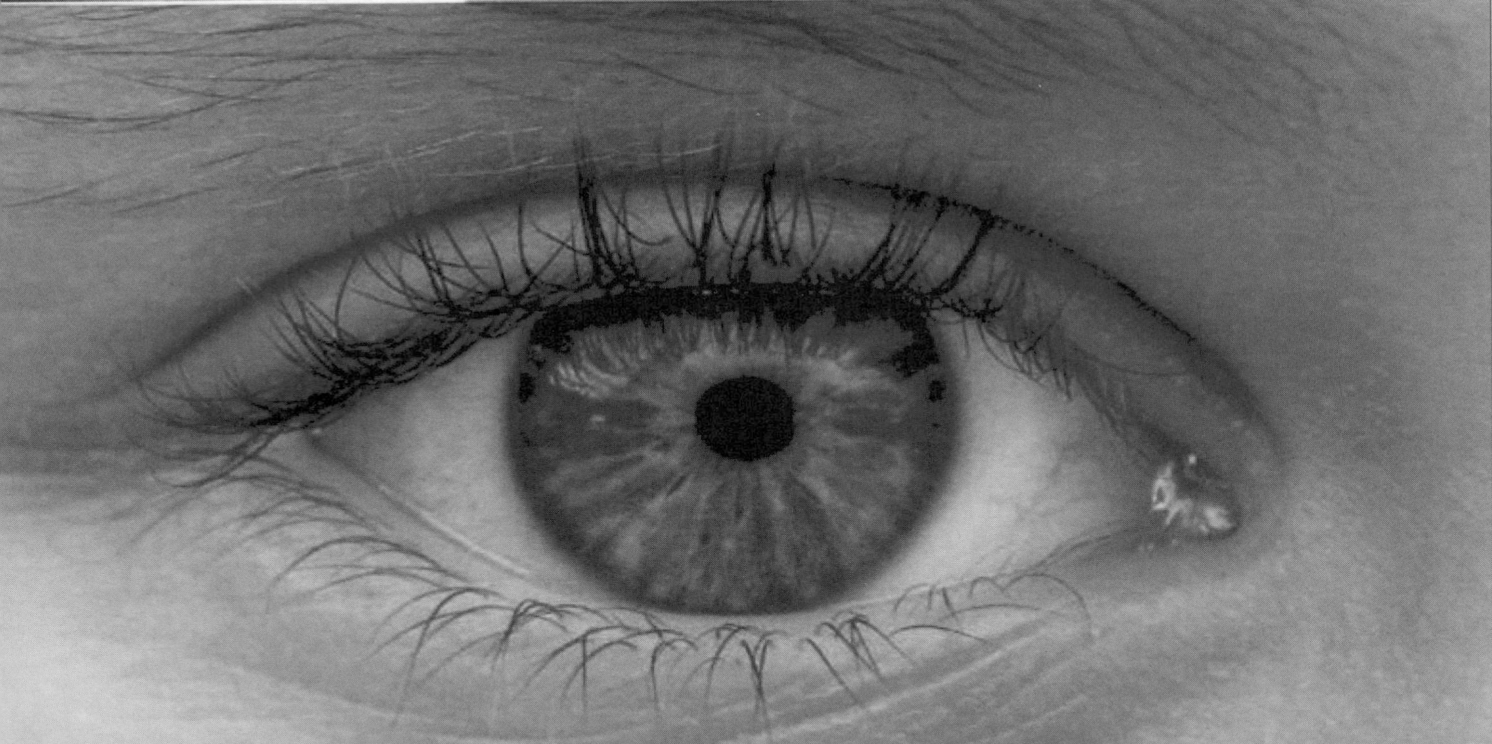

Biometric Enterprise Core Capability (BECC)

FOREIGN MILITARY SALES
None

CONTRACTORS
Program Management Support Services:
L-3 Communications (Canton, MA)
CACI (Arlington, VA)
General Dynamics (Falls Church, VA)
The Research Associates (New York, NY)
System Development and Integration:
To be determined pending Milestone B

Biometric Family of Capabilities for Full Spectrum Operations (BFCFSO)

INVESTMENT COMPONENT

Modernization

Recapitalization

Maintenance

MISSION

To provide tactical biometrics collection capability configurable for multiple operational mission environments, enabling identity superiority.

DESCRIPTION

Biometric Family of Capabilities for Full Spectrum Operations (BFCFSO) will provide the capability to capture, transmit, store, share, retrieve, exploit, and display biometrics data from multiple targets.

SYSTEM INTERDEPENDENCIES

Joint Biometrics Identity Intelligence Program, Identity Dominance System, Biometric Enterprise Core Capability, Distributed Common Ground System–Army

PROGRAM STATUS

- **4QFY08:** DoD Biometrics Acquisition Decision Memorandum directs Milestone B no later than FY10
- **1QFY09:** Biometrics in Support of Identity Management Initial Capabilities Document approved by Joint Requirements Oversight Council

PROJECTED ACTIVITIES

- **2–3QFY09:** Biometric analysis of alternatives
- **1QFY10:** Biometrics Capability Development Document(s) approved
- **3QFY10:** Milestone B, i.e. permission to enter system development and demonstration

ACQUISITION PHASE

Technology Development | Engineering & Manufacturing Development | Production & Deployment | Operations & Support

Biometric Family of Capabilities for Full Spectrum Operations (BFCFSO)

FOREIGN MILITARY SALES
None

CONTRACTORS
Program Management Support Services:
L-3 Communications (Canton, MA)
CACI (Arlington, VA)
General Dynamics (Falls Church, VA)
The Research Associates (New York, NY)
System Development and Integration:
To be determined pending Milestone B

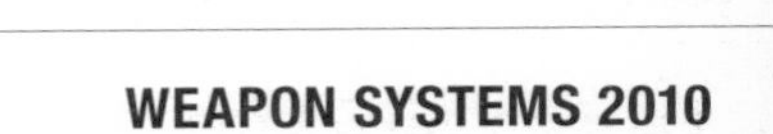

Black Hawk/UH-60

INVESTMENT COMPONENT

Modernization (UH-60M)

Recapitalization (UH-60 A-A/L)

Maintenance

MISSION

To provide air assault, general support, aero-medical evacuation, command and control, and special operations support to combat, stability, and support operations.

DESCRIPTION

The Black Hawk (UH-60) is the Army's utility tactical transport helicopter. The versatile Black Hawk has enhanced the overall mobility of the Army due to dramatic improvements in troop capacity and cargo lift capability. It will serve as the Army's utility helicopter in the Future Force. There are four basic versions of the UH-60: the original UH-60A; the UH-60L, which has greater gross weight capability, higher cruise speed, rate of climb, and external load; the UH-60M, which includes the improved GE-701D engine and provides greater cruising speed, rate of climb, and internal load than the UH-60A and L versions; and the UH-60M P3I Upgrade, which includes the Common Avionics Architecture System, and fly-by-wire and Full Authority Digital Engine Control upgrade to the GE-701D Engine. On the asymmetric battlefield, the Black Hawk enables the commander to get to the fight quicker and to mass effects throughout the battlespace across the full spectrum of conflict. A single Black Hawk can transport an entire 11-person, fully equipped infantry squad faster than predecessor systems and in most weather conditions. The Black Hawk can reposition a 105mm howitzer, its crew of six, and up to 30 rounds of 105mm ammunition in a single lift. The aircraft's critical components and systems are armored or redundant, and its airframe is designed to crush progressively on impact, thus protecting crew and passengers. The Army has put programs into place to extend the life of the UH-60 by providing it with the capabilities needed on the future battlefield. The UH-60M upgrade program will incorporate a digitized cockpit and improved handling characteristics, and will extend the system service life. The UH-60A-A/L recapitalization and rebuild program will be applied to a number of older aircraft while awaiting introduction of UH-60M aircraft into the fleet.

SYSTEM INTERDEPENDENCIES

Advanced Threat Infrared Countermeasures (ATRICM), Common Missile Warning System (CMWS), Air Warrior, Blue Force Tracker (BFT), and Joint Tactical Radio System (JTRS)

PROGRAM STATUS

- **1QFY08:** UH-60M multiyear VII award
- **2QFY08:** UH-60M first-unit equipped
- **4QFY08:** UH-60M upgrade first flight
- **4QFY08:** UH-60M upgrade customer test

PROJECTED ACTIVITIES

- **1QFY10:** UH-60M upgrade limited user test
- **2QFY10:** UH-60M upgrade low-rate production contract award

ACQUISITION PHASE

Technology Development | Engineering & Manufacturing Development | Production & Deployment | Operations & Support

Black Hawk/UH-60

FOREIGN MILITARY SALES
UH-60M: Bahrain, UAE

CONTRACTORS
UH-60M:
Sikorsky (Stratford, CT)
UH-60M Upgrade Development:
Sikorsky (Stratford, CT)
701D Engine:
General Electric (Lynn, MA)
CAAS Software:
Rockwell Collins (Cedar Rapids, IA)
Flight Control:
Hamilton Sundstrand (Windsor Locks, CT)

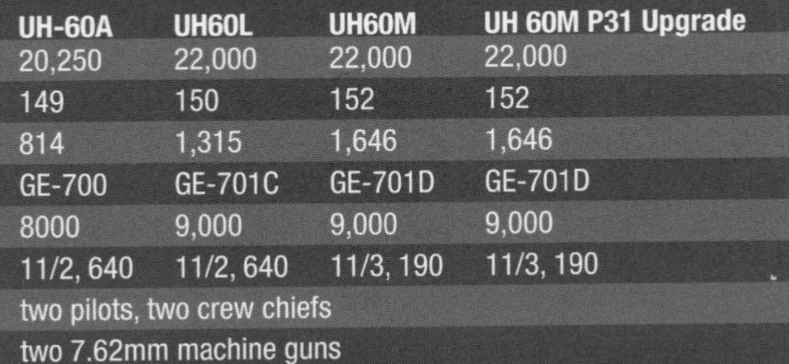

	UH-60A	UH60L	UH60M	UH 60M P31 Upgrade
MAX GROSS WEIGHT (pounds):	20,250	22,000	22,000	22,000
CRUISE SPEED (knots):	149	150	152	152
RATE CLIMB (feet per minute):	814	1,315	1,646	1,646
ENGINES (2 each):	GE-700	GE-701C	GE-701D	GE-701D
EXTERNAL LOAD (pounds):	8000	9,000	9,000	9,000
INTERNAL LOAD (troops/pounds):	11/2, 640	11/2, 640	11/3, 190	11/3, 190
CREW:	two pilots, two crew chiefs			
ARMAMENT:	two 7.62mm machine guns			

Bradley Upgrade

INVESTMENT COMPONENT

Modernization
Recapitalization
Maintenance

MISSION

To provide infantry and cavalry fighting vehicles with digital command and control capabilities, significantly increased situational awareness, enhanced lethality and survivability, and improved sustainability and supportability.

DESCRIPTION

The Bradley M2A3 Infantry/M3A3 Cavalry Fighting Vehicle (IFV/CFV) features two second-generation, forward-looking infrared (FLIR) sensors—one in the Improved Bradley Acquisition Subsystem (IBAS), the other in the Commander's Independent Viewer (CIV). These systems provide "hunter-killer target handoff" capability with ballistic fire control. The Bradley A3 also has embedded diagnostics and an Integrated Combat Command and Control (IC3) digital communications suite hosting a Force XXI Battle Command Brigade-and-Below (FBCB2) package with digital maps, messages, and friend/foe situational awareness. The Bradley's position navigation with GPS, inertial navigation, and enhanced squad situational awareness includes a squad leader display integrated into vehicle digital images and IC3.

SYSTEM INTERDEPENDENCIES

None

PROGRAM STATUS

- **1QFY09:** Bradley A3 fielded to 1st Armored Division; Bradley Operation Desert Storm (ODS) fielded to the 155th MS Army National Guard (ARNG)
- **1QFY09:** Bradley A3 fielded to 3rd Brigade, 3rd Infantry Division

PROJECTED ACTIVITIES

- **1QFY10:** Bradley A3 fielded to 1st Brigade, 4th Infantry Division; and ODS fielded to 170th Separate Infantry Brigade
- **2QFY10:** Bradley A3 fielded to 4th Brigade, 1st Cavalry Division and 2nd Brigade, 4th Infantry Division; and ODS fielded to 1st Brigade, 1st Infantry Division and 2nd Brigade, 1st Infantry Division
- **3QFY10:** Bradley A3 fielded to Army Prepositioned Stock 5, 3rd Brigade, 1st Cavalry Division; 2nd Brigade, 1st Cavalry Division; and ODS fielded to 172nd Separate Infantry Brigade
- **4QFY10:** Bradley A3 fielded to 1st Brigade, 1st Cavalry Division; and ODS Situational Awareness (ODS SA) fielded to 81st Washington Army National Guard
- **Current:** Bradley conversions continue for both the Active Army and the ARNG to meet the Army's modularity goals; A3 Bradley is in full-rate production through 3QFY11.

ACQUISITION PHASE

Technology Development | Engineering & Manufacturing Development | Production & Deployment | Operations & Support

Bradley Upgrade

FOREIGN MILITARY SALES
None

CONTRACTORS
BAE Systems (San Jose, CA)
Raytheon (McKinney, TX)
DRS Technologies (Melbourne, FL)
Elbit Systems of America (Fort Worth, TX)
L-3 Communications (Muskegon, MI)

LENGTH: 21.5 feet
WIDTH: 9.75 feet without armor tiles; 10.83 feet with armor tiles
HEIGHT: 11.8 feet
WEIGHT: 67,000 pounds combat loaded; 78,925 pounds with add-on armor, Bradley reactive tiles, and skirt armor to protect from explosively formed penetrators
POWER TRAIN: 600 hp Cummins VTA-903T diesel engine with L-3 Communications HMPT-500-3EC hydro-mechanical automatic transmission
CRUISING RANGE: 250 miles
ROAD SPEED: 38 miles per hour
CREW, PASSENGERS: M2A3: 10 (3 crew; 7 dismounts); M3A3: 5 (3 crew; 2 dismounts)
VEHICLE ARMAMENT: 25mm Bushmaster cannon; TOW II missile system; 7.62mm M240C machine gun

CURRENT MODELS/VARIANTS:
- M2/M3 A2
- M2/M3 A2 Operation Desert Storm (ODS)
- M2/M3 A3
- M2/M3 A2 ODS–E (Engineer Vehicle)
- Bradley Commander's Vehicle (BCV)
- M7 ODS Bradley Fire Support Team (BFIST)
- M3A3 Bradley Fire Support Team (BFIST)

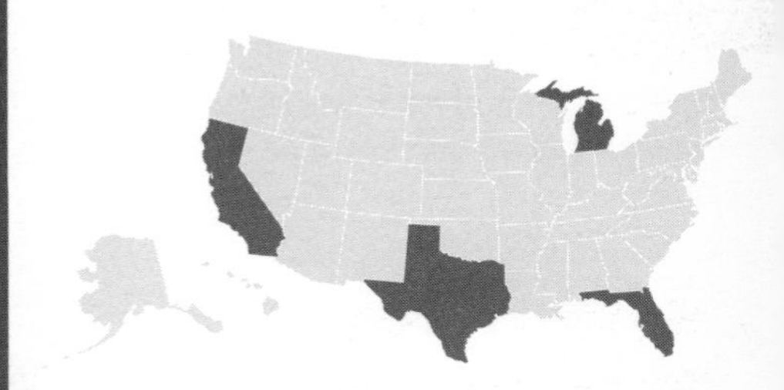

Calibration Sets Equipment (CALSETS)

INVESTMENT COMPONENT

Modernization

Recapitalization

Maintenance

MISSION

To provide the capability to test, adjust, synchronize, repair, and verify accuracy of Army test, measurement, and diagnostic equipment across all measurement parameters.

DESCRIPTION

Calibration Sets Equipment (CALSETS) consist of fixed and tactical shelters that house all instrumentation, components, and power generation equipment constituting a set. Calibration sets are capable of providing support to maintenance units or area support from brigade to multi-theater sustainment operations. The calibration sets are designed to calibrate 90 percent of the Army's test, measurement, and diagnostic equipment workload with an objective of 98 percent. The calibration sets Secondary Transfer Standards are deployed worldwide and are used to verify and transfer precision accuracy to the current and Future Force.

Secondary Transfer Standards Basic, AN/GSM-286
This set consists of baseline instruments and components capable of supporting precision maintenance equipment in the physical, dimensional, electrical, and electronic parameters.

Secondary Transfer Standards Augmented, AN/GSM-287
This set consists of baseline instruments and components with expanded capability to support a wider variety of precision maintenance equipment. It is capable of supporting precision maintenance equipment in the physical, dimensional, electrical, electronic, radiological, electro-optical, and microwave frequency parameters.

Secondary Transfer Standards, AN/GSM-421
This calibration platform consists of a M1152 High Mobility Multipurpose Wheeled Vehicle (HMMWV) with a mounted shelter and integrated 10-kilowatt power generator. This platform is equipped with a basic set of precision maintenance calibration standards designed to support up to 75 percent of the Army's high density precision measurement equipment. This system is modular and configurable to meet mission requirements and can operate in a true split-based mission posture. Designed for rapid deployment by surface or air, AN/GSM-21 will not radiate or be disrupted by electromagnetic interference.

Secondary Transfer Standards, AN/GSM-705
This calibration platform consists of a M1088A1 Medium Tactical Vehicle Tractor with a 35-foottrailer and integrated 15-kilowatt generator. Outfitted as a tactical mobile calibration system, it contains the baseline and expanded-issue instruments and components. The platform includes battlefield communication and applies a network-centric approach to precision maintenance support operations and data handling.

SYSTEM INTERDEPENDENCIES

None

PROGRAM STATUS

- **Current:** Sustainment of CALSETS Secondary Transfer Standards Basic, AN/GSM-286; Secondary Transfer Standards Augmented, AN/GSM-287; Secondary Transfer Standards, AN/GSM-421; Secondary Transfer Standards, AN/GSM-705
- **Current:** Production and fielding of CALSETS Secondary Transfer Standards, AN/GSM-705 (National Guard)
- **Current:** System development and demonstration of CALSETS Secondary Transfer Standards, AN/GSM-421(V2)

PROJECTED ACTIVITIES

- **2QFY09:** Total-package fielding to National Guard of CALSETS Secondary Transfer Standards, AN/GSM-705
- **3QFY10:** Production and fielding of CALSETS Secondary Transfer Standards, AN/GSM-421(V2)

ACQUISITION PHASE

Technology Development | Engineering & Manufacturing Development | Production & Deployment | Operations & Support

Calibration Sets Equipment (CALSETS)

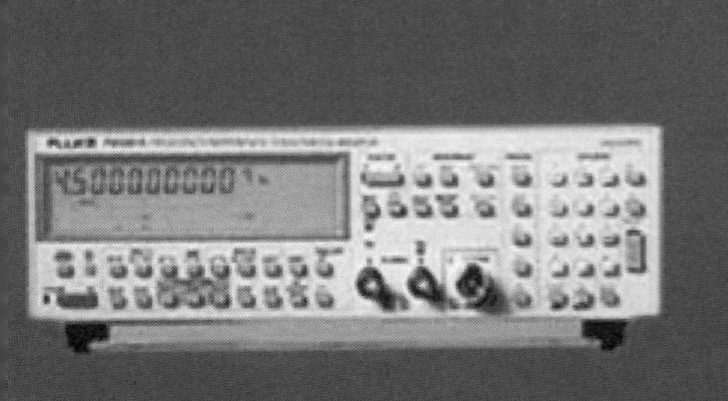

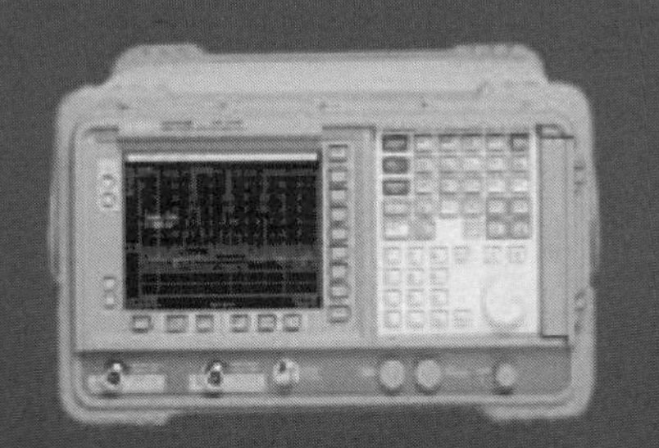

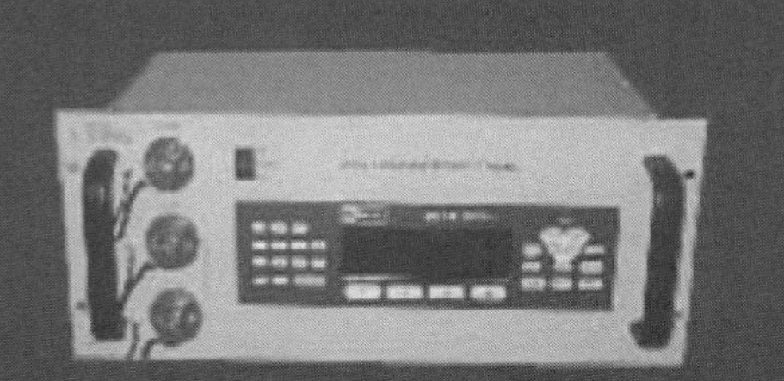

Calibration Instruments

FOREIGN MILITARY SALES

Afghanistan, Egypt, Japan, Lithuania, Saudi Arabia, Taiwan, United Arab Emirates

CONTRACTORS

Dynetics, Inc. (Huntsville, AL)
Agilent Technologies, Inc. (Santa Clara, CA)
Science Applications International Corp. (SAIC), (Huntsville, AL)
Fluke Corp. (Everett, WA)

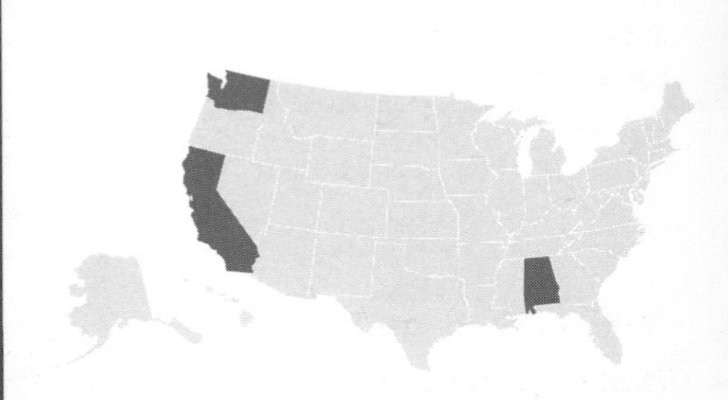

Chemical Biological Medical Systems–Diagnostics

INVESTMENT COMPONENT

Modernization

Recapitalization

Maintenance

MISSION

To provide the warfighter with safe, robust, affordable medical countermeasures against a broad spectrum of chemical, biological, radiological, and nuclear (CBRN) threats and use government and commercial best practices to acquire Food and Drug Administration (FDA)-approved CBRN medical countermeasures and diagnostics.

DESCRIPTION

The Joint Biological Agent Identification and Diagnostic System (JBAIDS) is a reusable, portable, modifiable biological agent identification and diagnostic system capable of rapid, reliable and simultaneous identification of multiple biological agents and other pathogens of operational concern. The JBAIDS anthrax, tularemia, brucellosis, and plague detection systems are FDA cleared for diagnostic use.

SYSTEM INTERDEPENDENCIES

None

PROGRAM STATUS

- **2QFY09:** National Guard fielding complete (26 systems)
- **4QFY09:** Emergency use authorization for Swine Flu (H1N1)
- **1QFY10:** FDA clearance for Avian Flu (H5N1) in vitro diagnostic (IVD) kit

PROJECTED ACTIVITIES

- **4QFY10:** FDA clearance for Q-fever IVD kit
- **1QFY11:** FDA clearance for typhus IVD kit
- **4QFY11:** Complete Navy fielding (31 systems)

ACQUISITION PHASE

Technology Development | Engineering & Manufacturing Development | Production & Deployment | Operations & Support

Chemical Biological Medical Systems–Diagnostics

FOREIGN MILITARY SALES
None

CONTRACTORS
Idaho Technologies (Salt Lake City, UT)

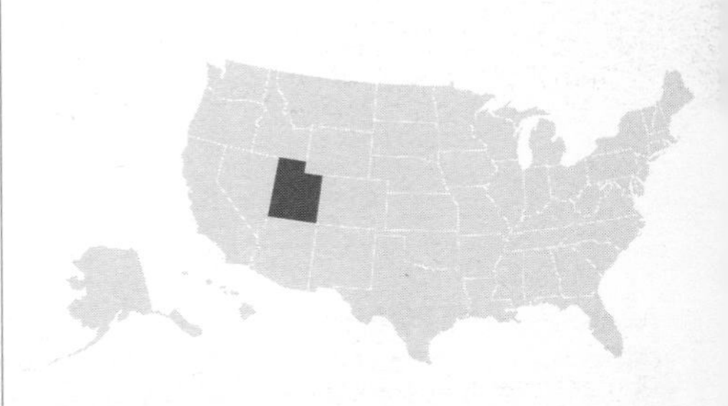

Chemical Biological Medical Systems–Prophylaxis

INVESTMENT COMPONENT

Modernization

Recapitalization

Maintenance

MISSION

To provide the warfighter with safe, robust, affordable medical countermeasures against a broad spectrum of chemical, biological, radiological, and nuclear (CBRN) threats; use government and commercial practices to acquire FDA-approved CBRN medical countermeasures and diagnostics.

DESCRIPTION

Plague Vaccine:

The Plague Vaccine is a highly purified polypeptide produced from non sporeforming bacterial cells transfected with a recombinant vector from Yersinia pestis.

Recombinant Botulinum Vaccine (rBV):

The Recombinant Botulinum Bivalent Vaccine (rBV) A/B comprises nontoxic botulinum toxin heavy chain (Hc) fragments of serotypes A and B formulated with an aluminum hydroxide adjuvant and delivered intramuscularly prior to potential exposure to botulinum toxins.

Bioscavenger II:

The Bioscavenger program fills an urgent capability gap in the warfighter's defense against nerve agents by development of a nerve agent prophylactic by rendering protective equipment, in theory, unnecessary. Bioscavenger Increment II consists of Protexia®, recombinant human butyrylcholinesterase produced in the milk of transgenic goats and modified with polyethylene glycol.

SYSTEM INTERDEPENDENCIES

None

PROGRAM STATUS

- **1QFY09:** Plague Vaccine Phase 2a clinical study complete
- **1QFY09:** Plague Vaccine manufacture scale-up and validation initiated
- **1QFY09:** rBV Phase 1b clinical study complete
- **1QFY09:** rBV Phase 2 clinical study initiated
- **1QFY10:** Bioscavenger Phase 1 clinical study complete
- **1QFY10:** Bioscavenger large-scale manufacturing, process qualification, and validation begins

PROJECTED ACTIVITIES

- **3QFY10:** Plague Vaccine Phase 2b clinical study begins
- **4QFY10:** rBV Phase 2 clinical study complete
- **4QFY10:** rBV large-scale manufacturing process validation complete

ACQUISITION PHASE

Technology Development | Engineering & Manufacturing Development | **Production & Deployment** | **Operations & Support**

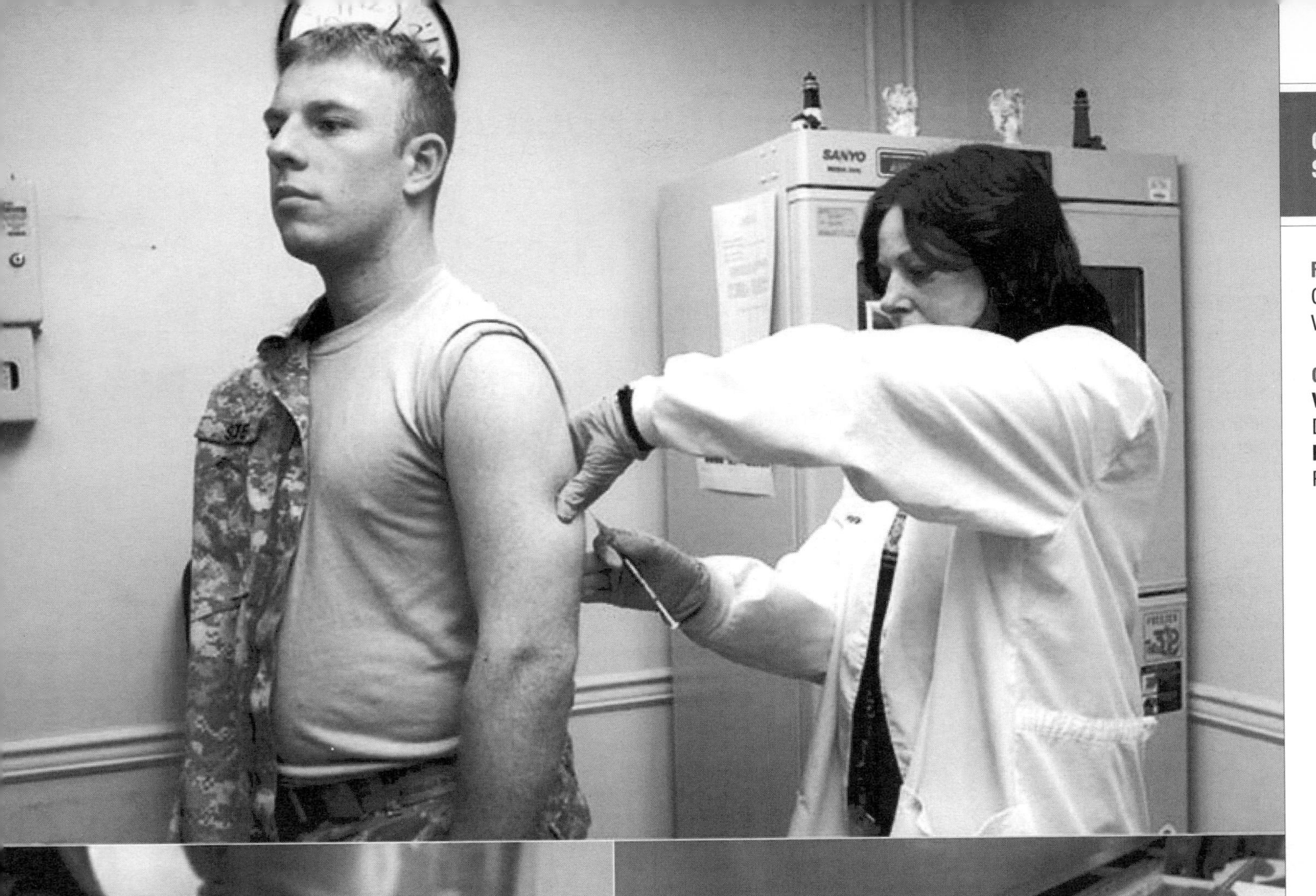

Chemical Biological Medical Systems–Prophylaxis

FOREIGN MILITARY SALES
Canada and United Kingdom: Plague Vaccine

CONTRACTORS
Vaccines:
DynPort Vaccine (Frederick, MD)
Bioscavenger:
PharmAthene (Annapolis, MD)

Chemical Biological Medical Systems–Therapeutics

INVESTMENT COMPONENT

- **Modernization**
- Recapitalization
- Maintenance

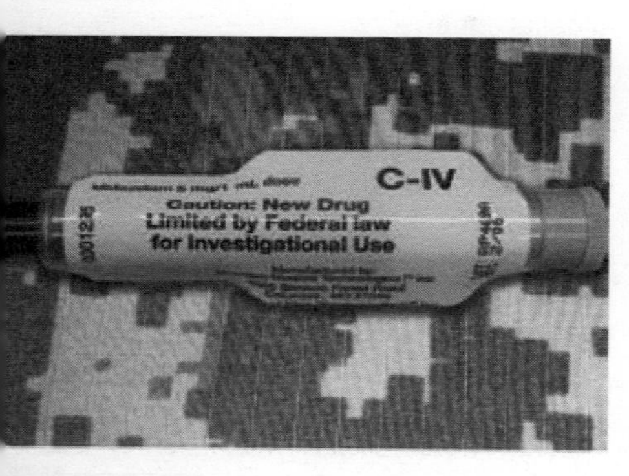

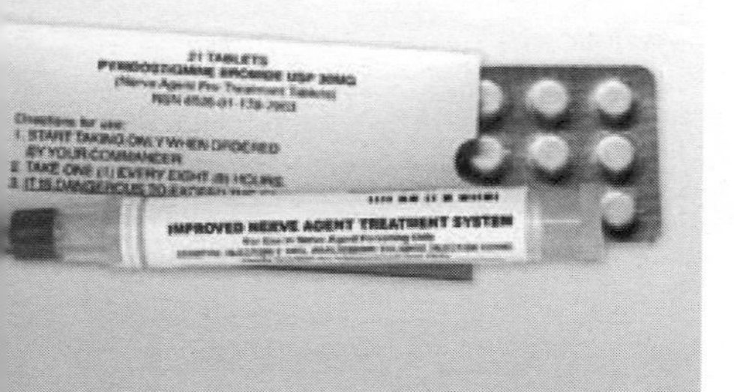

MISSION

To provide the warfighter with safe, robust, affordable medical countermeasures against a broad spectrum of CBRN threats; use government and commercial best practices to acquire FDA-approved CBRN medical countermeasures and diagnostics.

DESCRIPTION

Chemical Biological Medical Systems–Therapeutics consists of the following components:

Advanced Anticonvulsant System (AAS):
The AAS will consist of the drug midazolam in an autoinjector. The midazolam-filled autoinjector will replace the fielded Convulsant Antidote for Nerve Agents (CANA) that contains diazepam. Midazolam, injected intramuscularly, will treat seizures and prevent subsequent neurological damage caused by exposure to nerve agents. AAS will not eliminate the need for other protective and therapeutic systems.

Improved Nerve Agent Treatment System (INATS):
The INATS is an enhanced treatment regimen against the effects of nerve agent poisoning. The new oxime component of INATS will replace 2-PAM in the Antidote Treatment Nerve Agent Autoinjector (ATNAA).

Medical Radiation Countermeasure (MRADC):
Acute radiation syndrome (ARS) manifests as hematopoietic (bone marrow), gastrointestinal, and cerebrovascular subsyndromes depending on the dose of radiation received. The lead MRADC is adult-derived mesenchymal stem cells (Prochymal™) that will treat the gastrointestinal subsyndrome of ARS. The portfolio of MRADC will, when used as a system, provide a robust capability to the warfighter.

SYSTEM INTERDEPENDENCIES

None

PROGRAM STATUS

- **1QFY09:** AAS large-scale manufacturing and validation completed
- **1QFY09:** INATS pre-clinical safety studies complete
- **2QFY09:** INATS Investigational New Drug (IND) application submission to FDA
- **4QFY09:** MRADC pilot non-human primate (NHP) studies complete
- **1QFY10:** AAS Phase 2 clinical study complete
- **1QFY10:** MRADC pivotal NHP studies begin

PROJECTED ACTIVITIES

- **2QFY10:** AAS definitive NHP efficacy study complete
- **2QFY10:** INATS Phase 1 clinical study begins
- **3QFY10:** MRADC pivotal NHP studies complete
- **4QFY10:** INATS Phase 1 clinical study begins
- **4QFY09:** MDRAC Biologics License Application (BLA) submission to FDA
- **1QFY11:** AAS New Drug Application submission to FDA
- **1QFY11:** INATS Phase 2 clinical study begins

ACQUISITION PHASE

- Technology Development
- Engineering & Manufacturing Development
- **Production & Deployment**
- **Operations & Support**

Chemical Biological Medical Systems–Therapeutics

FOREIGN MILITARY SALES
None

CONTRACTORS
AAS:
Meridian Medical Technologies (Columbia, MD)
INATS:
Southwest Research Institute (San Antonio, TX)
MRADC:
Osiris Therapeutics (Columbia, MD)

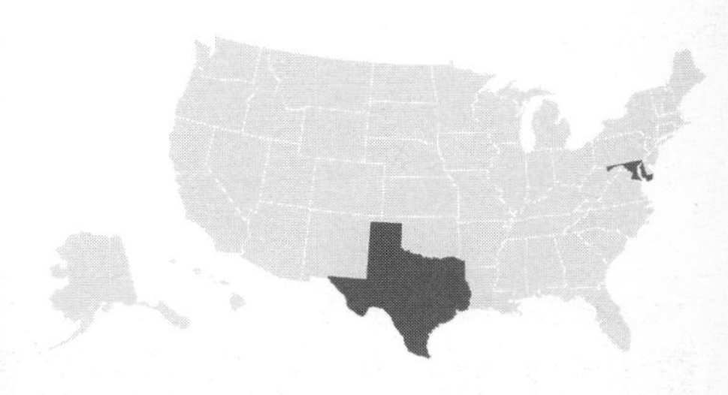

Chemical Biological Protective Shelter (CBPS)

INVESTMENT COMPONENT

Modernization

Recapitalization

Maintenance

MISSION

To enable medical personnel to treat casualties without the encumbrance of individual protective clothing and equipment in a highly mobile, easy-to-use, and self-contained, chemical biological (CB) hardened facility.

DESCRIPTION

The Chemical Biological Protective Shelter (CBPS) is a highly mobile, self-contained system designed to replace the M51 Collective Protection Shelter. CBPS consists of a Lightweight Multipurpose Shelter (LMS) mounted on an armored M1085A1 Medium Tactical Vehicle (MTV) and a 400-square-foot, airbeam-supported soft shelter. CBPS provides a contamination-free, environmentally controlled working area for medical, combat service, and combat service support personnel to obtain relief from the need to wear CB protective clothing for 72 hours of operation.

All ancillary equipment required to provide protection, except the generator, is mounted within the shelter. Medical equipment and crew gear are transported inside the LMS.

CBPS will be assigned to trauma treatment teams/squads of maneuver battalions, medical companies of forward and division support battalions, nondivisional medical treatment teams/squads, division and corps medical companies, and forward surgical teams.

SYSTEM INTERDEPENDENCIES

The shelter system is integrated onto an armored MTV.

PROGRAM STATUS

- **3QFY06:** Conduct first article testing

PROJECTED ACTIVITIES

- **2QFY10:** Conduct limited user testing
- **3QFY10:** Continue production

ACQUISITION PHASE

Technology Development | Engineering & Manufacturing Development | Production & Deployment | Operations & Support

Chemical Biological Protective Shelter (CBPS)

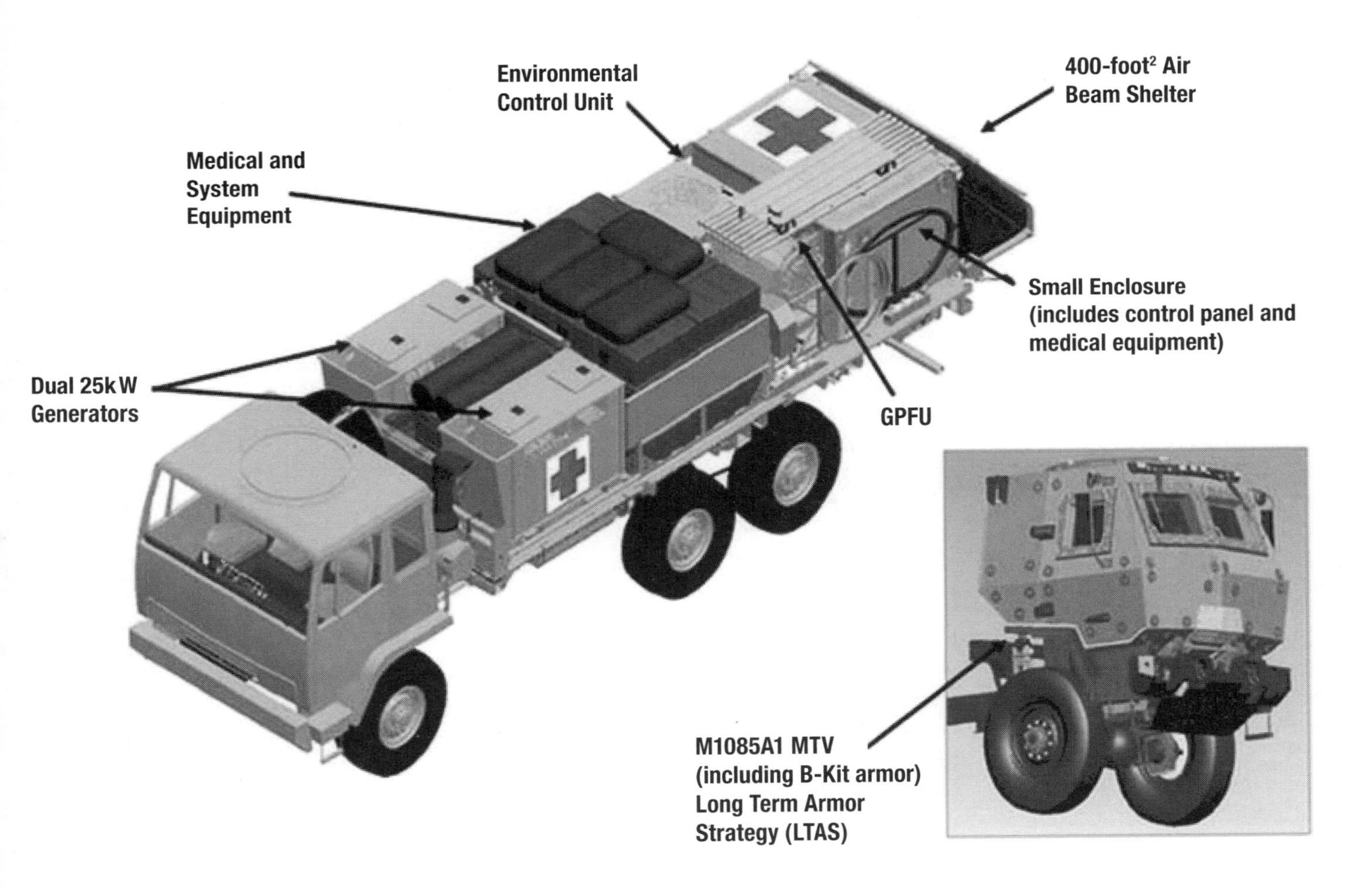

FOREIGN MILITARY SALES
None

CONTRACTORS
DRS Technologies (Parsippany, NJ)
Smiths Detection, Inc. (Edgewood, MD)

Chemical Demilitarization

INVESTMENT COMPONENT

Modernization

Recapitalization

Maintenance

MISSION

To safely destroy United States chemical warfare and related materiel, while ensuring maximum protection for the public, workers, and environment.

DESCRIPTION

The Chemical Materials Agency (CMA) mission includes the design, construction, systemization, operations and closure of chemical agent disposal facilities in Alabama, Arkansas, Indiana, Maryland, Oregon, Utah, and the Johnston Atoll in the South Pacific. Demilitarization operations have been completed in Indiana, Maryland, and the Johnston Atoll. Stockpile disposal at locations in Colorado and Kentucky is the responsibility of the Assembled Chemical Weapons Alternatives Program, which reports directly to the Office of the Secretary of Defense.

CMA is also responsible for emergency preparedness activities at chemical weapons storage depots; disposal of binary chemical munitions and non-stockpile chemical materiel; destruction of former chemical weapons production facilities; and assessment and destruction of recovered chemical materiel.

SYSTEM INTERDEPENDENCIES

None

PROGRAM STATUS

- **1QFY09:** Began final chemical agent campaign at the chemical disposal facility at Pine Bluff, AR
- **2QFY09:** Began final chemical agent campaign at the chemical disposal facility at Umatilla, OR
- **3QFY09:** Began final chemical agent campaign at the chemical disposal facility at Anniston, AL
- **4QFY09:** Completion of all Resource Conservation and Recovery Act (RCRA) permit requirements and release from future RCRA-required monitoring activities at Johnston Atoll
- **Current:** As of August 2, 2009, the Army has destroyed 63.5 percent of the chemical weapons stockpile

PROJECTED ACTIVITIES

- **3QFY10:** Complete closure operations at Newport, IN
- **3QFY11:** Complete final agent disposal campaign at Pine Bluff, AR

ACQUISITION PHASE

Technology Development | Engineering & Manufacturing Development | Production & Deployment | Operations & Support

Chemical Demilitarization

FOREIGN MILITARY SALES
None

CONTRACTORS
URS Corp. (Anniston, AL; Pine Bluff, AR; Umatilla, OR; Tooele, UT)

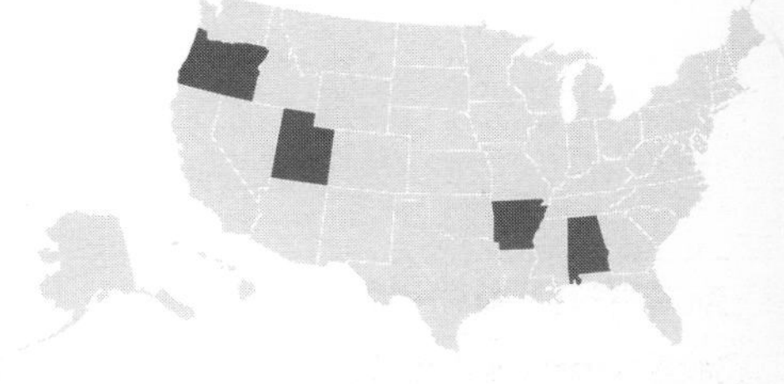

Chinook/CH-47 Improved Cargo Helicopter (ICH)

INVESTMENT COMPONENT

Modernization

Recapitalization

Maintenance

MISSION

To conduct air assault, air movement, mass casualty evacuation, aerial recovery, and aerial resupply across the full spectrum of operations.

DESCRIPTION

The Chinook/CH-47F Improved Combat Helicopter (ICH) upgrade program for the current CH-47D fleet will extend the service life of the current cargo helicopter fleet by an additional 20 years. The program includes the production of new aircraft and the remanufacture of CH-47Ds in the current fleet to meet the total Chinook fielding requirement of 513 aircraft. Both new and remanufactured CH-47F ICHs incorporate new monolithic airframes, a digital cockpit, a digital automatic flight control system, and improvements to reliability and maintainability. They provide an avionics architecture compliant with the DoD Information Technology Standards and Profile Registry (DISR), interoperability with DoD systems, and compliance with emerging Global Air Traffic Management requirements.

SYSTEM INTERDEPENDENCIES

Advanced Threat Infrared Countermeasures (ATRICM), Common Missile Warning System (CMWS), Air Warrior, Blue Force Tracker (BFT), and Joint Tactical Radio System (JTRS)

PROGRAM STATUS

- **2QFY07:** Complete initial operational testing
- **4QFY07:** First-unit equipped
- **1QFY08:** Multi-year procurement contract award

PROJECTED ACTIVITIES

- **4QFY19:** CH-47F fielding complete

ACQUISITION PHASE

Technology Development | Engineering & Manufacturing Development | Production & Deployment | Operations & Support

Chinook/CH-47 Improved Cargo Helicopter (ICH)

FOREIGN MILITARY SALES
None

CONTRACTORS
Aircraft:
Boeing (Philadelphia, PA)
Cockpit upgrade:
Rockwell Collins (Cedar Rapids, IA)
Engine upgrade:
Honeywell (Phoenix, AZ)
Extended range fuel system:
Robertson Aviation (Tempe, AZ)

MAX GROSS WEIGHT:	50,000 pounds
MAX CRUISE SPEED:	170 knots/184 miles per hour
TROOP CAPACITY:	36 (33 troops plus 3 crew members)
LITTER CAPACITY:	24
SLING-LOAD CAPACITY:	26,000 pounds center hook 17,000 pounds forward/aft hook 25,000 pounds tandem
MINIMUM CREW:	3 (pilot, copilot, and flight engineer)

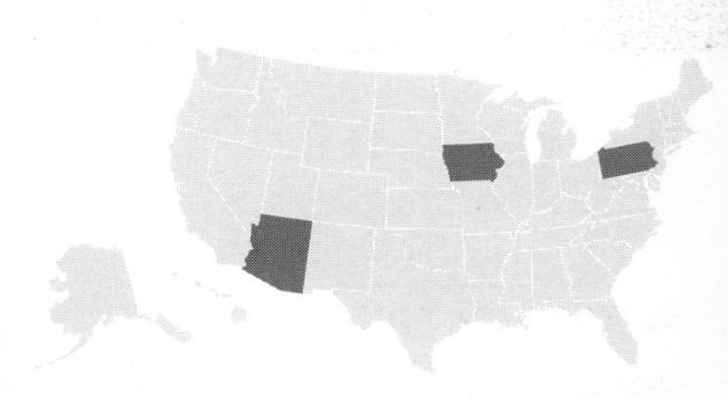

Close Combat Tactical Trainer (CCTT)

INVESTMENT COMPONENT

Modernization

Recapitalization

Maintenance

MISSION

To provide training of infantry, armor, mechanized infantry, and cavalry units from squad through battalion/squadron level, including battle staffs, using high-fidelity simulators within a virtual and collective training environment.

DESCRIPTION

The Close Combat Tactical Trainer (CCTT) is a virtual, collective training simulator that is fully interoperable with the Aviation Combined Arms Tactical Trainer. Soldiers operate from full-crew simulators and real or mockup command posts. Crewed simulators, such as the Abrams Main Battle Tank family, the Bradley Fighting Vehicle family, the High Mobility Multipurpose Wheeled Vehicle, the Heavy Expanded Mobility Tactical Truck and the M113A3 Armored Personnel Carrier, offer sufficient fidelity for collective mission accomplishment. Soldiers use command and control equipment to direct artillery, mortar, combat engineers, and logistics units to support the training mission. Semi-automated forces workstations provide additional supporting units (such as aviation and air defense artillery) and all opposing forces. All battlefield operating systems are represented, ensuring an effective simulation of a combat environment that encompasses daylight, night, and fog conditions. CCTT supports training of both Active Army and Army National Guard units at installations and posts in the United States, Europe, Korea, and Southwest Asia.

SYSTEM INTERDEPENDENCIES

CCTT requires Synthetic Environment Core (SE Core) to provide terrain databases and virtual models. The One Semi-Automated Forces (OneSAF) will provide a common SAF through SE Core in the future.

PROGRAM STATUS

- **3QFY08:** Fielded upgrades to the Bradley Fighting Vehicle for M2A3 chassis modernization and embedded diagnostics at Fort Bliss, TX; Fort Hood, TX; Fort Carson, CO; and Fort Knox, KY
- **4QFY08:** Obtained System Information Assurance Authority to Operate
- **1QFY09:** Successfully completed testing for Software Block II

PROJECTED ACTIVITIES

- **4QFY09:** Production and fielding of the Reconfigurable Vehicle Tactical Trainer (RVTT) to Fort Eustis, VA; United States Army Europe; and Central Command (Kuwait)
- **1QFY10:** Production and fielding of the RVTT to Fort Drum NY; Fort Bragg, NC; Fort Lewis, WA; Fort Wainwright, AK; and Fort Richardson, AK
- **2QFY10:** Production and fielding of the RVTT to Fort Dix, NJ; Fort Polk, LA; and Fort Campbell, KY
- **3QFY10:** Production and fielding of the RVTT to U.S. Army Europe; Schofield Barracks, HI; and Fort Sill, OK
- **4QFY10:** Production and fielding of the RVTT to Fort Bragg, NC; Fort McCoy, WI; and Fort Leonard Wood, MO
- **1QFY11:** Production and fielding of the RVTT to Fort Lee, VA; Camp Shelby, MS; and Gowen Field, ID

ACQUISITION PHASE

Technology Development | Engineering & Manufacturing Development | Production & Deployment | Operations & Support

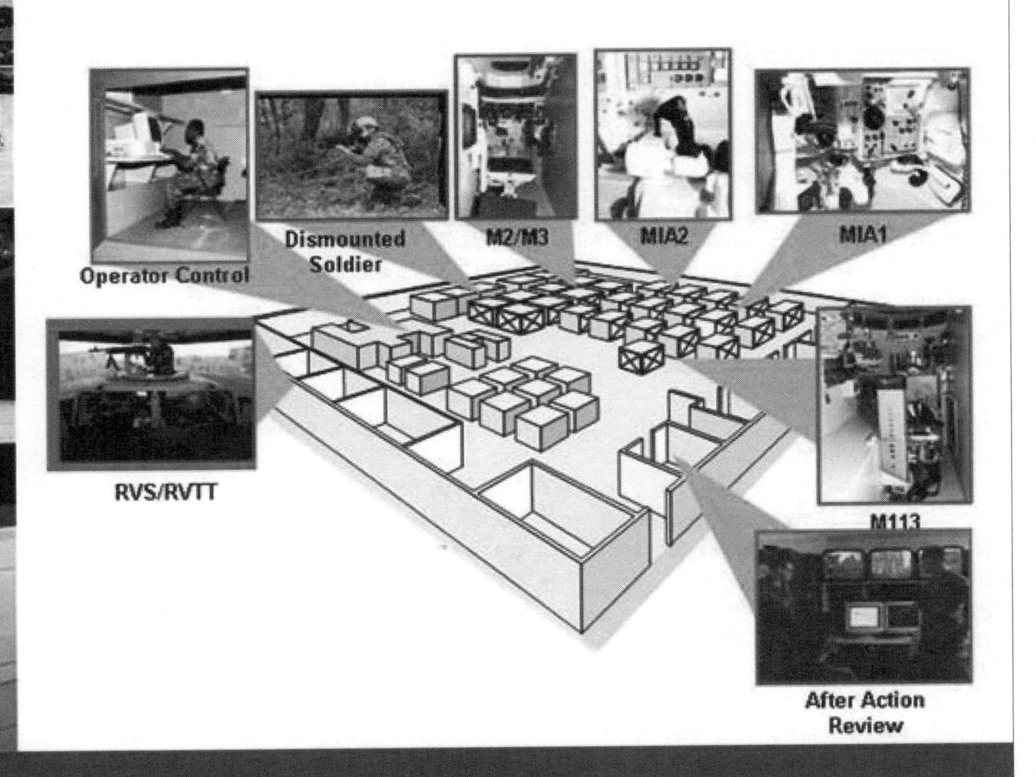

Close Combat Tactical Trainer (CCTT)

FOREIGN MILITARY SALES
None

CONTRACTORS
Lockheed Martin Simulation, Training and Support (Orlando, FL)
Rockwell Collins (Salt Lake City, UT)
Kaegan Corp. (Orlando, FL)
DRS Mobile Environmental Systems (Cincinnati, OH)
Meggitt Defense Systems (Suwanee, GA)

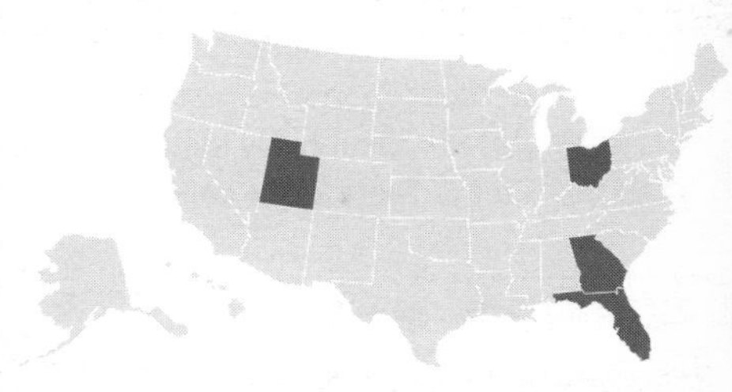

Combat Service Support Communications (CSS Comms)

INVESTMENT COMPONENT

Modernization

Recapitalization

Maintenance

MISSION

To provide a worldwide commercial satellite communications network, engineering services, Integrated Logistics Support, infrastructure, and portable remote terminal units in support of Army Combat Service Support (CSS) Logistics Management Information Systems operating from garrison or while deployed.

DESCRIPTION

Combat Service Support Communications (CSS Comms) includes the Combat Service Support Automated Information Systems Interface (CAISI) and the Combat Service Support Satellite Communications (CSS SATCOM) system. CAISI allows deployed Soldiers to connect CSS automation devices to a secure wireless network and electronically exchange information via tactical or commercial communications. CSS SATCOM complements CAISI by providing an easy-to-use, transportable SATCOM link to extend broadband information exchange worldwide.

CAISI employs a deployable wireless LAN infrastructure linking up to 92 tents, vans or shelters in a seven square-kilometer area. It includes Federal Information Processing Standards (FIPS) security requirements 140-2 Level 2-approved encryption for sensitive information.

CSS SATCOM includes commercial off-the-shelf Ku-band auto-acquire satellite terminals, called Combat Service Support Very Small Aperture Terminals (CSS VSATs), repackaged in fly-away transit cases, along with a contractor-operated fixed infrastructure of four teleports and high-speed terrestrial links that are connected to the unclassified segment of the Global Information Grid. CSS SATCOM supports operations at quick halt and rapid displacement within the Area of Operation (AO) and eliminates the often-dangerous need for Soldiers to hand-deliver requisitions via convoys in combat areas.

SYSTEM INTERDEPENDENCIES

CAISI, CSS SATCOM

PROGRAM STATUS

- **4QFY08–4QFY09:** Full-rate production and deployment of CAISI 2.0 along with CSS VSATs. Trained and equipped units in accordance with the Army Resourcing Priority List (ARPL)
- **3QFY08:** Conducted CAISI 2.0 User Assessment at Fort Drum, NY (10th Sustainment Bde) and Fort Hood, TX (15th Sustainment Bde)
- **4QFY08:** Army Acquisition Objective increased to 29,135 CAISI systems and 3,300 CSS VSATs
- **4QFY09:** CSS SATCOM supporting 2,400 systems on 16 networks with expanded coverage to Pacific Rim
- **4QFY09:** Authority to Operate (ATO) for two years granted for CSS SATCOM
- **4QFY08:** Secured host nation agreements for CSS SATCOM operations in support of 170 assigned users
- **4QFY09:** Completed reset/overhaul operations of 171 CSS SATCOM systems from Operation Enduring Freedom/Operation Iraqi Freedom (OEF/OIF)
- **4QFY08:** ATO granted for three years for CAISI 2.0
- **4QFY09:** Type Classification (TC) and assignment of Standard Line Item Numbers (SLIN) completed for CAISI 2.0
- **4QFY08:** First-unit equipped (FUE) with CAISI 2.0: 1st Brigade Combat Team (BCT)/1st Armored Division, Fort Bliss, TX
- **2QFY09:** Completed fielding of 266 CSS VSAT systems in support of National Guard Homeland Defense and civil support missions

PROJECTED ACTIVITIES

- **2QFY09–2QFY11:** Continue to field CAISI and CSS VSAT systems and train units an accordance with the ARPL
- **2QFY09–2QFY11:** Achieve pure fleet CAISI 2.0 and dispose of CAISI 1.1 and Natural Disaster relief missions
- **4QFY09–4QFY10:** Conduct reset/overhaul operations for 185 CSS VSAT systems redeploying from OEF/OIF

ACQUISITION PHASE

Technology Development | Engineering & Manufacturing Development | Production & Deployment | Operations & Support

Combat Service Support Communications (CSS Comms)

FOREIGN MILITARY SALES
None

CONTRACTORS
Equipment:
Telos Corp. (Ashburn, VA)
LTI DataComm, Inc. (Reston, VA)
L-3 Global Communications Solutions, Inc. (Victor, NY)
Segovia Global IP Services (Herndon, VA)
Project support/training:
Systems Technologies (Systek), Inc. (West Long Branch, NJ)
Tobyhanna Army Depot (Tobyhanna, PA)
CACI (Eatontown, NJ)

Command Post Systems and Integration (CPS&I)

INVESTMENT COMPONENT

Modernization

Recapitalization

Maintenance

MISSION

To provide commanders a standardized and mobile command post with a tactical, fully integrated, and digitized physical infrastructure to execute battle command and achieve information dominance.

DESCRIPTION

The Command Post Systems and Integration (CPS&I) program provides commanders with standardized, mobile, and fully integrated command posts for the modular expeditionary force, including support for Future Force capabilities as well as Joint and Coalition Forces. The command post is where commanders and their staffs collaborate, plan, and execute net-centric battle command, maintain situational awareness using the common operational picture (COP), and make decisions from available information. Based on the Standardized Integrated Command Post System (SICPS) Capabilities Production Document, a family of Command Post Platforms (CPPs) with standardized shelters, Command Center Systems (CCS), Command Post Communications Systems (CPCS), and Trailer Mounted Support Systems (TMSS) is currently being fielded to Army Active Component, National Guard, and Reserve units. SICPS provides the integrated Battle Command platform and infrastructure to allow shared situational understanding of the COP based on the various Army and Joint command and control communications and network systems in the command post. SICPS is modular and supports echelons from Battalion through Army Service Component Command, providing tactical flexibility to support all phases of operations. Integrating the Tactical Internet and the latest networking and Battle Command capability, command post operations are revolutionized through a combination of state-of-the-art data processing, communications, and information transport methods to achieve information dominance.

SYSTEM INTERDEPENDENCIES

JNN/WIN–T, BCCS Server, DCGS, MEP, CPOF

PROGRAM STATUS

- **2–3QFY08:** Executed Tactical Operations Centers (TOCFEST) engineering and logistical field study of the current Command Post System-of-Systems (SoS) at Fort Indiantown Gap, PA
- **3QFY08:** Implemented Battle Command SoS Integration training to train units on use of command post equipment in a SoS environment
- **3QFY08:** Initiated first delivery order for Trailer Mounted Support System (TMSS) under competitive contract awarded to Northrop Grumman Corporation
- **4QFY08:** Awarded contract option for CPP–Light development and integration effort
- **1QFY09:** Completed SICPS training/ fielding to 34th ID Div HQ; 43rd SUST; 155th HBCT; 32nd IBCT; Signal Center; 20th Support Command
- **1QFY09:** Completed CPP–Light system requirements review/ preliminary design review

PROJECTED ACTIVITIES

- **2–4QFY09:** Continue SICPS training/fielding in accordance with Unit Set Fielding schedule and reset support to units returning from deployment.
- **2QFY09:** Conduct CPP–Light critical design review
- **3QFY09:** Execute TOCFEST II in an operational environment with focus on command post mobility
- **4QFY09:** Conduct CPP–Light developmental test/operational test
- **1QFY10:** Conduct CPP–Light Milestone C low rate initial production decision review
- **1QFY10–1QFY11:** Continue SICPS training/fielding in accordance with Unit Set Fielding schedule and reset support to units returning from deployment
- **4QFY10:** Conduct CPP–Light Initial operational test and evaluation
- **2QFY11:** conduct CPP–Light Full-rate production decision review

ACQUISITION PHASE

Technology Development | Engineering & Manufacturing Development | Production & Deployment | Operations & Support

Command Post Systems and Integration (CPS&I)

FOREIGN MILITARY SALES
None

CONTRACTORS
Hardware Design, Integration, and Production (CPP, CCS, CPCS, TMSS):
Northrop Grumman (Huntsville, AL)

Common Hardware Systems (CHS)

INVESTMENT COMPONENT

Modernization

Recapitalization

Maintenance

MISSION

Provides state-of-the-art computer and networking equipment for the warfighter that improves connectivity, interoperability, logistics and worldwide repair, maintenance, and logistics support on the C4ISR battlefield.

DESCRIPTION

The Common Hardware Systems (CHS) program provides state-of-the-art, fully qualified, interoperable, compatible, deployable, and survivable hardware and computer networking equipment for command, control, and communications at all echelons of command for the Army and other DoD services. The CHS contract includes a technology insertion capability to continuously refresh the network-centric architectural building blocks, add new technology, and prevent hardware obsolescence. New products compliant with technology advances such as Internet Protocol Version Six (IPv6) can be easily added to the CHS offerings. CHS products include a spectrum of computer processors such as personal digital assistants (PDAs), high-end tactical computers, networking equipment, peripherals, displays, installation kits, and miscellaneous hardware needed for system integration. Four standardized environmental categories (Version 1, Version 1+, Version 2, and Version 3) are used to define hardware ruggedization and qualification test certification for the customers. Version 2 and Version 3 equipment items go through government-witnessed first article tests (FAT). Technical assistance and support services are also available.

CHS also provides worldwide repair, maintenance, logistics, and technical support through strategically located contractor-operated regional support centers (RSC) for tactical military units and management of a comprehensive five-year warranty. CHS hardware Version 1 includes commercial workstations, peripherals, and networking products. Version 1+ is Version 1 hardware that is modified for better survivability in the field with a minimal increase in cost. Version 2 includes ruggedized workstations, peripherals, and networking products. Version 3 includes near-military specification rugged handheld units.

SYSTEM INTERDEPENDENCIES

None

PROGRAM STATUS

- **1QFY08:** Acquired Battlefield Video Teleconferencing (BVTC) program
- **2QFY08:** Completed high-altitude electromagnetic pulse test
- **4QFY08:** Tactical Switching Requirement-3 (TSR-3) contract was signed

PROJECTED ACTIVITIES

- **2QFY09–2QFY11:** Continue CHS-3 hardware and software deliveries
- **2QFY09–2QFY11:** CHS-3 hardware and software; CHS Southwest Asia repair facility
- **3QFY09:** Joint User Interoperability Communications Exercise (JUICE)
- **3QFY09–4QFY11:** BVTC fieldings
- **4QFY09:** Award Total Asset Visibility (TAV) contract
- **2QFY10:** CHS-4 contract award

ACQUISITION PHASE

Technology Development | Engineering & Manufacturing Development | Production & Deployment | Operations & Support

Common Hardware Systems (CHS)

CHS Hardware

WIN–T Transit Cases

Antenna Interface Case (AIC)

Multi-processor Ethernet-switched Combat Chassis

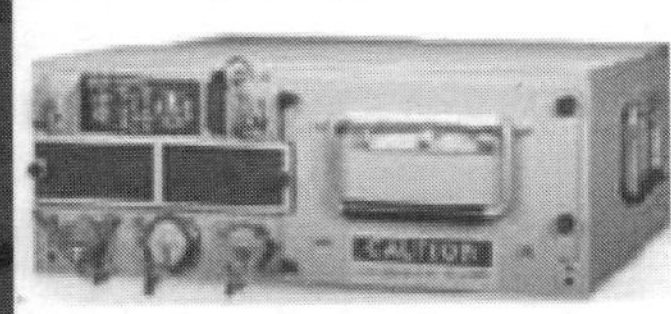
Uninterruptible Power Supply (UPS)

CHS hardware includes:

- BCCS V3 Operational Transit Cases (OTCs) (Battle Command Common Services–Version 3)
- Standalone Computer Unit (SCU-2)
- Miltope TSC-V3-GM45 RLC (Rugged Laptop Computer) Army and USMC AFATDS configurations
- Antenna Interface Case (AIC)
- CISC-2 Servers
- Multi processor Ethernet-switched Combat Chassis–5 Slice (MECC–5), 7 Slice (MECC–7)
- Modular Four-Slice Multiple Processor Unit (M4S MPU-2)
- Laptops and Servers
- Routers, Switches, and Firewalls
- Network and Communications
- Displays (LCDs)
- Peripherals
- Tablets and Handheld
- Uninterruptible Power Supplies (UPS) and Power Converter/Conditioners (PCC)
- Storage/RAID

FOREIGN MILITARY SALES
None

CONTRACTORS
General Dynamics C4 Systems (Taunton, MA)
Sun MicroSystems (Santa Clara, CA)
Cisco (San Jose, CA)
DRS Technologies (Parsippany, NJ)
Dell (Austin, TX)
Hewlett Packard (Palo Alto, NM)
Elbit Systems (Tallahassee, FL)
ECS (Lanham, MD)

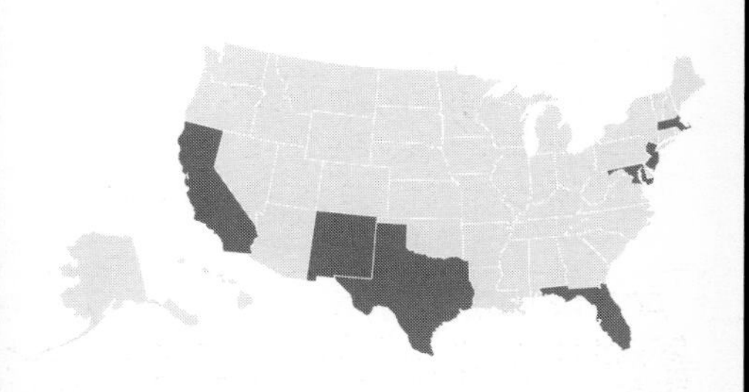

Common Remotely Operated Weapon Station (CROWS)

INVESTMENT COMPONENT
Modernization
Recapitalization
Maintenance

MISSION

To protect the gunner inside various armored vehicles, including the up-armored High Mobility Multipurpose Wheeled Vehicle, while providing mobile, first-burst engagement of targets day or night.

DESCRIPTION

The XM153 Common Remotely Operated Weapon Station (CROWS) consists of a weapon mount, display, and a joystick controller. Within the mount are a day camera, thermal camera, laser rangefinder, and fiberoptic gyroscopes. CROWS uses input from these sensors to calculate a ballistic solution to a target seen on the display. The Soldier uses the joystick controller to operate CROWS and engage the enemy from within the safety of the vehicle. CROWS operates with current weapons. It features:

- Three-axis stabilized mount that allows firing on the move.
- Auto Target Tracking to help the operator lock on target.

SYSTEM INTERDEPENDENCIES

CROWS mounts the MK19, M2, M240B, or M249 machine guns

PROGRAM STATUS

- **4QFY07:** Full and open competitive five-year indefinite delivery/ indefinite quantity contract to Kongsberg Defence and Aerospace for up to 6,500 systems
- **Ongoing:** Deliveries; Over 3,500 systems on order
- **Current:** Fielding in support of several urgent materiel releases on various platforms.
- **Ongoing:** Additional vehicle integrations and testing.

PROJECTED ACTIVITIES

- **Continue:** fielding and sustainment of systems with urgent materiel release
- **3QFY09:** Production verification test start
- **FY10:** Type classification standard

ACQUISITION PHASE
Technology Development | Engineering & Manufacturing Development | Production & Deployment | Operations & Support

Common Remotely Operated Weapon Station (CROWS)

FOREIGN MILITARY SALES
None

CONTRACTORS
Kongsberg Defence & Aerospace (Johnstown, PA)
MICOR Industries, Inc. (Decatur, AL)

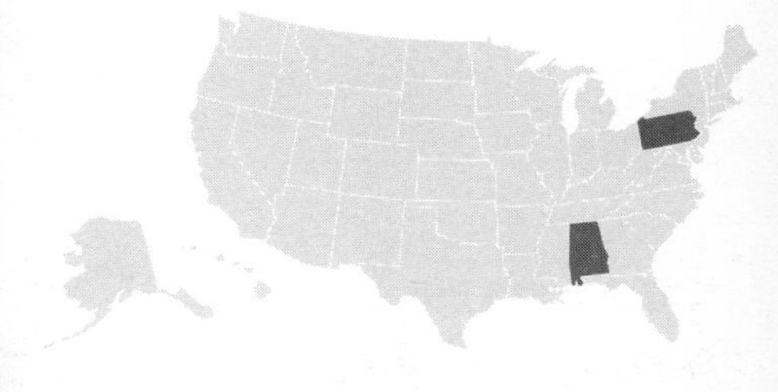

Counter-Rocket, Artillery and Mortar (C-RAM)

INVESTMENT COMPONENT

Modernization

Recapitalization

Maintenance

MISSION

To integrate multiple Army- and DoD-managed systems and commercial off-the-shelf systems with a command and control (C2) system to provide protection of fixed and semi-fixed sites from rockets and mortar rounds.

DESCRIPTION

The Counter-Rocket, Artillery, and Mortar (C-RAM) system was developed in response to a Multi-National Force–Iraq Operational Needs Statement (ONS) that was validated in September 2004. An innovative system-of-systems approach was implemented in which multiple DoD Program of Record systems were integrated with two commercial off-the-shelf (COTS) items to provide seven C-RAM functions: sense, warn, respond, intercept, command and control (C2), shape, and protect.

The C-RAM component systems are the Forward Area Air Defense command and control system; the Air and Missile Defense Workstation for C2; the Lightweight Counter Mortar Radar and Firefinder radars as sensors; Landbased Phalanx Weapon System to intercept; and wireless audio/visual emergency system and a wireless local area network for warning. Response is provided through C-RAM integration with Army and Marine Corps battle command systems and the Air Force Tactical Automated Security System.

Using this system-of-systems approach, C-RAM completed development, integration, and testing in April 2005, meeting the requirements of the ONS. The C-RAM system was then fielded five months after initial funding and just eight months after ONS validation.

In transition to a program of record, the C-RAM warning capability will be fielded to all Army Brigade Combat Teams (BCTs) as Increment 1 of the indirect fire protection capability.

SYSTEM INTERDEPENDENCIES

Army and Marine Corps Battle Command Systems, Sentinel radar

PROGRAM STATUS

Sustainment

PROJECTED ACTIVITIES

- **2QFY09:** Increment 1 CPD validation
- **1QFY10:** Increment 1 low-rate initial production milestone

ACQUISITION PHASE

Technology Development | Engineering & Manufacturing Development | Production & Deployment | Operations & Support

Counter-Rocket, Artillery and Mortar (C-RAM)

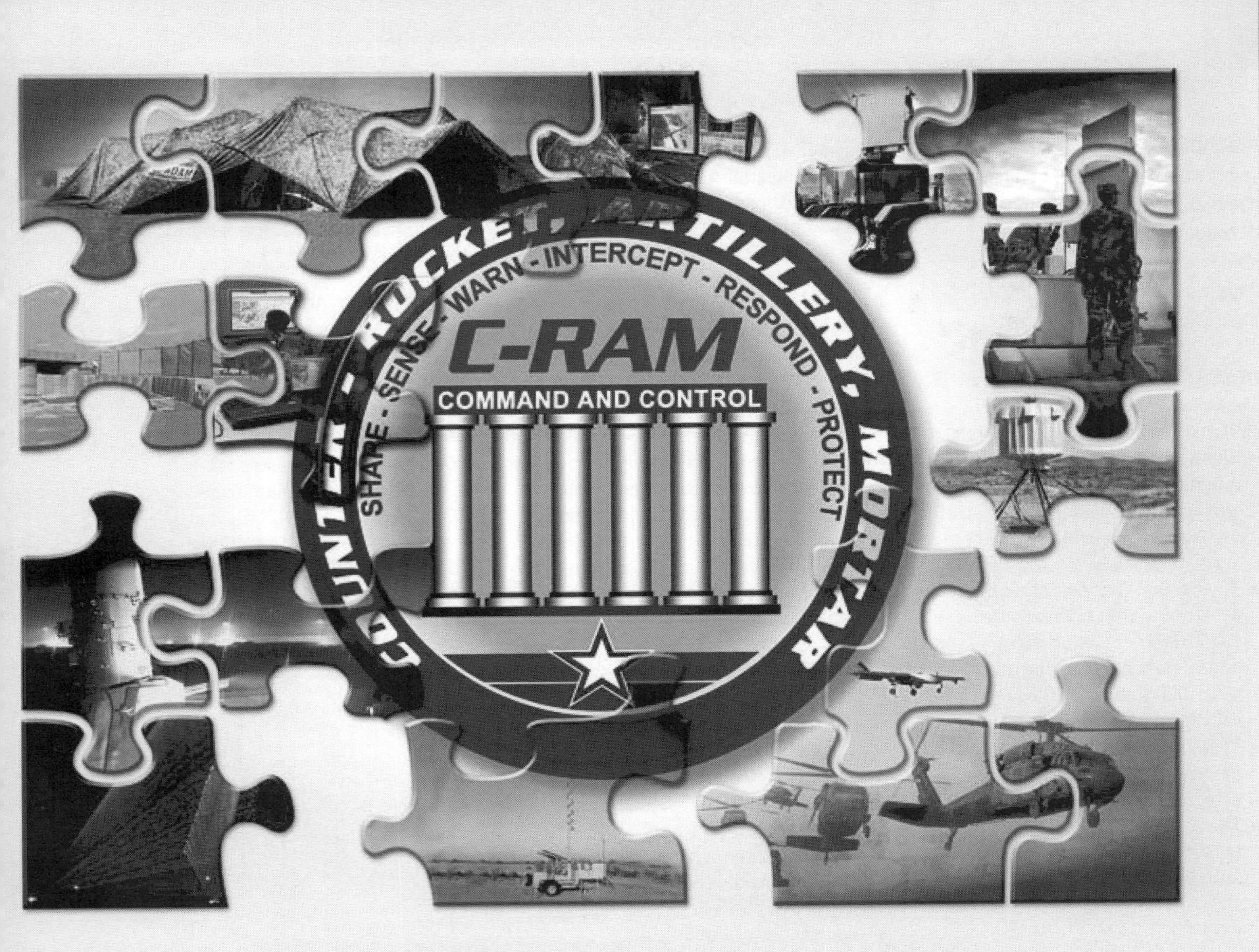

FOREIGN MILITARY SALES
United Kingdom (UK)

CONTRACTORS
Software/Hardware:
Northrop Grumman Mission Systems (Redondo Beach, CA; Huntsville, AL)
SETA:
ITT-CAS, Inc. (Huntsville, AL)

Countermine

INVESTMENT COMPONENT

Modernization

Recapitalization

Maintenance

MISSION

To provide Soldiers and maneuver commanders with a full range of countermine capabilities, plus immediate solutions to counter improvised explosive devices (IEDs) and other explosive hazards.

DESCRIPTION

The Countermine product line comprises several different systems:

- The Airborne Surveillance, Target Acquisition and Minefield Detection System (ASTAMIDS) puts a small, multi-spectral sensor payload on aerial platforms.
- The Ground Standoff Mine Detection System Future Combat Systems (GSTAMIDS FCS) provides mine detection, marking and neutralization for FCS countermine vehicles.
- The AN/PSS-14 Mine Detecting Set is a handheld multisensor mine detector.
- The Area Mine Clearance System (AMCS) is a flail system that destroys all types of landmines
- The HMDS upgrades the Husky mine detection vehicle with a ground penetrating radar
- The Autonomous Mine Detection System (AMDS) will detect, mark and neutralize hazards from a small, robotic platform.
- The Vehicle Optics Sensor System (VOSS) is a multisensor camera system for route clearance and explosive ordnance disposal (EOD) operations.
- The Improvised Explosive Device (IED) Interrogation Arms provides greater capabilities for investigating suspected IEDs.
- The EOD equipment product line provides capabilities such as the Man Transportable Robotic System (MTRS) and dual use blasting machines.

SYSTEM INTERDEPENDENCIES

ASTAMIDS, GSTAMIDS FCS, AMDS

PROGRAM STATUS

- **3QFY09:** AN/PSS-14 full-rate production and Army-wide fielding continues through FY15
- **3QFY09:** VOSS Phase 2 production contract award and fielding
- **3QFY09:** HMDS production and fielding
- **4QFY09:** ASTAMIDS flight testing on manned UH-1 and Fire Scout Unmanned Aerial System (UAS)

PROJECTED ACTIVITIES

- **1QFY10:** ASTAMIDS combined flight test (CFT) and CDR
- **1QFY10:** MTRS full rate production and Army-wide fielding continues through FY15
- **2QFY10:** AMCS Milestone C
- **2QFY10:** MTRS production
- **4QFY10:** ASTAMIDS Milestone C and low-rate initial production (LRIP)
- **4QFY10:** AMDS technology development contract award

ACQUISITION PHASE

Technology Development | Engineering & Manufacturing Development | Production & Deployment | **Operations & Support**

FOREIGN MILITARY SALES
4QFY09: IED Interrogation Arm to the Royal Netherlands Army; VOSS to Canadian Army

CONTRACTORS
AN/PSS-14:
L-3 CyTerra Corp. (Waltham, MA; Orlando, FL)
ASTAMIDS:
Northrop Grumman Integrated Systems (Melbourne, FL)
GSTAMIDS FCS:
BAE Systems (Austin, TX)
VOSS:
Gyrocam Systems LLC (Sarasota, FL)
IED Interrogation Arm:
FASCAN International (Baltimore, MD)
HMDS:
NIITEK (Sterling, VA)

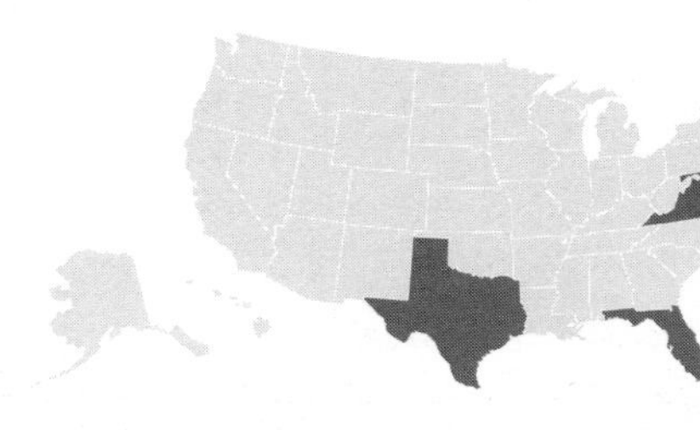

Defense Enterprise Wideband SATCOM Systems (DEWSS)

INVESTMENT COMPONENT

Modernization

Recapitalization

Maintenance

MISSION

To provide combatant commanders and deployed warfighters secure, high-capacity satellite connectivity enabling reachback for voice, video, and data communications and transfer of intelligence information.

DESCRIPTION

The Defense Enterprise Wideband SATCOM System (DEWSS) program is modernizing the enterprise satellite terminals and baseband, payload, and network control systems required to support combatant commander and warfighter use of high-capacity Wideband Global SATCOM (Satellite Communications) (WGS) satellites, which DoD began launching in October 2007. DEWSS consists of a geosynchronous orbiting satellite network, fixed enterprise military satellite terminals, and baseband, payload control, and network control systems. DEWSS provides superhigh-frequency, beyond-line-of-sight communications; reachback, via DoD Teleport and Standard Tactical Entry Point (STEP) sites; a critical conduit for intelligence information transfer; survivable communications for critical nuclear command and control; and an anti-jam and anti-scintillation capability for key strategic forces.

SYSTEM INTERDEPENDENCIES

None

PROGRAM STATUS

- **3QFY09:** Modernization of Enterprise (MET) contract awarded
- **3QFY09:** 350 EBEMs fielded worldwide as well as Automatic Uplink Power Control (AUPC)
- **3QFY09:** KaSTARS completed four terminal installations (Landstuhl, Camp Roberts, and 2 Lago)
- **3Q–4QFY09:** KaSTARS support toWGS-2 launch
- **4QFY09:** Wideband Global SATCOM (WGS) KaSTARS terminal performance certification awarded
- **4QFY09:** DSCS Integrated Management System (DIMS) V5.2 material release
- **4QFY09:** CNPS V2.1 material release
- **1QFY10:** MET Preliminary Implementation Review (IPR)

PROJECTED ACTIVITIES

- **2QFY10:** Critical Implementation Review (CIR)
- **2QFY10:** Provide IP capability to EBEM
- Major DCSS technology refresh/ modernization to coincide with MET terminal installations
- **4QFY10:** Conduct installation of and training for Wideband Global Spectrum Monitoring System (WGSMS) V2.0
- **4QFY10:CNPS** V3.1 material release
- **1QFY11:** Complete installation and checkout of Wahiawa, HI starter kit

ACQUISITION PHASE

Technology Development | Engineering & Manufacturing Development | Production & Deployment | Operations & Support

Defense Enterprise Wideband SATCOM Systems (DEWSS)

FOREIGN MILITARY SALES

None

CONTRACTORS

Johns Hopkins University Applied Physics Laboratory (Laurel, MD)
Northrop Grumman (Winter Park, FL)
U.S. Army Information Systems Engineering Command (Fort Huachuca, AZ)
ITT Industries (Colorado Springs, CO)
Harris Corp. (Melbourne, FL)

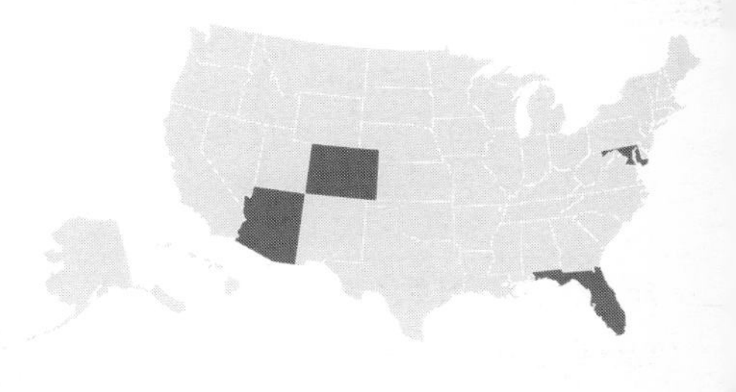

Distributed Common Ground System (DCGS–Army)

INVESTMENT COMPONENT

Modernization

Recapitalization

Maintenance

MISSION

To provide timely, multi-intelligence battle management and targeting information to field commanders at all echelons, improve data access, and reduce the forward footprint.

DESCRIPTION

Distributed Common Ground System–Army (DCGS–A) provides an integrated intelligence, surveillance, and reconnaissance (ISR) ground processing system, operating in a secure distributed and collaborative environment, enabled by networks. DCGS–A will serve as the primary ground system-of-systems for airborne and ground sensor platforms. DCGS–A enables the commander to achieve situational understanding by leveraging multiple sources of data, information, intelligence, and to synchronize the elements of Joint and Combined Arms combat power to See First, Understand First, Act First and Finish Decisively. DCGS–A consolidates/replaces nine systems. The core functions of DCGS–A are receipt and processing of select ISR sensor data, control of select Army sensor systems, intelligence synchronization, ISR planning, reconnaissance and surveillance (R and S) integration, fusion of sensor information, and direction and distribution of relevant threat, nonaligned, friendly, and environmental (weather and geospatial) information. DCGS–A emphasizes the use of reach- and split-based operations to improve data access, reduce forward footprint, and increase interoperability via a network-enabled modular, tailorable system in fixed, mobile, and embedded configurations.

DCGS–A will support three primary roles: As an analyst tool set, DCGS–A enables the user to collaborate, synchronize, and integrate organic and non-organic direct and general-support collection elements with operations; as the ISR component of the Army Battle Command, DCGS–A can discover and use all relevant threat, noncombatant, weather, and geospatial data and evaluate technical data and information on behalf of a Commander; DCGS–A provides organizational elements the ability to control select sensor platforms/payloads and process the collected data.

SYSTEM INTERDEPENDENCIES

DCGS, ACS, Battle Command System (BCS)–Army, Network Enabled Command Capability (NECC), Global Information Grid (GIG), Warfighter Information Network–Tactical (WIN–T), and Joint Tactical Radio System (JTRS).

PROGRAM STATUS

- **1QFY09:** Version 3.1 (V3.1) Joint Certification received from JITC on October 3, 2008.
- **1QFY09:** V3.1 limited user test (LUT) was completed on November 21, 2008. V3.1 provides system improvements such as the DCGS–A Application Framework (DAF) for seamless user experience, the Tactical Entity Database (TED), a persistent local store that facilitates interoperability with Battle Command and Joint systems, and DIB enhancements such as the Dynamic DIB Node Acquisition (DNA) which simplifies configuration.
- **2QFY09:** Field DCGS–A Version 3.1 to OIF and OEF
- **3QFY09:** Begin worldwide fielding of V3.1. V3.1 displaces All Source Analysis System–Light (ASAS–L).
- **3QFY09:** DCGS–A Mobile Basic (MB) Design Update Review 4-5 June 09
- **4QFY09:** DCGS–A MB IPR 2
- **4QFY09:** DCGS–A was a key system in JFCOM Empire Challenge 09 which demonstrated technology enhancements in collection and sharing of real-time ISR data
- **1QFY10:** DCGS-A V3.1 transition to post-production software support

PROJECTED ACTIVITIES

- **1QFY11:** DCGS–A MB Maintenance Demo
- **1QFY11:** DCGS–A MB Logistics Demo
- **3QFY11:** DCGS–A MB LUT
- **1QFY11:** DCGS-A MB FCA/PCA
- **1QFY12:** DCGS–A MB MS C / LRIP

ACQUISITION PHASE

Technology Development | Engineering & Manufacturing Development | Production & Deployment | **Operations & Support**

Distributed Common Ground System (DCGS–Army)

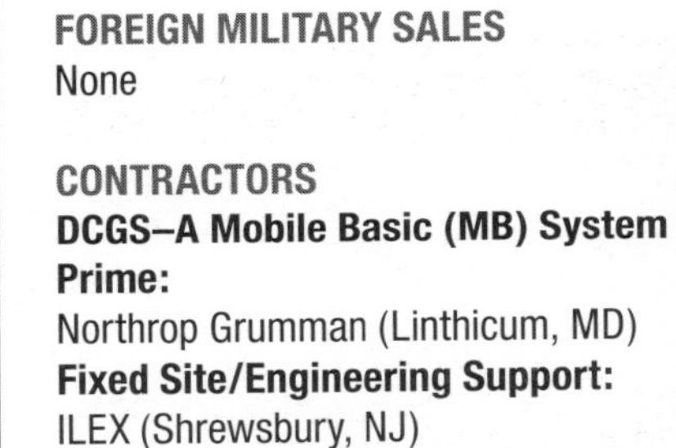

FOREIGN MILITARY SALES
None

CONTRACTORS
DCGS–A Mobile Basic (MB) System Prime:
Northrop Grumman (Linthicum, MD)
Fixed Site/Engineering Support:
ILEX (Shrewsbury, NJ)
Science Applications International Corp. (SAIC) (Alexandria, VA)
Program Support, System Engineering & Architecture:
Booz Allen Hamilton (Eatontown, NJ)
MITRE (Eatontown, NJ)
Battle Command Interoperability:
Overwatch Systems (Austin, TX)
DIB:
Raytheon (Garland, TX)

Distributed Learning System (DLS)

INVESTMENT COMPONENT

Modernization

Recapitalization

Maintenance

MISSION

To ensure that Soldiers receive critical mission training for mission success.

DESCRIPTION

The Distributed Learning System (DLS) provides digital training facilities (DTFs) equipped with computers and video equipment, enabling Soldiers to take digital training anywhere in the world at any time. Currently, 226 digital training facilities are operational at 93 sites worldwide.

DLS provides:

- Digital training facilities capable of delivering courseware for individual or group training
- Enterprise management of the DLS system
- A web-based learning management system for centralizing training management and delivery
- Deployed Digital Training Campuses (DDTCs) (currently in testing) to deliver multimedia courseware to deployed Soldiers
- Army e-Learning: commercial web-based training for business, information technology, or language skills

SYSTEM INTERDEPENDENCIES

AKO is used for identification and authorization and to gain access to the Army Learning Management System (ALMS). Student training results are transmitted via the ALMS to the Army Training Requirements and Resources System (ATRRS) as the system of record for Army training.

PROGRAM STATUS

- **2QFY08–1QFY09:** Sustained a centrally managed global training enterprise; electronically delivered training in military occupational specialties and self-development; fielded ALMS to Army schools; awarded contract to develop and build the DDTCs and completed preliminary testing for DDTC; increased Army e-Learning and Rosetta Stone (foreign language training) enrollments.

PROJECTED ACTIVITIES

- **2QFY09:** Activate new DLS disaster recovery site; sustain operation of DLS
- **4QFY09:** Complete DDTC testing
- **4QFY09:** Conduct full-rate production decision review; begin fielding DDTC
- **2QFY09–2QFY11:** Continue to sustain all DLS increments; continue to produce and deploy the DDTC

ACQUISITION PHASE

Technology Development | Engineering & Manufacturing Development | Production & Deployment | Operations & Support

Distributed Learning System (DLS)

FOREIGN MILITARY SALES
None

CONTRACTORS
Army Learning Management System (ALMS); Enterprise Management Center (EMC) Operations:
IBM (Fairfax, VA)
Army e-learning courseware:
Skillsoft (Nashua, NH)
DDTC development:
Lockheed Martin (Bethesda, MD)
Rosetta Stone foreign language training courseware:
Fairfield (Harrisonburg, VA)
Program management support:
MPRI (An L-3 Company) (Arlington, VA)

Dry Support Bridge (DSB)

INVESTMENT COMPONENT

Modernization

Recapitalization

Maintenance

MISSION

To support military load classification 100 (wheeled)/80 (tracked) vehicles over 40-meter gaps via a mobile, rapidly erected, modular military bridge.

DESCRIPTION

The Dry Support Bridge (DSB) system is fielded to Multi-Role Bridge Companies (MRBC) and requires a crew of eight Soldiers to deploy a 40-meter bridge in fewer than 90 minutes (daytime). The bridge modules are palletized onto seven flat racks and transported by equipment organic to the MRBC. DSB uses a launcher mounted on a dedicated Palletized Load System (PLS) chassis to deploy the modular bridge sections, which have a 4.3-meter road width and can span up to 40 meters. DSB is designed to replace the M3 Medium Girder Bridge.

SYSTEM INTERDEPENDENCIES

DSB operations rely and are interdependent upon fully mission capable M1977 CBTs and M1076 PLS Trailer assets within a fully MTOE equipped MRBC.

PROGRAM STATUS

- **4QFY07:** Fielded to 652nd Multi-Role Bridge Company
- **3QFY08:** Fielded to 1437th Multi-Role Bridge Company
- **1QFY09:** Fielded to 35th Engineer Company
- **1QFY09:** Fielded to 739th Multi-Role Bridge Company

PROJECTED ACTIVITIES

- **3QFY09:** Fielding to 671st Multi-Role Bridge Company
- **4QFY09:** Fielding to 74th Multi-Role Bridge Company
- **2QFY10:** Fielding to 1438th Multi-Role Bridge Company
- **3QFY10:** Fielding to 957th Multi-Role Bridge Company
- **4QFY10:** Fielding to 1041st Multi-Role Bridge Company
- **2QFY11:** Fielding to 502nd Multi-Role Bridge Company

ACQUISITION PHASE

Technology Development | Engineering & Manufacturing Development | Production & Deployment | Operations & Support

Dry Support Bridge (DSB)

FOREIGN MILITARY SALES
None

CONTRACTORS
Manufacturer:
Williams Fairey Engineering, Ltd. (Stockport, UK)
PLS chassis:
Oshkosh Truck Corp. (Oshkosh, WI)
Logistics:
XMCO (Warren, MI)

Early Infantry Brigade Combat Team (E-IBCT) Capabilities

INVESTMENT COMPONENT

- **Modernization**
- Recapitalization
- Maintenance

MISSION

To empower Soldiers with increased intelligence, surveillance, and reconnaissance (ISR) and lethality capabilities by incrementally modernizing the Army's Brigade Combat Teams (BCTs)

DESCRIPTION

The Army will build a versatile mix of mobile, networked Brigades that will leverage mobility, protection, information, and precision fires to conduct effective operations across the spectrum of conflict. Starting with the fielding of Early Infantry Brigade Combat Team (E-IBCT) capabilities in 2011, Soldiers in Infantry BCTs will incrementally receive capabilities that will increase their warfighting effectiveness.

The E-IBCT package contains:

- Network Integration Kit (NIK) (for the High Mobility Multipurpose Wheeled Vehicle [HMMWV] platform)
- XM501 Non Line of Sight–Launch System (NLOS–LS)
- XM156 Class I Unmanned Aerial Vehicle (UAV)
- AN/GSR 9 & AN/GSR 10 Unattended Ground Sensors (tactical and urban)
- XM1216 Small Unattended Ground Vehicle (SUGV)
- Interceptor Body Armor (See page 148)
- Ground Soldier System (See page 116)

SYSTEM INTERDEPENDENCIES

None

PROGRAM STATUS

The Army is addressing the capability gaps in our current force by accelerating delivery of advanced warfighting capabilities to all 73 Army BCTs. Starting with the E-IBCT capabilities, the Army is developing platforms and equipment to meet emerging Soldier requirements, and, as capabilities mature, they will be fielded incrementally according to the most urgent needs of the Army.

PROJECTED ACTIVITIES

CAPABILITIES

Fielding by capabilities is a key element of the Army's transition to a broader modernization strategy to build a versatile mix of mobile, networked, and combat effective BCTs. Accelerating proven solutions, these capabilities will provide planned and integrated upgrades to the Force every few years. These sets of capabilities include doctrine, organization, and training in conjunction with materiel to fill the highest priority shortfalls and mitigate risk for Soldiers.

Capability sets allow the Army to meet the evolving needs of the operating environment by providing units with the latest materiel and non-materiel solutions. The best capabilities available go to the Soldiers who need them most, based on the continually evolving combat environment. By fielding capabilities in alignment with the way BCTs are structured and

ACQUISITION PHASE

- Technology Development
- **Engineering & Manufacturing Development**
- Production & Deployment
- Operations & Support

trained, the Army is ensuring that Soldiers have the right capabilities to fight effectively as a system in the environments they are facing. The incremental deliveries will build upon one another as the Army continually adapts and modernizes.

EARLY INFANTRY BRIGADE COMBAT TEAM CAPABILITIES

E-IBCT fielding will provide enhanced warfighter capabilities to the force. The Army's priority is to provide Soldiers with enhanced situational awareness, force protection, and lethality through the use of unattended and attended sensors and munitions. In addition, the Soldier is provided improved communications and data sharing through the NIK. The E-IBCT package will consist of the following systems: the Non Line of Sight–Launch System (NLOS-LS), Urban and Tactical Unattended Ground Sensors (U/T UGS), Class 1 (Block 0) Unmanned Aerial Vehicle (UAV), and Small Unmanned Ground Vehicle (SUGV) Block 1. The E-IBCT will be fully integrated and networked through fielding of the ground tactical network and the NIK.

THE NETWORK

The Army will continue development and fielding of an incremental ground tactical network capability, fielded to all Army BCTs. This network is a layered system of interconnected computers and software, radios, and sensors within the BCT.

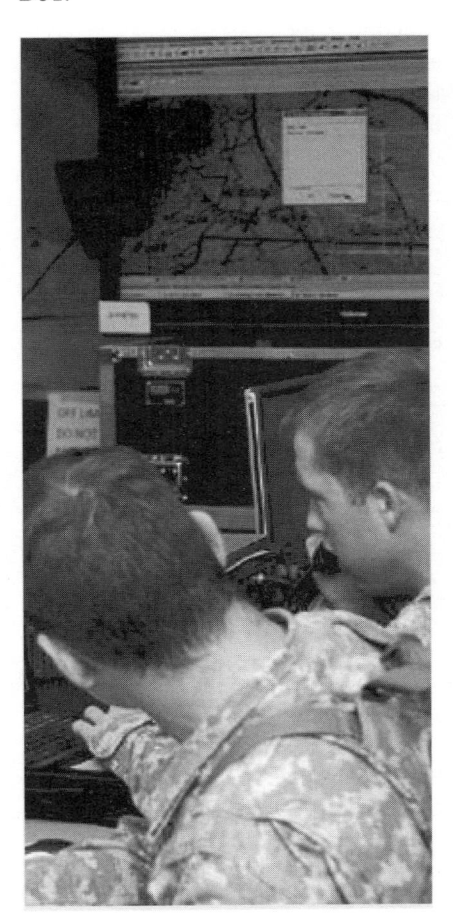

The network is essential to enable Unified Battle Command and will be delivered to the BCTs in increments of increasing capability. The first increment is currently completing System Development and Demonstration testing and will be delivered to Infantry BCTs in the form of NIK (B-kits).

THE NETWORK INTEGRATION KIT (NIK)

The NIK is an integrated suite of equipment on a HMMWV that provides network connectivity and battle command software to integrate and fuse sensor data into the commander's common operational picture (COP). The NIK consists of an integrated computer system (ICS) that hosts Battle Command software and the Systems of Systems Common Operating Environment (SOSCOE) software, along with a Joint Tactical Radio System (JTRS) Ground Mobile Radio (GMR) radio to provide the interface to selected sensors and unmanned systems, as well as voice and data communications with other vehicles and tactical operations centers.

XM501 NON LINE OF SIGHT–LAUNCH SYSTEM (NLOS–LS)

The XM501 NLOS–LS consists of a platform-independent Container Launch Unit (CLU) with self-contained technical fire control electronics and software for remote and unmanned operations. Each CLU consists of a computer and communications system and 15 Precision Attack Missiles (PAM). The NLOS–LS provides a rapidly deployable and network-linked precision-guided munitions launch capability that is currently not available within the Army.

XM156 CLASS I UNMANNED AERIAL VEHICLE (UAV)

The XM156 Class I UAV is a platoon-level asset that provides the dismounted Soldier with Reconnaissance, Surveillance, and Target Acquisition (RSTA) and laser designation capabilities. The air vehicle operates in open, rolling, complex, and urban terrains with a vertical take-off and landing capability. It is interoperable with select ground and air platforms and controlled by mounted or dismounted Soldiers.

The Class I uses autonomous flight and navigation, but it will interact with the network and Soldier to dynamically update routes and target information. It provides dedicated reconnaissance support and early warning to the lowest echelons of the BCT in environments not suited to larger assets.

AN/GSR-9 & AN/GSR-10 UNATTENDED GROUND SENSORS (UGS)

The UGS program is divided into two major subgroups of sensing systems: AN/GSR-9 (V) 1 Tactical-UGS (T-UGS), which includes Intelligence, Surveillance and Reconnaissance (ISR)-UGS and Radiological and Nuclear UGS; and AN/GSR-10 (V) 1 Urban-UGS (U-UGS), also known as Urban Military Operations on Urban Terrain (MOUT) Advanced Sensor System (UMASS). The UGS are used to perform mission tasks such as perimeter defense, surveillance, target acquisition, and situational awareness, including radiological, nuclear, and early warning. Soldiers involved in the recent testing of the UGS provided invaluable feedback, which was incorporated into new versions (form factors) that are now in testing.

XM1216 SMALL UNMANNED GROUND VEHICLE (SUGV)

The XM1216 SUGV is a lightweight, Soldier-portable UGV capable of conducting military operations in urban terrain, tunnels, sewers, and caves. The SUGV aids in the performance of urban ISR missions, chemical/Toxic Industrial Chemicals (TIC), and Toxic Industrial Materials (TIM) reconnaissance and inspecting suspected booby traps and improvised explosive devices without exposing Soldiers to these hazards. The SUGV's modular design allows multiple payloads to be integrated in a plug-and-play fashion that will minimize the Soldier's exposure to hazards. Payloads to be fielded are the manipulator arm, tether capability, chemical/radiation detection, and a laser target designator. Weighing 32 pounds, the SUGV is capable of carrying up to four pounds of payload weight.

Early Infantry Brigade Combat Team (E-IBCT) Capabilities

FOREIGN MILITARY SALES
None

CONTRACTORS
Boeing Corp.
Science Applications International Corp. (SAIC)
Network Integration Kit:
Boeing Corp. (Huntington Beach, CA)
General Dynamics C4 Systems, Inc. (Bloomington, MN)
Overwatch Systems (Austin, TX)
XM501 Non Line of Sight-Launch System:
Raytheon Company (Plano, TX)
Lockheed Martin Missiles & Fire Control (Grand Prairie, TX)
XM156 Class I Unmanned Aerial Vehicle:
Honeywell (Albuquerque, NM)
AN/GSR 9 & AN/GSR 10 Unattended Ground Sensors:
Textron Defense Systems (Wilmington, MA)
XM1216 Small Unmanned Ground Vehicle:
iRobot (Burlington, MA)

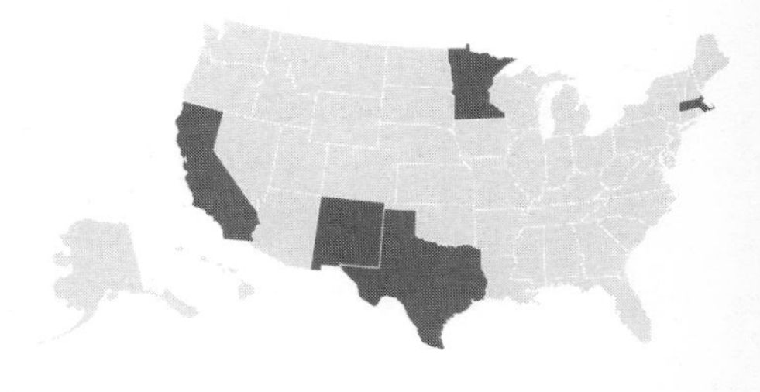

Excalibur (XM982)

INVESTMENT COMPONENT

Modernization

Recapitalization

Maintenance

MISSION

To provide improved fire support to the maneuver force commander through a precision-guided, extended range-artillery projectile that increases lethality and reduces collateral damage.

DESCRIPTION

The Excalibur (XM982) is a 155mm, Global Positioning System (GPS)-guided, fire-and-forget projectile, in use today in Operation Iraqi Freedom and Operation Enduring Freedom as the Army's next-generation cannon artillery precision munition. The target, platform location, and GPS-specific data are entered into the projectile's mission computer through an enhanced portable inductive artillery fuze setter.

Excalibur uses a jam-resistant internal GPS receiver to update the inertial navigation system, providing precision guidance and dramatically improving accuracy regardless of range. Excalibur has three fuze options: height-of-burst, point-detonating, and delay/penetration; and is effective in all weather conditions and terrain.

The program is using an incremental approach to provide a combat capability to the Soldier as quickly as possible, and to deliver advanced capabilities and lower costs as technology matures. The initial variant (Increment Ia1) was fielded in 2007 to provide an urgently needed capability. It includes a unitary high-explosive warhead capable of penetrating urban structures and is also effective against personnel and light materiel targets. Increment Ia2 will provide increased range (up to 40 kilometers) and reliability improvements. The third variant (Increment Ib) will maintain performance and capabilities while significantly reducing unit cost and increasing reliability.

Excalibur is designed for fielding to the digitized Lightweight 155mm Howitzer (LW155), the 155mm M109A6 selfpropelled howitzer (Paladin), and the Swedish Archer howitzer. Excalibur is an international cooperative program with Sweden, which contributes resources toward the development in accordance with an established project agreement and plans to join in procurement.

SYSTEM INTERDEPENDENCIES

None

PROGRAM STATUS

- **Current:** Army and Marine Corps units in Afghanistan and Iraq are now Excalibur capable.

PROJECTED ACTIVITIES

- **FY10:** Initial operational test and evaluation for Increment Ia2.
- **FY10:** Full materiel release and full-rate production of Increment Ia-2.
- **FY10:** Conduct competition between Increment Ib competitors and down-select to one contractor team for Phase 2 (Qualification and Production).
- **FY11:** Milestone C decision for Ib
- **FY12:** Operational test for Increment Ib
- **FY13:** Full material release for Ib

ACQUISITION PHASE

Technology Development | Engineering & Manufacturing Development | Production & Deployment | Operations & Support

Excalibur (XM982)

FOREIGN MILITARY SALES
Canada, Australia, Sweden, United Kingdom (compatibility testing with AS90 howitzer)

CONTRACTORS
Excalibur Increment Ia (Systems Integration):
Raytheon (Tucson, AZ)
Atlantic Inertial Units (Plymouth, England)
BAE Systems Bofors Defense (teamed with Raytheon) (Karlskoga, Sweden)
General Dynamics Ordnance and Tactical Systems (Healdsburg, CA; Niceville, FL)
Excalibur Ib Phase 1 (Design Maturation):
Raytheon (Tucson, AZ)
Alliant Techsystems (Minneapolis, MN)

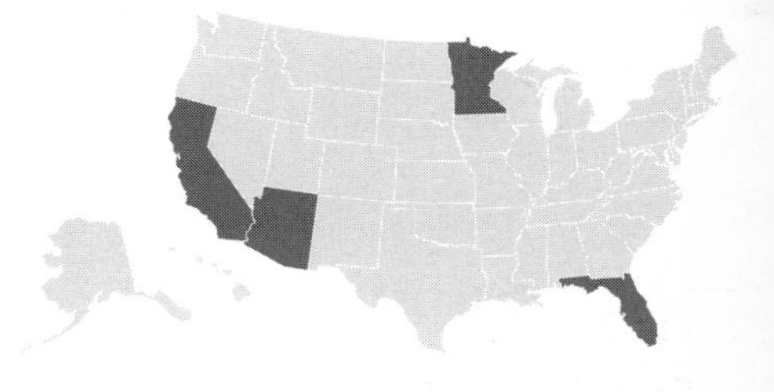

Extended Range Multipurpose (ERMP) Unmanned Aircraft System (UAS)

INVESTMENT COMPONENT

Modernization

Recapitalization

Maintenance

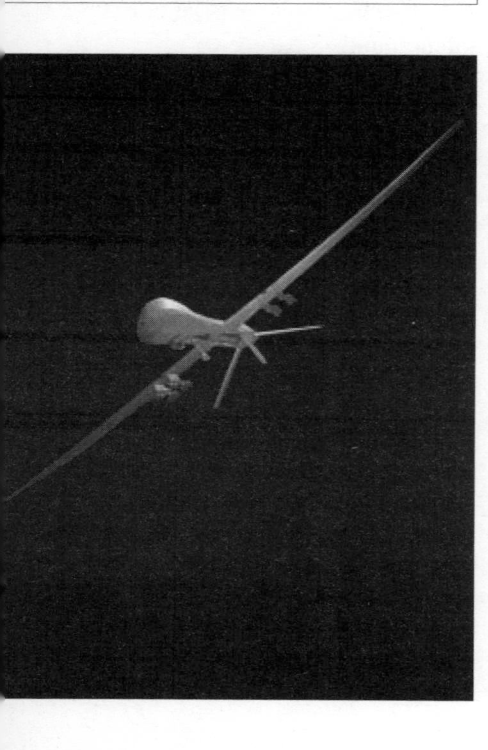

MISSION

To provide combatant commanders a real-time responsive capability to conduct long-dwell, persistent stare, wide-area reconnaissance, surveillance, target acquisition, communications relay, and attack missions.

DESCRIPTION

The Extended Range Multipurpose (ER/MP) Unmanned Aircraft System (UAS) addresses the need for a long endurance, armed, unmanned aircraft system that offers greater range, altitude, and payload flexibility.

The ER/MP is powered by a heavy fuel engine (HFE) for higher performance, better fuel efficiency, common fuel on the battlefield, and a longer lifetime. Its specifications include the following:

Length: 28 feet
Wingspan: 56 feet
Gross take off weight: 3,200 pounds (Growth to 3,600 pounds)
Maximum speed: 150 knots
Ceiling: 25,000 feet
Range: 1,200 nautical miles via satellite communications (SATCOM)
Endurance: 30+ hours

The ER/MP UAS is fielded in company sets, consists of 12 multi-role aircraft (six with SATCOM); five Universal Ground Control Stations (UGCS); two portable ground control stations; five Tactical Common Data Link (TCDL) ground data terminals; two TCDL portable ground data terminals; one ground SATCOM system; four automatic takeoff and landing systems; 12 Electro-Optical/Infrared/Laser Designator (EO/IR/LD) payloads; 12 Synthetic Aperture Radar/Ground Moving Target Indicator (SAR/GMTI) payloads; and ground support equipment.

SYSTEM INTERDEPENDENCIES

Payloads: PM-Robotic Unmanned Sensors (PM-RUS) provides the EO/IR and SAR/GMTI payloads; **Weapons:** PM-Joint Attack Munition Systems (PM-JAMS) provides Hellfire missiles; **Communications:** PM-Warfighter Information Network Terrestrial (PMWINT) provides communications relay payload

PROGRAM STATUS

- **Current:** System development and demonstration

PROJECTED ACTIVITIES

- **3QFY09:** Operational assessment
- **1QFY10:** Milestone C acquisition decision
- **4QFY11:** Initial operational test and evaluation

ACQUISITION PHASE

Technology Development | Engineering & Manufacturing Development | Production & Deployment | Operations & Support

Extended Range Multipurpose (ERMP) Sky Warrior Unmanned Aircraft System (UAS)

FOREIGN MILITARY SALES
None

CONTRACTORS
Aircraft:
General Atomics, Aeronautical Systems Inc. (San Diego, CA)
Ground Control Station:
AAI (Hunt Valley, MD)
Tactical Common Data Link:
L-3 Communications (Salt Lake City, UT)

Family of Medium Tactical Vehicles (FMTV)

INVESTMENT COMPONENT
- Modernization
- Recapitalization
- Maintenance

MISSION

To provides unit mobility/resupply, equipment/personnel transportation, and key ammunition distribution, using a family of vehicles based on a common chassis.

DESCRIPTION

The Family of Medium Tactical Vehicles (FMTV) is a system of strategically deployable vehicles that perform general resupply, ammunition resupply, maintenance and recovery, engineer support missions, and serve as weapon systems platforms for combat, combat support, and combat service support units in a tactical environment.

The Light Medium Tactical Vehicle (LMTV) has a 2.5-ton capacity (cargo and van models).

The Medium Tactical Vehicle (MTV) has a 5-ton capacity (cargo, long-wheelbase-cargo with and without materiel handling equipment, tractor, van, wrecker, 8.8-ton Load Handling System (LHS), 8.8-ton LHS trailer, and 10-ton dump truck models). Three truck variants and two companion trailers, with the same cube and payload capacity as their prime movers, provide air drop capability. MTV also serves as the platform for the High Mobility Artillery Rocket System (HIMARS) and resupply vehicle for PATRIOT and HIMARS. MTV operates worldwide in all weather and terrain conditions.

FMTV enhances crew survivability through the use of hard cabs, three-point seat belts, and central tire inflation capability. FMTV enhances tactical mobility and is strategically deployable in C5, C17, C141, and C130 aircraft. It reduces the Army's logistical footprint by providing commonality of parts and components, reduced maintenance downtime, high reliability, and high operational readiness rate (more than 90 percent). FMTV incorporates a vehicle data bus and class V interactive electronic technical manual, significantly lowering operating and support costs compared with older trucks. Units are equipped with FMTVs at more than 68 locations worldwide, 39,663 trucks and 8,332 trailers are in field units as of June 30, 2009. The Army developed, tested, and installed add-on-armor and enhanced add-on-armor kits, and a Low Signature Armored Cab (LSAC) for Southwest Asia. The newest armored version, the Long Term Armor Strategy (LTAS) A1P2 cabs are now in production and are being fielded. Approximately 4,000 FMTVs have been armored in Southwest Asia in support of Operation Iraqi Freedom.

SYSTEM INTERDEPENDENCIES

None

PROGRAM STATUS

- **Current:** Installation of enhancements for the LSAC equipped FMTVs operating in Southwest Asia
- **1QFY09:** Long Term Armor Strategy (LTAS) deliveries begin

PROJECTED ACTIVITIES

- **Ongoing:** Continue full production and fielding to support Army transformation
- **3QFY09:** Competitive rebuy multi-year contract award
- **3QFY09:** First fieldings of LTAS vehicles
- **4QFY09:** Competitive rebuy multiyear contract award

ACQUISITION PHASE

Technology Development	Engineering & Manufacturing Development	Production & Deployment	Operations & Support

Family of Medium Tactical Vehicles (FMTV)

FOREIGN MILITARY SALES
Jordan

CONTRACTORS
BAE Systems (Sealy, TX)
Meritor (Troy, MI)
Allison (Indianapolis, IN)
Caterpillar (Greenville, SC)
AAR Mobility Systems (Cadillac, MI)

FMTV with Armor Kit

	LMTV A1 Cargo	MTV A1 Cargo
Payload:	5,000 pounds	10,000 pounds
Towed load:	12,000 pounds	21,000 pounds
Engine:	Caterpillar 6-cylinder diesel	Caterpillar 6-cylinder diesel
Transmission:	Allison Automatic	Allison Automatic
Horsepower:	275	330
Drive:	4 x 4	6 x 6

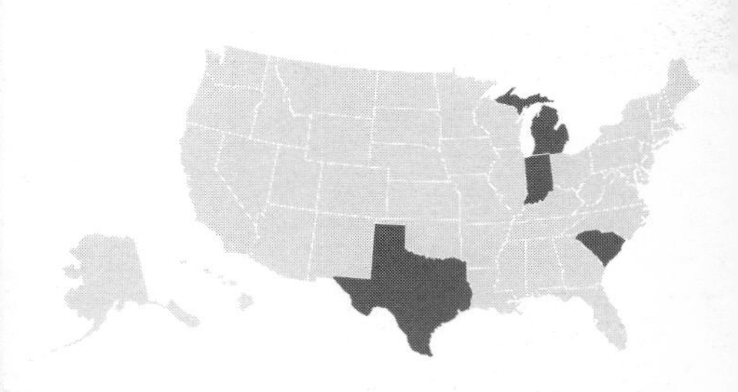

Fixed Wing

INVESTMENT COMPONENT

Modernization

Recapitalization

Maintenance

UC-35

MISSION

To perform operational support and focused logistics missions for the Army, joint services, national agencies, and multinational users in support of intelligence and electronic warfare, transporting key personnel, and providing logistical support for battle missions and homeland security.

DESCRIPTION

The Fixed Wing fleet consists of eight aircraft platforms and 256 aircraft that allow the Army to perform day-to-day operations in a more timely and cost-efficient manner without reliance on commercial transportation. The fleet provides timely movement of key personnel to critical locations throughout the theater of operations, and transports time-sensitive and mission-critical supply items and repair parts needed to continue the warfight. Special electronic-mission aircraft provide commanders with critical intelligence and targeting information, enhancing lethality and survivability on the battlefield.

All Army fixed-wing aircraft are commercial off-the-shelf products or are developed from those products. The fleet includes:

- C-12 Utility
- C-20/C-37 Long range transport
- C-23 Cargo
- C-26 Utility
- EO-5 Airborne Reconnaissance Low (ARL)
- RC-12 Guardrail Common Sensor (GR/CS)
- UC-35 Utility

The EO-5 and RC-12 are classified as special electronic mission aircraft and provide real-time intelligence collection in peace and wartime environments. The C-12, C-23, C-26, and UC-35 are classified as operational support aircraft and provide direct fixed-wing support to warfighting combatant commanders worldwide. The C-20 and C-37 are assigned to Andrews Air Force Base and are classified as senior support aircraft for the chief of staff and service secretary.

SYSTEM INTERDEPENDENCIES

None

PROGRAM STATUS

- C-12, RC-12, and UC-35 aircraft are sustained using a Life Cycle Contractor Support (LCCS) maintenance contract (DynCorp)
- C-23 aircraft are sustained using an LCCS maintenance contract (M7 Aerospace)
- C-37 and UC-35 aircraft were purchased with Congressional plus-up funding
- C-37, C-20, and C-26 aircraft are sustained using Air Force LCCS maintenance contracts (Gulfstream and M7 Aerospace)
- EO-5 aircraft are sustained using an LCCS maintenance contract (King Aerospace)

PROJECTED ACTIVITIES

- Acquire 10 C-12 replacement aircraft for the Army Reserve
- Re-compete the C12/RC-12/UC-35 aircraft Life Cycle Contractor Support contract
- Re-compete and assume responsibility from the Air Force for C-26 aircraft Life Cycle Contractor Support

ACQUISITION PHASE

Technology Development | Engineering & Manufacturing Development | Production & Deployment | Operations & Support

Fixed Wing

C-20/37

C-26

RC-12

EO-5 (ARL)

C-12

C23

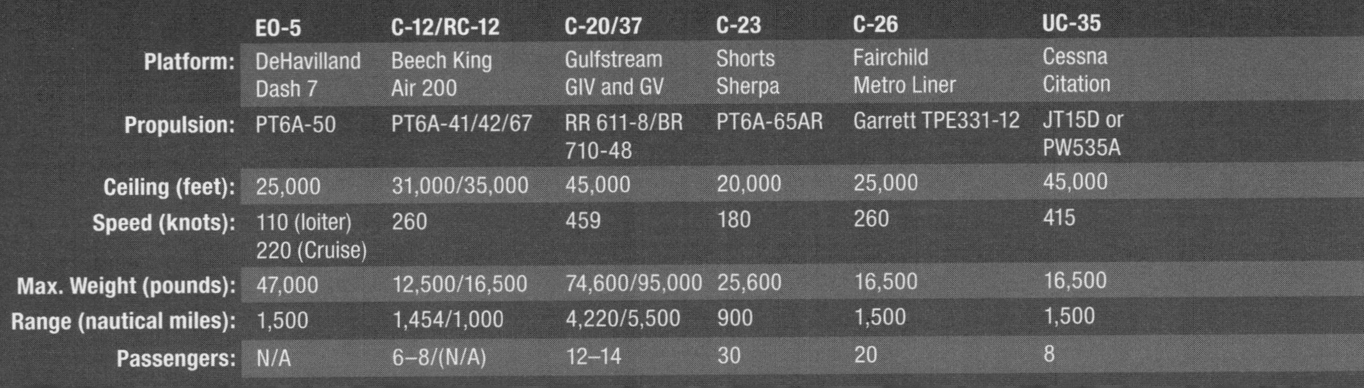

	EO-5	C-12/RC-12	C-20/37	C-23	C-26	UC-35
Platform:	DeHavilland Dash 7	Beech King Air 200	Gulfstream GIV and GV	Shorts Sherpa	Fairchild Metro Liner	Cessna Citation
Propulsion:	PT6A-50	PT6A-41/42/67	RR 611-8/BR 710-48	PT6A-65AR	Garrett TPE331-12	JT15D or PW535A
Ceiling (feet):	25,000	31,000/35,000	45,000	20,000	25,000	45,000
Speed (knots):	110 (loiter) 220 (Cruise)	260	459	180	260	415
Max. Weight (pounds):	47,000	12,500/16,500	74,600/95,000	25,600	16,500	16,500
Range (nautical miles):	1,500	1,454/1,000	4,220/5,500	900	1,500	1,500
Passengers:	N/A	6–8/(N/A)	12–14	30	20	8

FOREIGN MILITARY SALES
None

CONTRACTORS
DynCorp (Fort Worth, TX)
Gulfstream (Savannah, GA)
King Aerospace (Addison, TX)
M7 Aerospace (San Antonio, TX)

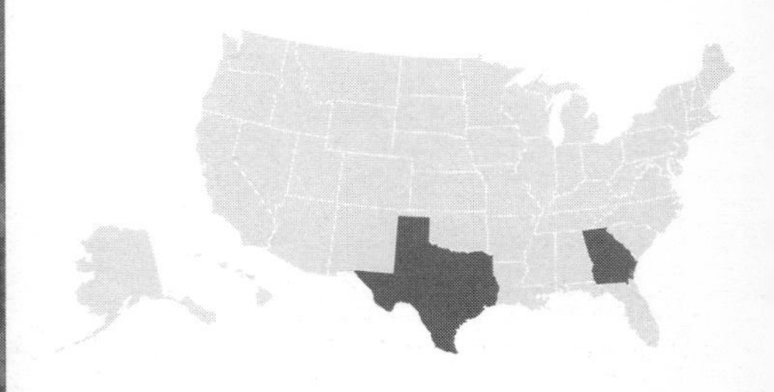

Force Protection Systems

INVESTMENT COMPONENT

Modernization
Recapitalization
Maintenance

MISSION

To detect, assess, and respond to unauthorized entry or attempted intrusion into installation/facilities.

DESCRIPTION

Force Protection Systems consist of the following components:

Automated Installation Entry (AIE) is a software and hardware system designed to read and compare vehicles and personnel identification media. The results of the comparison are used to permit or deny access to installation in accordance with installation commanders' criteria. AIE will use a database of personnel and vehicles that have been authorized entry onto an Army installation and appropriate entry lane hardware to permit/deny access to the installation. AIE will validate the authenticity of credentials presented by a person with data available from defense personnel and vehicle registration databases. AIE will have the capability to process permanent personnel and enrolled visitors, and to present a denial barrier to restrict unauthorized personnel. AIE will be capable of adapting to immediate changes in threat conditions and apply restrictive entrance criteria consistent with the force protection condition.

The Battlefield Anti-Intrusion System (BAIS) is a compact, modular, sensor-based warning system that can be used as a tactical stand-alone system. The system consists of a handheld monitor and three seismic/acoustic sensors and provides coverage across a platoon's defensive front (450 meters). It delivers early warning and situational awareness information, classifying detections as personnel, vehicle, wheeled, or tracked intrusions.

The Lighting Kit, Motion Detector (LKMD) is a simple, compact, modular, sensor-based early-warning system providing programmable responses of illumination and sound. The LKMD enhances unit awareness during all types of operations and environments, including those in urban terrain.

SYSTEM INTERDEPENDENCIES

None

PROGRAM STATUS

- **FY02–10:** BAIS procurement and fielding
- **FY06–11:** LKMD operational testing and procurement

PROJECTED ACTIVITIES

- **1QFY10:** BAIS fielding
- **2QFY10:** LKMD procurement and fielding

ACQUISITION PHASE

Technology Development | Engineering & Manufacturing Development | Production & Deployment | Operations & Support

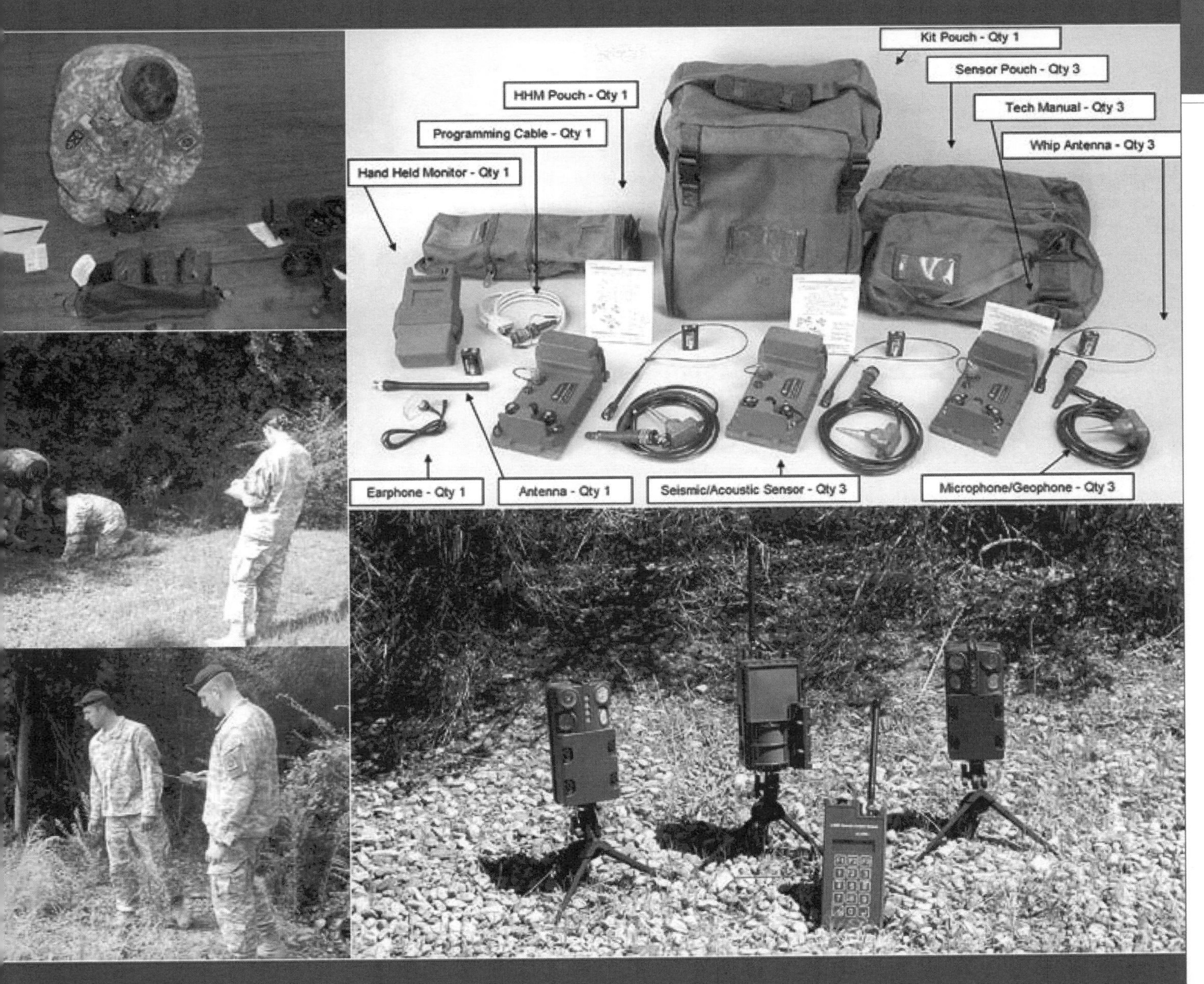

FOREIGN MILITARY SALES
None

CONTRACTORS
BAIS:
L-3 Communications (Camden, NJ)
LKMD:
EG&G (Albuquerque, NM)

Force Provider (FP)

INVESTMENT COMPONENT

- **Modernization**
- Recapitalization
- Maintenance

MISSION

To provide the Army, joint U.S. Military, host nation, and coalition forces personnel with a high-quality deployable base camp to support the expeditionary missions, develop, integrate, acquire, field, sustain, and modernize base camp support systems to improve the warfighters fighting capabilities, performance, and quality of life.

DESCRIPTION

Each Force Provider (FP) includes 71 deployable triple container (TRICON) systems, including eight latrine systems, eight shower systems, four kitchen system, containerized batch laundry system, four TRICON refrigerated containers, 26 60-kilowatt tactical quiet generators, 26 modular personnel tents (air supported), four 400,000 BTU water heaters, four improved fuel distribution systems, two wastewater evacuation tank/ trailers, 26 mobile electric power distribution replacement systems, 56 environmental control units, and eight air compressors that are diesel engine driven and operate at 17 cubic feet per minute. Additional operational add-on kits include a cold-weather modification system, prime-power modification system, electric kitchen, and shower water reuse system.

SYSTEM INTERDEPENDENCIES

60-kilowatt tactical quiet generator

PROGRAM STATUS

- **3QFY08:** Deployment of FP 600-man base camp, 150-man subset through urgent materiel release to support an operational needs statement (ONS)
- **4QFY08:** Deployment of two FP 600-man base camps through urgent materiel release to support ONS
- **1QFY09:** Module number 53 production initiated

PROJECTED ACTIVITIES

- **4QYF09:** Projected delivery of FP Module number 53
- **1QFY10:** Capabilities Production Document approval supporting improved capabilities
- **3QFY10:** Integration of shower water reuse system into FP baseline

ACQUISITION PHASE

- Technology Development
- Engineering & Manufacturing Development
- **Production & Deployment**
- Operations & Support

Force Provider (FP)

FOREIGN MILITARY SALES
None

CONTRACTORS
Force Provider Assembly:
Global Defense Engineering (Easton, MD)
Letterkenny Army Depot (Chambersburg, PA)
Expeditionary TRICON Kitchen System and FP Electric Kitchen:
Tri-Tech USA Inc. (South Burlington, VT)
Airbeam TEMPER Tent:
Vertigo Inc. (Lake Elsinore, CA)
Environmental Control:
Hunter Manufacturing (Solon, OH)
TRICON Container:
Charleston Marine Containers (Charleston, SC)
Waste Water Evacuation Tank/Trailer:
Marsh Industrial (Kalkaska, MI)
Cold Weather Kit Assembly:
Berg Companies, Inc. (Spokane, WA)
Mobile Electric Power Distribution System Replacement:
Lex Products Corp. (Stamford, CT)
Expeditionary TRICON Systems (shower, laundry, latrine):
To be determined

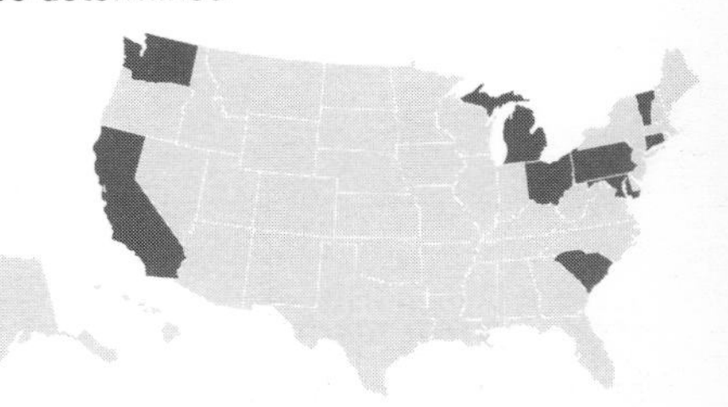

Force XXI Battle Command Brigade-and-Below (FBCB2)

INVESTMENT COMPONENT

- Modernization
- Recapitalization
- Maintenance

MISSION

To provide enhanced situational awareness to the lowest tactical level—the individual Soldier—and a seamless flow of command and control information across the battlefield.

DESCRIPTION

The Force XXI Battle Command Brigade-and-Below (FBCB2) forms the principal digital command and control system for the Army at brigade levels and below. It provides increased situational awareness on the battlefield by automatically disseminating throughout the network timely friendly force locations, reported enemy locations, and graphics to visualize the commander's intent and scheme of maneuver.

FBCB2 is a key component of the Army Battle Command System (ABCS). Applique hardware and software are integrated into the various platforms at brigade and below, as well as at appropriate division and corps slices necessary to support brigade operations. The system features the interconnection of platforms through two communication systems: FBCB2–Enhanced Position Location Reporting System (EPLRS), supported by the tactical Internet; and FBCB2–Blue Force Tracking, supported by L-band satellite. The Joint Capabilities Release (JCR) is the next software release and addresses joint requirements, database simplification, Type 1 encryption, a product line software approach and enables transition to the Blue Force Tracking II (BFT II) transceiver allowing a tenfold increase in data throughput.

SYSTEM INTERDEPENDENCIES

Enhanced Position Location Reporting System (EPLRS)

PROGRAM STATUS

- **2QFY08:** Completed 10,000th installation in theater of FBCB2 on Up-Armored High Mobility Multipurpose Wheeled Vehicles (UAH). New production installations have been completed, although the Program Manager continues to assist theater units with support.
- **4QFY08:** Completed 8,000th installation of FBCB2 on Mine Resistant Ambush Protected (MRAP) vehicles, based on the Army's current requirement for 12,000 MRAP vehicles.
- **1–4QFY08:** Completed fielding of 1,942 FBCB2 systems to the Army National Guard (ARNG).
- **1QFY09:** Began system software acceptance testing (SSAT) for Joint Capabilities Release (JCR) software. JCR upgrades FBCB2 version 6.4.4.2 and 6.5 providing Type 1 encryption capability, simplified database builds, use by the Marine Corps for command and control, and transition to the new Blue Force Tracking II network (reduces latency from minutes to seconds).

PROJECTED ACTIVITIES

- **2QFY10:** Begin AIC testing of JCR Software.
- **1QFY10:** Conduct field test of JCR Software
- **4QFY10:** Conduct operational test of JCR Software

ACQUISITION PHASE

- Technology Development
- Engineering & Manufacturing Development
- Production & Deployment
- Operations & Support

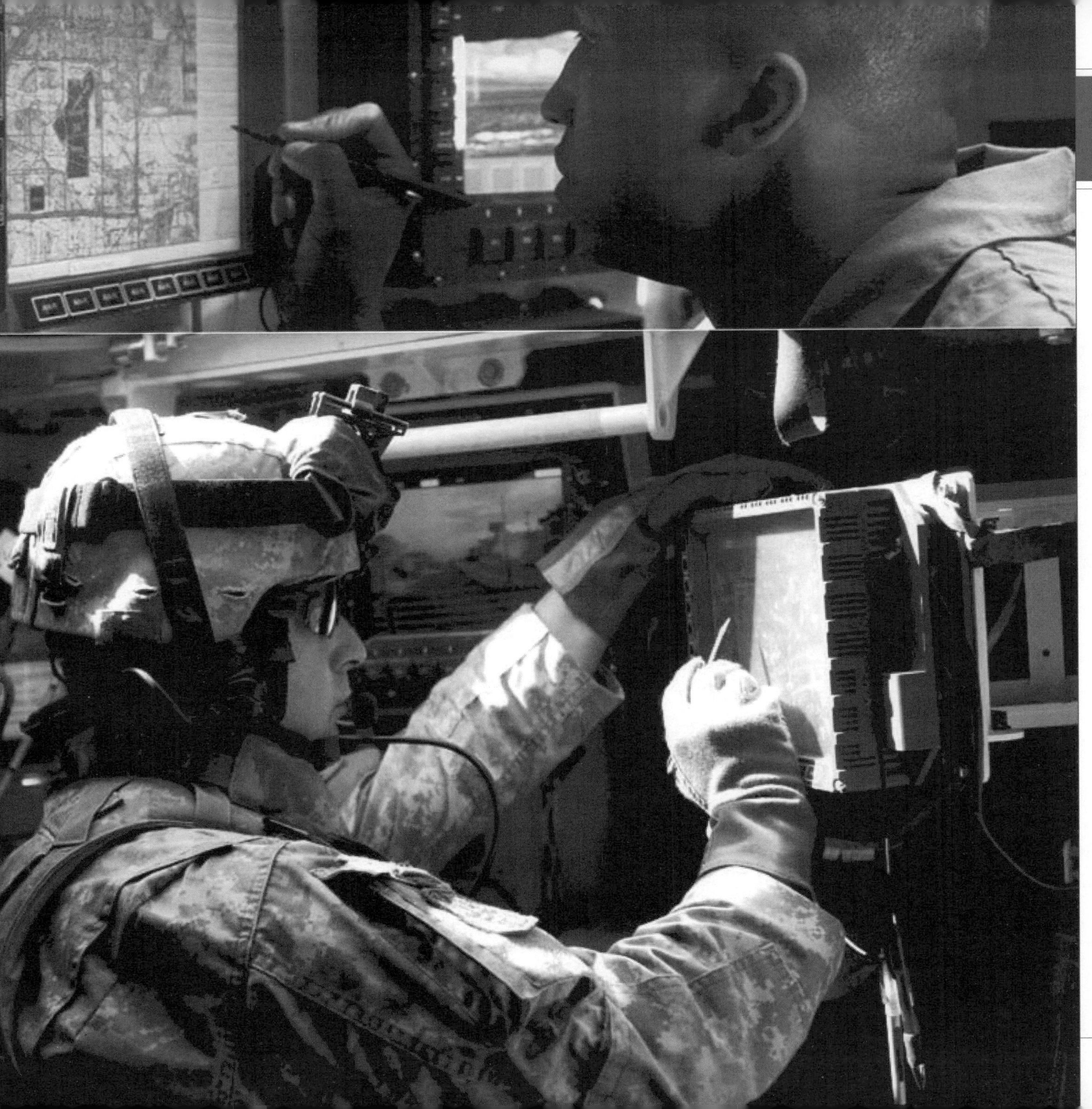

Force XXI Battle Command Brigade-and-Below (FBCB2)

FOREIGN MILITARY SALES
Australia

CONTRACTORS

Software/systems engineering:
Northrop Grumman Space & Mission Systems Corp. (Redondo Beach, CA)

Hardware:
DRS Technologies (Palm Bay, FL)

Installation kits:
Northrop Grumman Space & Mission Systems Corp. (Redondo Beach, CA)

Satellite services:
COMTECH (Germantown, MD)

Field support:
Engineering Solutions and Products, Inc. (Eatontown, NJ)

Forward Area Air Defense Command and Control (FAAD C2)

INVESTMENT COMPONENT
- Modernization
- Recapitalization
- Maintenance

MISSION

To collect, process, and disseminate real-time target tracking and cuing information to all short-range air defense weapons and provide command and control for the Counter-Rocket, Mortar and Artillery (C-RAM) program.

DESCRIPTION

Forward Area Air Defense Command and Control (FAAD C2) software provides critical C2, situational awareness, and automated air track information by integrating engagement operations software for multiple systems, including:

- Avenger
- Sentinel
- Army Battle Command System (ABCS)
- C-RAM Program

FAAD C2 supports air defense and C-RAM weapon systems engagement operations by tracking friendly and enemy aircraft, cruise missiles, unmanned aerial vehicles, mortar and rocket rounds as identified by radar systems, and by performing C2 for Avenger and the C-RAM system.

FAAD C2 uses the following communication systems:

- Enhanced Position Location Reporting System (EPLRS)
- Multifunctional Information Distribution System (MIDS)
- Single Channel Ground and Airborne Radio System (SINCGARS)

FAAD C2 provides joint C2 interoperability and horizontal integration with all Army C2 and air defense artillery system, including, but not limited to:

- Surface Launched Advanced Medium Range Air-to-Air Missile (SLAMRAAM)
- PATRIOT
- Avenger
- Theater High Altitude Area Defense (THAAD)
- Airborne Warning and Control System (AWACS)
- C-RAM
- ABCS

SYSTEM INTERDEPENDENCIES

Radar systems providing input data such as Sentinel, Firefinder, Lightweight Counter-Mortar Radar (LCMR), and AWACS.

PROGRAM STATUS

- **1QFY08:** C-RAM forward operating base fieldings completed
- **4QFY08:** complete Active Army transformation

PROJECTED ACTIVITIES

- **Continuing:** In-country reset of Operation Iraqi Freedom/Operation Enduring Freedom (OIF/OEF) assets
- **3QFY09:** FAAD C2 Version 5.4B materiel release
- **3QFY10:** FAAD C2 fielded to final National Guard unit
- **4QFY10:** FAAD C2 Complete CHS-3 upgrades

ACQUISITION PHASE

Technology Development | Engineering & Manufacturing Development | Production & Deployment | Operations & Support

Forward Area Air Defense Command and Control (FAAD C2)

FOREIGN MILITARY SALES
Egypt

CONTRACTORS
Software:
Northrop Grumman Space & Mission Systems Corp. (Redondo Beach, CA)
Hardware:
Tobyhanna Army Depot (Scranton, PA)
PKMM (Las Vegas, NV)
SETA:
ITT–CAS, Inc. (Huntsville, AL)
CHS 3:
General Dynamics (Taunton, MA)

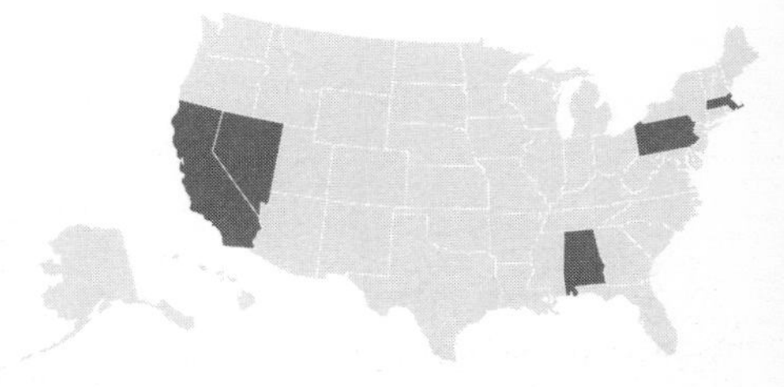

Future Tank Main Gun Ammunition

INVESTMENT COMPONENT

- Modernization
- Recapitalization
- Maintenance

MISSION

To provide overwhelming lethality overmatch to the heavy armor fleet.

DESCRIPTION

The Future Tank Main Gun Ammunition Suite consists of two cartridges and will provide enhanced lethality and increased capability to the Heavy Brigade Combat Team.

The Advanced Kinetic Energy (AKE) cartridge, designated M829E4, will use an advanced penetrator to defeat future heavy armor targets equipped with explosive reactive armor and active protection systems. This will increase survivability of the Abrams tank in the 0–4 kilometer range.

The Advanced Multi-Purpose (AMP) cartridge will combine the capabilities of a number of existing munitions into one cartridge. This cartridge will utilize air bursting warhead and multimode fuze technology to combine those capabilities and provide new capability against dismounted infantry at longer ranges. This cartridge will employ high-explosive, anti-personnel, obstacle reduction, and anti-helicopter capabilities into one munition, thus streamlining the logistical footprint associated with deploying heavy forces. This cartridge will further enhance survivability and lethality for Abrams tanks and Mounted Combat Systems vehicles in the 0–4 kilometer range.

SYSTEM INTERDEPENDENCIES

The Future Tank Main Gun Ammunition suite must be compatible with the Abrams tank fleet through the remainder of its service life.

PROGRAM STATUS

- **4QFY09:** Milestone B for AKE
- **FY10:** AKE Engineering and Manufacturing Development (EMD) initiation
- **Currently:** AKE TRL-6 demonstrated; AMP TRL-6 demonstrated

PROJECTED ACTIVITIES

- **FY10:** Award of two competing EMD contracts for AKE
- **FY11:** Milestone B for AMP

ACQUISITION PHASE

Technology Development | Engineering & Manufacturing Development | Production & Deployment | Operations & Support

Future Tank Main Gun Ammunition

FOREIGN MILITARY SALES
None

CONTRACTORS
To be determined

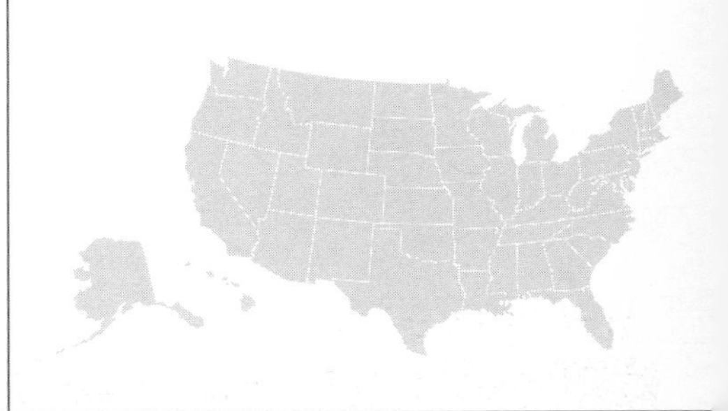

General Fund Enterprise Business Systems (GFEBS)

INVESTMENT COMPONENT

- **Modernization**
- Recapitalization
- Maintenance

MISSION

To acquire a new core financial management capability for administering the Army's General Fund to improve performance, standardize processes, and ensure that it can meet future needs.

DESCRIPTION

The Army will implement a commercial off-the-shelf Enterprise Resource Planning (ERP) system that meets the requirements of the Chief Financial Officers Act and the Federal Financial Management Improvement Act of 1996 (FFMIA), and that is capable of supporting DoD with accurate, reliable, and timely financial information.

The GFEBS implementation involves standardizing financial management, accounting functions, real property inventory, and management across the Army. As a result, Army financial and real property professionals will have access to timely, reliable, and accurate information. GFEBS will also improve cost management and control, allow more time to perform financial analysis, and facilitate a more accurate understanding of the value, location, and characteristics of all property. GFEBS will provide a comprehensive system for many of the Army's financial and accounting functions including general ledger, accounts payable, revenue and accounts receivable, cost management, financial reporting, and real property inventory and management.

SYSTEM INTERDEPENDENCIES

None

PROGRAM STATUS

- **3QFY08:** Developmental testing for release 1.2
- **3QFY08:** Begin build phase of second production release (Release 1.3)
- **1QFY09:** Go live of first production release
- **1QFY09:** Limited user test

PROJECTED ACTIVITIES

- **2QFY09:** Go live of release 1.3
- **2QFY09:** Milestone C and move to production and deployment/ operations and support phase
- **2QFY09:** Initial operational capability
- **3QFY09:** Full Army-wide deployment of release 1.3
- **1QFY10:** Limited user testing of release 1.4

ACQUISITION PHASE

Technology Development | **Engineering & Manufacturing Development** | Production & Deployment | Operations & Support

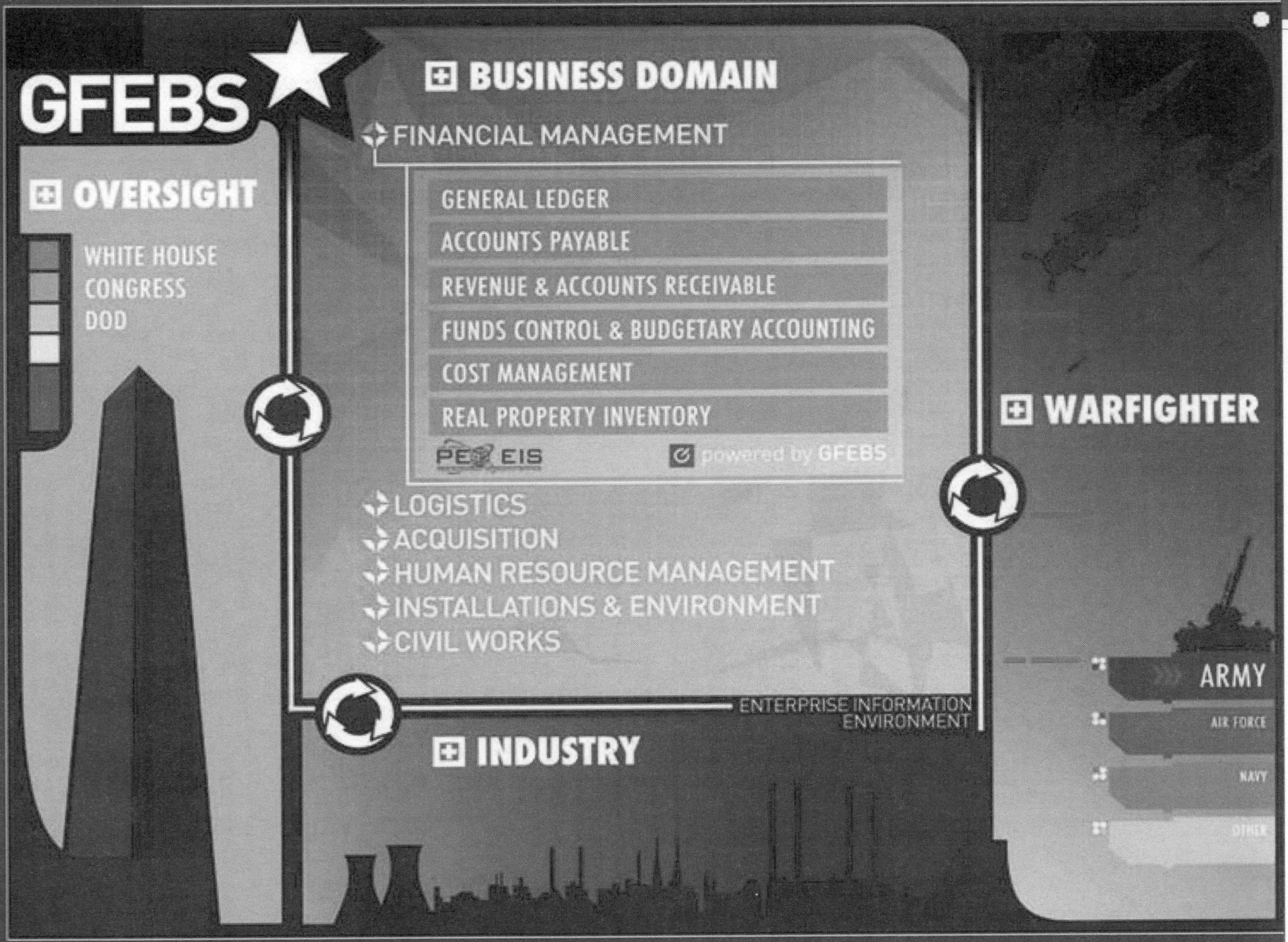

General Fund Enterprise Business Systems (GFEBS)

FOREIGN MILITARY SALES
None

CONTRACTORS
Systems Integration:
Accenture (Reston, VA)
Technical Program Management:
iLumina Solutions (California, MD)
Program Management:
Binary Group (Bethesda, MD)
IV&V:
SNVC (Fairfax, VA)

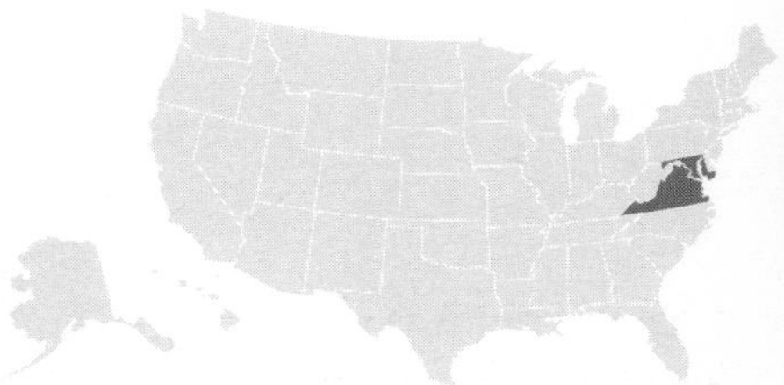

Global Combat Support System–Army (GCSS–Army)

INVESTMENT COMPONENT
- **Modernization**
- Recapitalization
- Maintenance

MISSION

To provide responsive and efficient logistical support by reengineering current business processes, and by developing and fielding modernized tactical automation systems to achieve a Single Army Logistics Enterprise (SALE).

DESCRIPTION

Global Combat Support System–Army (GCSS–Army) and Product Lifecycle Management Plus (PLM+) enable Army and joint transformation of combat support/combat service support (CS/CSS) using Enterprise Resource Planning (ERP) software products. GCSS–Army supports rapid force projection in the battlefield functional areas of arming, fixing, fueling, moving, sustaining, and tactical logistics financial processes. The GCSS–Army development includes the reengineering of 12 legacy Army logistics processes and the interface/integration with applicable command and control (C2) and joint systems. GCSS–Army is the primary enabler for the Army transformation vision of a technologically advanced ERP that manages the flow of logistics resources and information to satisfy the Army's modernization requirements. PLM+ integrates Army business functions by providing a single source for enterprise hub services, business intelligence and analytics, and centralized master data management. It will become the Army Enterprise Systems Integration Program (AESIP) as it evolves to support the Army's vision of ERP-based cross-domain business integration. GCSS–Army and PLM+ will meet the warfighter's need for responsive support at the right place and time.

SYSTEM INTERDEPENDENCIES

General Fund Enterprise Business System

PROGRAM STATUS

- **4QFY08:** Milestone B
- **1QFY09:** Acquisition program baseline signed

PROJECTED ACTIVITIES

- **2QFY09:** Material Master R.1.0 go-live (PLM+)
- **3QFY10:** Release 1.1 go-live (GCSS–Army)
- **4QFY10:** Release 1.1 developmental test and evaluation; initial government testing (GCSS–Army)

ACQUISITION PHASE
- Technology Development
- **Engineering & Manufacturing Development**
- Production & Deployment
- Operations & Support

What GCSS-Army Provides

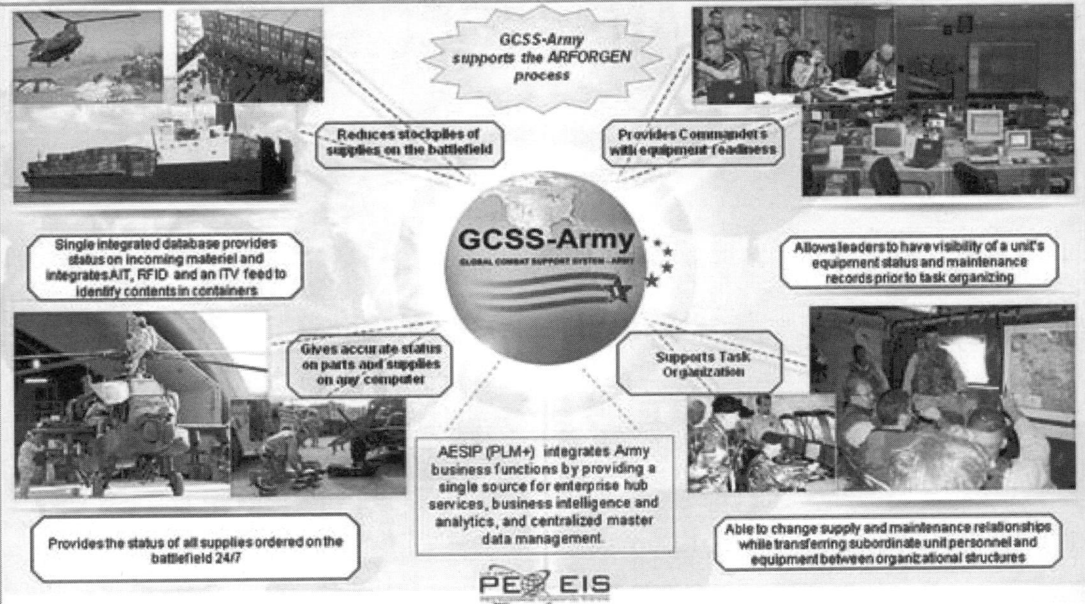

System Architecture

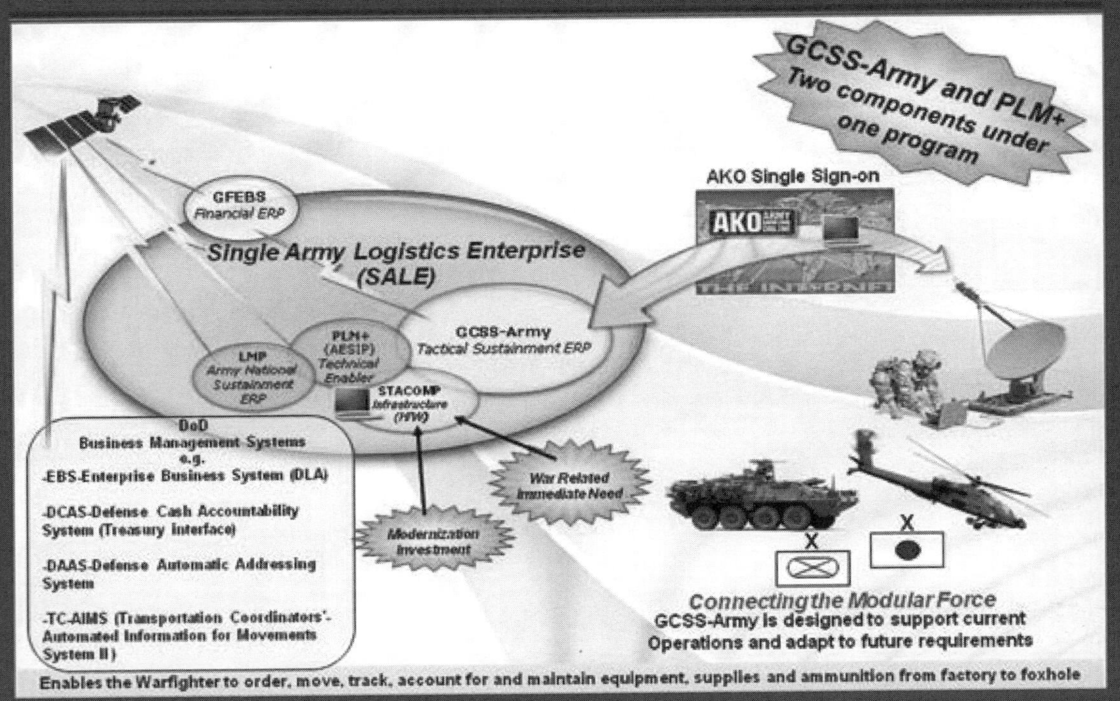

Global Combat Support System–Army (GCSS–Army)

FOREIGN MILITARY SALES

None

CONTRACTORS

GCSS–Army:
Northrop Grumman (Richmond, VA)
PLM+:
Computer Sciences Corp. (CSC) (Falls Church, VA)
LMI Consulting (McLean, VA)
MPRI (L-3 Communications Division) (Colonial Heights, VA)
SNVC (Fairfax, VA)

Global Command and Control System–Army (GCCS–A)

INVESTMENT COMPONENT

Modernization
Recapitalization
Maintenance

MISSION

To enhance warfighter capabilities during joint and combined operations, through automated command and control tools for strategic and operational commanders.

DESCRIPTION

The Global Command and Control System–Army (GCCS–A) is the Army's strategic and operational command and control (C2) system, providing readiness, planning, mobilization, and deployment capability information for strategic commanders. For theater commanders, GCCS–A provides the following:

- Common operational picture and associated friendly and enemy status information
- Force-employment planning and execution tools (receipt of forces, intra-theater planning, readiness, force tracking, onward movement, and execution status)
- Overall interoperability with joint, coalition, and the tactical Army Battle Command System (ABCS)

GCCS–A supports Army units from the strategic commanders and regional combatant commanders in theater, down through the joint task force commander. As part of ABCS, GCCS–A provides a seamless Army extension from the joint GCCS system to echelons corps and below. Compatibility and interoperability are achieved by building the GCCS–A applications to function on the common operating environment and through interfaces with other C2 systems within the Army and other services.

The common operating environment specifies a common system infrastructure for all C2 systems in accordance with the joint technical architecture guidelines. These provide a common support architecture and modular software for use by the services and agencies in developing mission-specific solutions to their C2 requirements. The hardware platform is based on commercial off-the-shelf hardware. The system users are linked via local area networks in client/server configurations with an interface to the Secret Internet Protocol Router Network for worldwide communication.

SYSTEM INTERDEPENDENCIES

ABCS, AFATDS, GCCS–J, MCS, ASAS, GSORTS, JOPES, DRRS, DTSS, BCS3

PROGRAM STATUS

- **3QFY08–1QFY09:** Release GCCS–A Block IV Version 4.1.b to the field
- **3QFY08–1QFY09:** Support Operations Enduring Freedom and Iraqi Freedom (OEF/OIF)
- **3QFY08–1QFY09:** Support Net-Enabled Command Capability (NECC) development of Army capability modules with unique Army requirements
- **3QFY08–1QFY09:** Release Defense Readiness Reporting System–Army (DRRS–A) Force Readiness Tool (Phase 3) to the field

PROJECTED ACTIVITIES

- **2–3QFY09:** Continue spiral development in support of GCCS–A Version 4.1.x and DRRS–A Phase 4 requirements
- **2QFY09–1QFY11:** Continue directed fieldings and required support for OEF/OIF
- **2QFY09–1QFY11:** Continue support to NECC technology demonstration phase and development of Army unique capability modules

ACQUISITION PHASE

Technology Development | Engineering & Manufacturing Development | Production & Deployment | Operations & Support

Global Command & Control System - Army

Global Command and Control System–Army (GCCS–A)

FOREIGN MILITARY SALES
None

CONTRACTORS
Lockheed Martin (Springfield, VA)
Software:
Lockheed Martin (Springfield, VA)
Accenture (Camden, NJ)
Hardware:
General Dynamics (Taunton, MA)
GTSI (Chantilly, VA)
Fielding support:
General Dynamics Information Technology (Springfield, VA)
Engineering Solutions & Products. Inc. (Eatontown, NJ)
Software training:
General Dynamics Information Technology (Atlanta, GA)

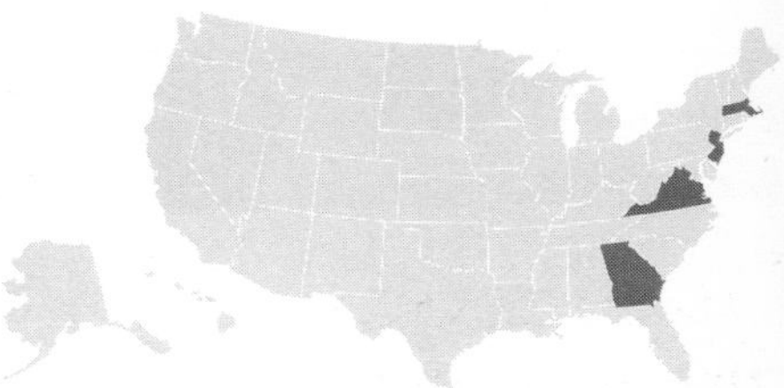

Ground Soldier System (GSS)

INVESTMENT COMPONENT

- Modernization
- Recapitalization
- Maintenance

MISSION

To provide unparalleled situational awareness and understanding to the dismounted leader (team leader and above) allowing for faster, more accurate decisions in the tactical fight and connecting the dismounted Soldier to the network.

DESCRIPTION

The Ground Soldier System (GSS) is an integrated dismounted Soldier situational awareness system for use during combat operations. It consists of a hands-free display to view information; a computer to process information and populate the screen; an interface device for user-screen interaction; a system power source; a software operating system for system functionality; a tactical applications and battle command; and a networked radio transmitter/receiver to send and receive information.

SYSTEM INTERDEPENDENCIES

Battle Command product line, Enhanced Position Location Reporting System Radio, Rifleman Radio, Core Soldier System equipment

PROGRAM STATUS

- **1QFY09:** Milestone A Defense Acquisition Board; release of request for proposal
- **2QFY09:** Award of three competitive protoyping contracts
- **4QFY09:** Preliminary design review

PROJECTED ACTIVITIES:

- **1QFY10:** Critical design review
- **3QFY10:** Capability Development Document joint validation
- **4QFY10:** Capability Production Document joint-validation
- **4QFY10:** Limited user tests (3 each)
- **2QFY11:** Milestone C

ACQUISITION PHASE

Technology Development | Engineering & Manufacturing Development | Production & Deployment | Operations & Support

Ground Soldier System (GSS)

FOREIGN MILITARY SALES
To be determined

CONTRACTORS
General Dynamics (Scottsdale, AZ)
Raytheon (Plano, TX)
Rockwell Collins (Cedar Rapids, IA)

Guardrail Common Sensor (GR/CS)

INVESTMENT COMPONENT

Modernization
Recapitalization
Maintenance

MISSION

To provide signal intelligence collection and precision targeting that intercepts, collects, and precisely locates hostile communications intelligence radio frequency emitters and electronic intelligence threat radar emitters.

DESCRIPTION

The Guardrail Common Sensor (GR/CS) is a fixed-wing, airborne, signals intelligence (SIGINT) collection and precision targeting location system. It provides near-real-time information to tactical commanders in the corps/joint task force area with emphasis on deep battle and follow-on forces attack support. It collects low-, mid-, and high-band radio signals and electronic intelligence (ELINT) signals; identifies and classifies them; determines source location; and provides near-real-time reporting, ensuring information dominance to commanders. GR/CS uses a Guardrail Ground Baseline (GGB) for the control, data processing, and message center for the system. It includes:

- Integrated communications intelligence (COMINT) and ELINT collection and reporting
- Enhanced signal classification and recognition and precision emitter geolocation
- Near-real-time direction finding
- Advanced integrated aircraft cockpit
- Tactical Satellite Remote Relay System (Systems 1, 2, 3, and 4)

A standard system has eight to 12 RC-12 aircraft flying operational missions in sets of two or three. Up to three airborne relay facilities simultaneously collect communications and noncommunications emitter transmissions and gather lines of bearing and time-difference-of-arrival data, which is transmitted to the GGB, correlated, and supplied to supported commands.

Planned improvements through Guardrail modernization efforts include an enhanced precision geolocation subsystem, the Communications High-Accuracy Location Subsystem–Compact (CHALS–C), with increased frequency coverage and a higher probability to collect targets; a modern COMINT infrastructure and core COMINT subsystem, providing a frequency extension, Enhanced Situational Awareness (ESA); a capability to process special high-priority signals through the high-end COMINT subsystem; and elimination of non-supportable hardware and software. Ground processing software and hardware are being upgraded for interoperability with the Distributed Common Ground System–Army (DCGS–A) architecture and Distributed Information Backbone.

SYSTEM INTERDEPENDENCIES

DCGS–A

PROGRAM STATUS

- **3QFY08:** Initial CHALS–C flight test
- **1QFY09:** ESA factory acceptance test

PROJECTED ACTIVITIES

- **3QFY09:** High band COMINT (HBC) factory acceptance test
- **4QFY09:** CHALS–C, ESA, HBC, and ELINT system flight test
- **3QFY10:** CHALS–C, ESA, HBC, and ELINT upgrades system assessment
- **3QFY10:** CHALS–C, ESA, HBC, and ELINT fieldings; GGB fieldings to the 224th Military Intelligence (MI), 3rd MI, 15th MI; Initiating new contracts for additional GGB hardware

ACQUISITION PHASE

Technology Development | Engineering & Manufacturing Development | Production & Deployment | Operations & Support

Guardrail Common Sensor (GR/CS)

FOREIGN MILITARY SALES
None

CONTRACTORS
System Integrator, ESA Subsystem, and GGB Software/System Support:
Northrop Grumman (Sacramento, CA)
Data links:
L-3 Communications (Salt Lake City, UT)
CHALS-C:
Lockheed Martin (Owego, NY)
X-MIDAS software:
ZETA (Fairfax, VA)
HBC Subsystem:
ArgonST Radix (Mountain View, CA)

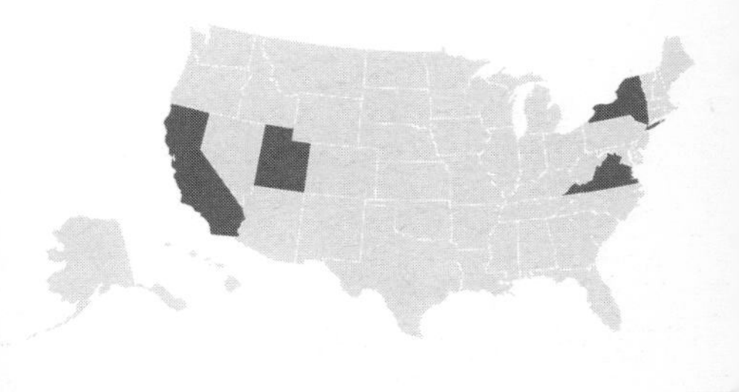

Guided Multiple Launch Rocket System (GMLRS)

INVESTMENT COMPONENT

Modernization

Recapitalization

Maintenance

MISSION

To provide responsive, long-range, precision fires against area and point targets in open/complex/urban terrain with effects matched to the target and rules of engagement.

DESCRIPTION

The Guided Multiple Launch Rocket System (GMLRS) is a major upgrade to the M26 rocket, producing precise destructive and shaping fires against a variety of target sets. GMLRS is employed with the M270A1 upgraded Multiple Launch Rocket System (MLRS) tracked launcher and the M142 High Mobility Artillery Rocket System (HIMARS) wheeled launchers. GMLRS munitions have greater accuracy with a resulting higher probability of kill, smaller logistics footprint, and minimized collateral damage. There are two variants of the GMLRS: the previously produced dual-purpose improved conventional munitions (DPICM) variant designed to service area targets; and the unitary variant with a single 200-pound class high-explosive charge to provide blast and fragmentation effects on, above, or in a specific point target.

The development of a third alternative warhead has been initiated. The Alternative Warhead Program (AWP) will service area target sets without producing unexploded ordnance and will be fielded in FY15. The original GMLRS development was an international cooperative program with the United Kingdom, Germany, France, and Italy. An urgent materiel release version of the GMLRS unitary variant has been produced and fielded in support of U.S. Central Command (CENTCOM) forces with nearly 1,000 rockets used in operations through November 2008.

Rocket Length: 3,937mm
Rocket Diameter: 227mm
Rocket Reliability: Threshold 92 percent; objective: 95 percent
Ballistic Range(s): 15 to 70+ kilometers

SYSTEM INTERDEPENDENCIES

Global Positioning System, Advanced Field Artillery Tactical Data System (AFATDS), M270 A1, and HIMARS Launchers

PROGRAM STATUS

- **2–3QFY08:** GMLRS unitary initial operational test
- **1QFY09:** GMLRS AWP Configuration Steering Board (CSB), Acquisition Decision Memorandum (ADM) approved for Technology Development Initiation of the AWP
- **1QFY09:** GMLRS AWP CSB ADM halts new DPICM procurements
- **1QFY09:** GMLRS Unitary full-rate production decision

PROJECTED ACTIVITIES

- **4QFY09:** GMLRS AWP Milestone A
- **1QFY11:** GMLRS AWP Warhead Prototype Technical Demonstrations
- **4QFY11:** GMLRS AWP Milestone B

ACQUISITION PHASE

Technology Development | Engineering & Manufacturing Development | Production & Deployment | Operations & Support

Guided Multiple Launch Rocket System (GMLRS)

FOREIGN MILITARY SALES
United Kingdom, UAE, Singapore, Bahrain, Japan, Germany, and France

CONTRACTORS
Prime munitions integrator:
Lockheed Martin (Dallas, TX)
Rocket assembly:
Lockheed Martin (Camden, AR)
Motor assembly:
Aerojet (Camden, AR)
G&C section:
Honeywell (Clearwater, FL)
Motor case/warhead skins:
Aerojet (Vernon, CA)

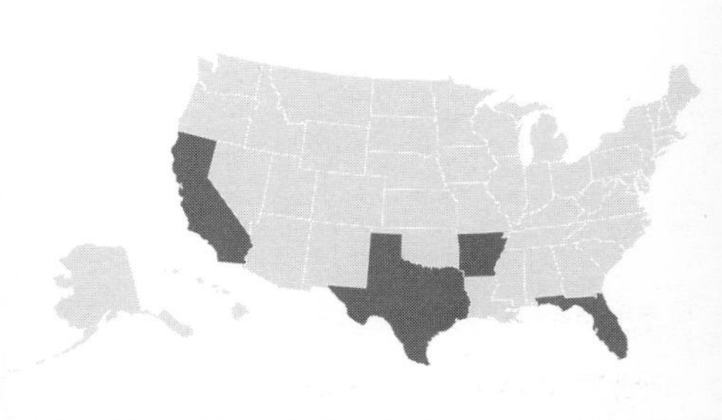

Heavy Expanded Mobility Tactical Truck (HEMTT)/ HEMTT Extended Service Program (ESP)

INVESTMENT COMPONENT

Modernization

Recapitalization

Maintenance

MISSION

To support combat units by performing line and local haul, unit resupply, helicopter and tactical vehicle refueling, and related missions in a tactical environment.

DESCRIPTION

The rapidly deployable Heavy Expanded Mobility Tactical Truck (HEMTT), developed for cross-country military missions, transports ammunition, petroleum, oils, and lubricants to currently equipped, digitized, and brigade/battalion areas of operation.

The HEMTT A4 model began fielding in December 2008. Enhancements include air-ride suspension, a new Caterpillar C-15 engine, the Allison 4500 transmission, anti-lock braking system and traction control, J-1939 data-bus, a larger common cab, which is common with the Palletized Load System (PLS) A1, and Long Term Armor Strategy (LTAS)-compliant.

The HEMTT ESP, also known as HEMTT RECAP, is a recapitalization program that converts high-mileage HEMTT trucks to 0 Miles/0 Hours and to the current A4 production configuration. The trucks are disassembled and rebuilt with improved technology found in the HEMTT A4.

HEMTT comes in six basic configurations:

- **M977:** Cargo truck with light materiel handling crane
- **M985:** Cargo truck with materiel-handling crane
- **M978:** 2,500-gallon fuel Tanker
- **M983:** Tractor
- **M984:** Wrecker
- **M1120:** Load-Handling System (LHS) transports palletized materiel and ISO containers

Truck payload: 11 tons
Trailer payload: 11 tons
Flatrack dimensions: 8-foot-by-20-foot (International Standards Organization (ISO) container standard)
Engine type: Diesel
Transmission: Automatic
Number of driven wheels: 8
Range: 300 miles
Air transportability: C-130, C-17, C-5

SYSTEM INTERDEPENDENCIES

M983 HEMTT tractor Light Equipment Transporter (LET), M1977 HEMTT Common Bridge Transporter, M18 Dry Support Bridge, and the Theatre High Altitude Area Defense Missile System (THAAD); M984A2 and M984A4 Tractors with Fifth Wheel Towing Device and High Mobility Recovery Trailer are designated Stryker and MRAP Interim Recovery Systems

PROGRAM STATUS

- **1QFY09:** HEMTT A4 Family of Vehicles was type classified standard and full materiel released (TC/MR), also received TC/MR and is being fielded to Multi-Role Bridging Company's (MRBC)

PROJECTED ACTIVITIES

- **FY10:** Distribute HEMTT A4s to next deployers in accordance with HQDA G8 distribution plan

ACQUISITION PHASE

Technology Development | Engineering & Manufacturing Development | Production & Deployment | Operations & Support

Heavy Expanded Mobility Tactical Truck (HEMTT)/HEMTT Extended Service Program (ESP)

FOREIGN MILITARY SALES
Turkey, Israel, and Jordan

CONTRACTORS
Oshkosh Truck Corp. (Oshkosh, WI; Kileen, TX)
Caterpillar (Peoria, IL)
Allison Transmissions (Indianapolis, IN)
Michelin (Greenville, SC)

Heavy Loader

INVESTMENT COMPONENT

Modernization

Recapitalization

Maintenance

MISSION

To support engineer construction missions by providing engineer units the capability to perform multiple operations with one piece of equipment supporting division, corps, and theater Army missions.

DESCRIPTION

The Heavy Loader is a commercial vehicle modified for military use. Modifications include Chemical Agent Resistant Coating (CARC) paint, blackout lights, rifle rack, and military standard (MIL-STD-209) lift and tiedown. The military version of the loader will be armored with an A kit (armored floor plate) on all loaders and C kit (armored cab) on selected loaders. The loader bucket is multifunctional with a capacity of 4.5 cubic yards (Type I–Quarry Teams) and 5 cubic yards (Type II–all other units for general use).

SYSTEM INTERDEPENDENCIES

M916/M870 truck trailer for highway transportability

PROGRAM STATUS

- All ballistics and vehicle performance testing completed. Logistics development activities well underway.

PROJECTED ACTIVITIES

- **4QFY09:** Type classification–standard and full material release scheduled.

ACQUISITION PHASE

Technology Development | Engineering & Manufacturing Development | Production & Deployment | Operations & Support

and phone cards are not refundable.

REFUNDS

Items in new condition may be exchanged or returned below within 90 days* except for the following:

* **15 days for:** Computers, GPS Units and **UNOPENED** peripherals, software, pre-recorded music, videos and video games

* **30 days for:** Jewelry and Watches, Camcorders, Televisions, Digital Cameras, Furniture, Mattresses, Major Appliances and Gas Powered Equipment

* **Sales receipt required**

If a sales receipt is not available the refund amount will be loaded on an AAFES Merchandise Card.

Gift Cards, pre-paid music, wireless, and phone cards are not refundable.

REFUNDS

Items in new condition may be exchanged or returned below within 90 days* except for the following:

* **15 days for:** Computers, GPS Units and **UNOPENED** peripherals, software, pre-recorded music, videos and video games

* **30 days for:** Jewelry and Watches, Camcorders, Televisions, Digital Cameras, Furniture, Mattresses, Major Appliances and Gas Powered Equipment

* **Sales receipt required**

06371 4079 103
SERVING THE BEST CUSTOMERS IN THE WORLD
SHOP US ONLINE AT WWW.SHOPMYEXCHANGE.COM
MGR JASON ROSENBERG,ROSENBERGJ@AAFES.COM

US ARMY WEAPONS SYS 978160239725
17.95

TOTAL $17.95
MASTERCARD $17.95
XXXXXXXXXXXX6881
EXPIRY: XX/XX SWIPED BANK
AUTH# 01842P
SEQ# 8681

ITEMS 1
03/18/2012 17:12 3162 53 000518 2220

532220120318316200

THE EXCHANGE VALUES YOUR OPINION!
TAKE A SURVEY AT:
WWW.INTOUCHINSIGHT.COM/EXCHANGE
STORE NUMBER 3162
WIN SWEEPSTAKES PRIZES

Heavy Loader

FOREIGN MILITARY SALES
None

CONTRACTORS
OEM:
Caterpillar Defense and Federal Products (Peoria, IL)
Armor:
BAE Systems (Rockville, MD)
Logistics:
XMCO (Warren, MI)

HELLFIRE Family of Missiles

INVESTMENT COMPONENT

Modernization

Recapitalization

Maintenance

MISSION

To engage and defeat individual moving or stationary advanced-armor, mechanized or vehicular targets, patrol craft, buildings or bunkers while increasing aircraft survivability.

DESCRIPTION

The HELLFIRE family of munitions, consisting of the AGM-114 A, C, F, K, L, M, N, and P model missiles, provides air-to-ground precision strikes and is designed to defeat individual hard-point targets. The Laser HELLFIRE (HELLFIRE II) comes with either a shaped-charge warhead for defeating armor targets or a penetrating-blast-fragmentation warhead for defeating buildings and bunkers. It uses semi-active laser terminal guidance and is the primary anti-tank armament for the AH-64 Apache, OH-58 Kiowa Warrior, Special Operations aircraft, the Marine Corps' AH-1W Super Cobra Helicopters, and the Army's Sky Warrior Unmanned Aircraft System (UAS).

The Longbow HELLFIRE (L model–no longer in production) uses millimeter wave technology for terminal guidance. The Longbow HELLFIRE ability to engage single or multiple targets directly or indirectly and to fire single, rapid, or ripple (salvo) rounds gives combined arms forces a decisive battlefield advantage.

Laser HELLFIRE (AGM-114K, M, and N models) and Longbow HELLFIRE incorporate many improvements over the basic HELLFIRE missile, including:

- Electro-optical countermeasure hardening
- Software-controlled digital seeker and autopilot electronics to adapt to changing threats and mission requirements
- Increased warhead lethality capable of defeating all projected armor threats into the 21st century

Laser HELLFIRE semi-active laser precision guidance and Longbow HELLFIRE fire-and-forget capability will provide the battlefield commander with fast battlefield response and flexibility across a wide range of mission scenarios.

Laser HELLFIRE
Diameter: 7 inches
Weight: 100 pounds
Length: 64 inches
Range: 0.50–8.0 kilometers
Longbow HELLFIRE:
Diameter: 7 inches
Weight: 108 pounds
Length: 69.2 inches
Range: 0.50–8.0 kilometers

SYSTEM INTERDEPENDENCIES

None

PROGRAM STATUS

Laser HELLFIRE

- **3QFY08–1QFY09:** 114 N (thermobaric) missiles were delivered to replace missiles expended in the Global War on Terrorism (GWOT).

PROJECTED ACTIVITIES

Laser HELLFIRE

- Continue production

Longbow HELLFIRE

- Continue sustainment activities

ACQUISITION PHASE

Technology Development | Engineering & Manufacturing Development | Production & Deployment | Operations & Support

HELLFIRE Family of Missiles

Family of HELLFIRE Munitions
All Variants

System Description	Production	Characteristics	Performance
AGM-114A, B, C, F, FA – HELLFIRE Weight = 45 kg Length = 163 cm	1982 – 1992	**A, B, C** have a Single Shaped-Charge Warhead; Analog Autopilot	• Not Capable Against Reactive Armor • Non-Programmable
		F has Tandem Warheads; Analog Autopilot	• **Reactive Armor Capable** • Non-Programmable • **FA** adds **Blast Frag Sleeve**
AGM-114K/K2/K2A – HELLFIRE II Weight = 45 kg Length = 163 cm	1993 – UTC	• Tandem Warheads • Electronic Safe & Arm Device • Digital Autopilot & Electronics • Improved Performance Software	• Capable Against 21st Century Armor • Hardened Against Countermeasures • **K-2** adds **Insensitive Munitions (IM)** • **K-2A** adds **Blast-Frag Sleeve**
AGM-114L – HELLFIRE LONGBOW Weight = 49 kg Length = 180 cm	1995 – 2005	• Tandem Warheads • Digital Autopilot & Electronics • Millimeter-Wave **(MMW)** Seeker • IM Warheads	• Initiate on Contact • Hardened Against Countermeasures • Programmable Software • All-Weather
AGM-114M – HELLFIRE II (Blast Frag) Weight = 49 kg Length = 180 cm	1998 – UTC	• Blast-Frag Warhead • 4 Operating Modes • Digital Autopilot & Electronics • Delayed-Fuse Capability	• For Buildings, Soft-Skin Vehicles • Optimized for Low Cloud Ceilings • Hardened Against Countermeasures • WH Penetrates Target Before Detonation
AGM-114N – HELLFIRE II (MAC) Weight = 49 kg Length = 180 cm	2003 – UTC	• Metal-Augmented Charge – Sustained Pressure Wave • 4 Operating Modes • Delayed-Fuse Capability	• For Buildings, Soft-Skin Vehicles • Optimized for Low Cloud Ceilings • Hardened Against Countermeasures • WH Penetrates Target Before Detonation
AGM-114R – HELLFIRE II (for UAS) Weight = 49kg Length = 180cm	2012 – UTC	• Multi-Purpose Warhead • Designed for UAV and and Rotary Wing • 21st Century Armor Capability	• Variable Delay Fuze • IMU & Improved Software • Fully HFII Backward Compatible

FOREIGN MILITARY SALES

Laser HELLFIRE:
Singapore, Israel, Kuwait, Netherlands, Greece, Egypt, Saudi Arabia, Taiwan, Australia, Spain
Direct commercial sale:
United Kingdom
Longbow HELLFIRE:
Singapore, Israel, Kuwait, Japan
Direct commercial sale:
United Kingdom

CONTRACTORS

Lockheed Martin (Troy, AL)
L-3 Communications (Chicago, IL)
Alliant Techsystems (Rocket City, WV)
Moog (Salt Lake City, UT)
Laser HELLFIRE Missile System, guidance section, sensor group:
HELLFIRE LLC (Orlando, FL)
Longbow HELLFIRE:
Longbow LLC (Orlando, FL)

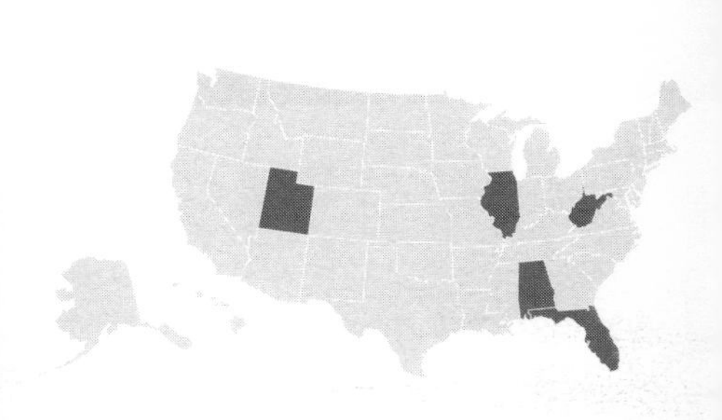

Helmet Mounted Enhanced Vision Devices

INVESTMENT COMPONENT

Modernization

Recapitalization

Maintenance

MISSION

To enhance the warfighter's visual ability and situational awareness while successfully engaging and executing operations day or night, whether in adverse weather or battlefield obscurant conditions.

DESCRIPTION

The AN/PVS-14 Monocular Night Vision Device (MNVD) is a lightweight, multipurpose, passive device used by the individual warfighter in close combat, combat support, and combat service support. It amplifies ambient light and very-near infrared energy for night operations. AN/PVS-14 can be mounted to the M16/M4 receiver rail.

PVS-14
Field of View: ≥ 40 degrees
Weight (maximum): 1.25 pounds
Magnification: 1x
Range: 150 meters

SYSTEM INTERDEPENDENCIES

None

PROGRAM STATUS

- **Current:** AN/PVS-14 in production and being fielded

PROJECTED ACTIVITIES

- **Continue:** Production and fielding in accordance with Headquarters Department of the Army (HQDA)-G8 priorities

ACQUISITION PHASE

Technology Development | Engineering & Manufacturing Development | Production & Deployment | Operations & Support

Helmet Mounted Enhanced Vision Devices

FOREIGN MILITARY SALES
None

CONTRACTORS
L-3 Communications Electro-Optic Systems (Tempe, AZ; Garland, TX; Huntsville, AL; West Springfield, MA)
ITT Industries (Roanoke, VA; West Springfield, MA)

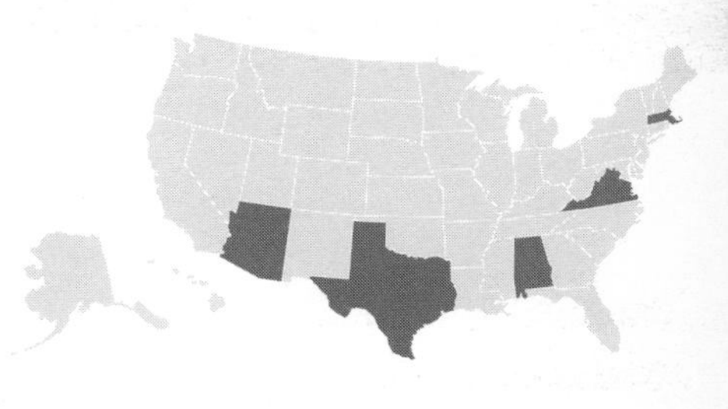

High Mobility Artillery Rocket System (HIMARS)

INVESTMENT COMPONENT

Modernization

Recapitalization

Maintenance

MISSION

To provide early entry and contingency forces with highly lethal, responsive, and precise long-range rocket and missile fires that defeat point and area targets in both urban/complex and open terrain, with minimal collateral damage, via a highly mobile and deployable multiple launch system.

DESCRIPTION

The M142 High Mobility Artillery Rocket System (HIMARS) is a wheeled artillery system that provides close- and long-range precision rocket and missile fire support for Army and Marine early-entry expeditionary forces, contingency forces, and modular fire support brigades supporting Brigade Combat Teams. The combat-proven HIMARS is rapidly deployable via C-130 and operates in all weather and visibility conditions. HIMARS is mounted on a five-ton modified Family of Medium Tactical Vehicles chassis. The wheeled chassis allows for faster road movement and lower operating costs, and requires far fewer strategic airlifts (via C-5 or C-17) to transport a firing battery than the current tracked M270 Multiple Launch Rocket System (MLRS) that it replaces. The M142 provides responsive, highly accurate, and extremely lethal surface-to-surface rocket and missile fires 15 to 300 kilometers. HIMARS can fire all current and planned suites of MLRS munitions, including Army Tactical Missile System missiles and guided MLRS rockets. HIMARS carries either six rockets or one missile, is self-loading and self-locating, and is operated by a three-man crew protected during firings in either a reinforced man-rated cab or an armored cab. It operates within the MLRS command, control, and communications structure.

SYSTEM INTERDEPENDENCIES

Family of Medium Tactical Vehicles (FMTV) chassis, Advanced Field Artillery Tactical Data System (AFATDS)

PROGRAM STATUS

- **1QFY09:** Full-rate production IV contract award
- **2QFY09:** Fielded 1st Battalion, 94th Field Artillery Regiment, 17th Fires Brigade, the seventh battalion to be fielded
- **3QFY09:** Fielded 3rd Battalion, 197th Field Artillery Regiment, 197th Fires Brigade, the eighth battalion fielded

PROJECTED ACTIVITIES

- Continue fielding to active and reserve components
- Continue retrofit of Increased Crew Protection Armored Cab to fleet
- Begin fielding to Foreign Military Sales customers
- Provide support to fielded units in combat

ACQUISITION PHASE

Technology Development | Engineering & Manufacturing Development | Production & Deployment | Operations & Support

High Mobility Artillery Rocket System (HIMARS)

FOREIGN MILITARY SALES
United Arab Emirates, Singapore

CONTRACTORS
Prime and launcher:
Lockheed Martin (Dallas, TX; Camden, AR)
Family of Medium Tactical Vehicles:
BAE Systems (Sealy, TX)
Improved Weapons Interface Unit:
Harris Corp. (Melbourne, FL)
Position Navigation Unit:
L-3 Communications Space & Navigation (Budd Lake, NJ)
Hydraulic pump and motor:
Vickers (Jackson, MS)

High Mobility Engineer Excavator (HMEE)

INVESTMENT COMPONENT
- **Modernization**
- Recapitalization
- Maintenance

MISSION
To provide the Army with self-deployability, mobility, and speed to keep pace with the Brigade Combat Teams within the Future Engineer Force.

DESCRIPTION
The High Mobility Engineer Excavator Type I (HMEE-I) is a non-developmental, military-unique vehicle that will be fielded to the Army's Brigade Combat Teams (BCTs) and other selected engineer units. The HMEE-I can travel up to 60 miles per hour on primary roads and up to 25 miles per hour on secondary roads. The high mobility of the HMEE-I provides earthmoving machines capable of maintaining pace with the Army's current and Future Combat Systems. All HMEE-Is will be capable of accepting armor.

SYSTEM INTERDEPENDENCIES
None

PROGRAM STATUS
- All ballistics and vehicle testing performance completed
- **1QFY09:** Type classification—standard and full material release granted

PROJECTED ACTIVITIES
- **2QFY09:** Fielding begins

ACQUISITION PHASE
- Technology Development
- Engineering & Manufacturing Development
- **Production & Deployment**
- Operations & Support

High Mobility Engineer Excavator (HMEE)

FOREIGN MILITARY SALES
None

CONTRACTORS
OEM:
JCB Inc. (Pooler, GA)
Armor:
ADSI (Hicksville, NY)
Logistics:
XMCO (Warren, MI)

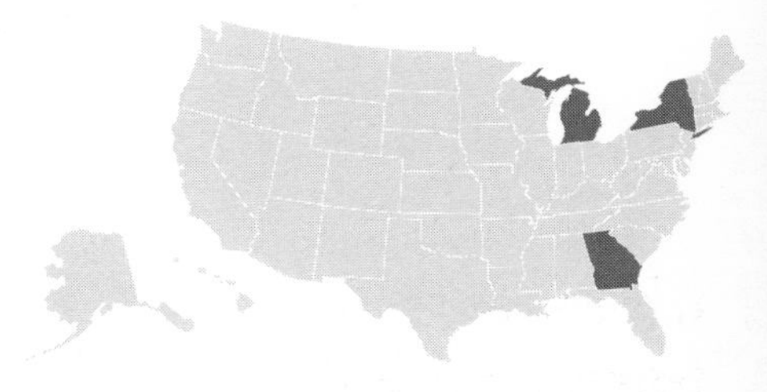

High Mobility Multipurpose Wheeled Vehicle (HMMWV)

INVESTMENT COMPONENT

- Modernization
- Recapitalization
- Maintenance

MISSION

To support combat and combat service support units with a versatile, light, mission-configurable, tactical wheeled vehicle.

DESCRIPTION

The High Mobility Multipurpose Wheeled Vehicle (HMMWV) is a tri-service program that provides light, highly mobile, diesel-powered, four-wheel-drive vehicles to satisfy Army, Marine Corps, and Air Force requirements. The HMMWV uses common components to enable its reconfiguration as a troop carrier, armament carrier, shelter carrier, ambulance, TOW missile carrier, and scout vehicle. Since its inception, the HMMWV has undergone numerous improvements, including: technological upgrades; higher payload capacity; radial tires; Environmental Protection Agency emissions updates; commercial bucket seats; three-point seat belts and other safety enhancements; four-speed transmissions; and, in some cases, turbocharged engines and air conditioning.

There are numerous HMMWV variants. The HMMWV A2 configuration incorporates a four-speed, electronic transmission, a 6.5-liter diesel engine, and improvements in transportability. It serves as a platform for other Army systems such as the Ground-Based Common Sensor. The heavy variant has a payload of 4,400 pounds and is the prime mover for the light howitzer and heavier shelters. The expanded capacity vehicle (ECV) has a payload capacity of 5,100 pounds, including crew and kits. The ECV chassis serves as a platform for mission payloads and for systems that exceed 4,400 pounds and is used for the M1114 Up-armored HMMWV. The Up-Armored HMMWV was developed to provide increased ballistic and blast protection, primarily for military police, special operations, and contingency force use.

The current production variants–M1151A1 Armament Carrier, the M1152A1 (2-door variant) Troop/Cargo/Shelter Carrier, and the M1165A1 (4-door variant) Command and Control Carrier–are built on an ECV chassis, providing additional carrying capacity for an integrated armor package (A-Kit) and the capability to accept add-on-armor kits (B-Kits). The M1151A1 is currently fielded with a gunner's protection kit.

The HMMWV recapitalization program reconfigures older base models to R1 vehicles with increased capability, reliability, and maintainability.

SYSTEM INTERDEPENDENCIES

The HMMWV supports numerous data interchange customers, who mount various shelters and other systems on it. The M1101/1102 Light Tactical Trailer is the designed trailer for this vehicle.

PROGRAM STATUS

- Fielding of ECV HMMWVs to Army, Marine Corps, Air Force, and foreign military sales customers
- Recapitalization of older model HMMWVs
- Continued product improvement in response to Army requirements

PROJECTED ACTIVITIES

- Continuous product improvements through the introduction of upgraded components in response to Army requirements

ACQUISITION PHASE

Technology Development | Engineering & Manufacturing Development | **Production & Deployment** | Operations & Support

High Mobility Multipurpose Wheeled Vehicle (HMMWV)

FOREIGN MILITARY SALES
Afghanistan, Argentina, Bahrain, Bolivia, Chad, Colombia, Djibouti, Ecuador, Egypt, Ethiopia, Honduras, Israel, Kuwait, Luxembourg, Mexico, Oman, Philippines, Saudi Arabia, Sudan, Taiwan, Tanzania, Tunisia, Uganda

CONTRACTORS
AM General (AMG) (South Bend, IN)
BAE Systems (Fairfield, OH)
GEP (Franklin, OH)
Defiance (Defiance, OH)
General Motors (Warren, MI)
Red River Army Depot (Red River, TX)
Letterkenny Army Depot (Chambersburg, PA)
Maine Military Authority (Limestone, ME)

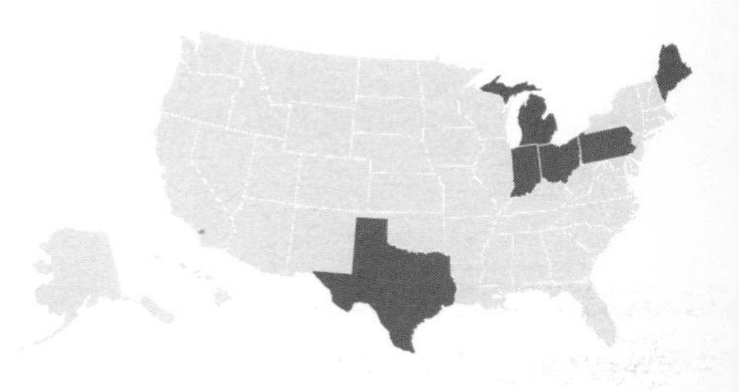

Improved Ribbon Bridge (IRB)

INVESTMENT COMPONENT

Modernization

Recapitalization

Maintenance

MISSION

To improve mobility by providing continuous roadway or raft capable of crossing military load classification 96 (wheeled)/80 (tracked) vehicles over non-fordable wet gaps.

DESCRIPTION

The Improved Ribbon Bridge (IRB) Float Ribbon Bridge System is issued to the Multi-Role Bridge Company (MRBC). US Army Modified Table of Organization and Equipment (MTOE) authorizes MRBCs consist of the 42ea IRB bays (30ea.M17 Interior and 12ea.M16 Ramps), 42ea Bridge Adapter Pallets (BAPs), 14ea Bridge Erection Boats (BEBs), 14ea Improved Boat Cradles (IBCs) and 56ea Common Bridge Transporters (CBTs). These assets collectively address Tactical Float Ribbon Bridge "Wet Gap" Bridging. These components are required to transport, launch, erect, and retrieve up to 210 meters of floating bridge. The IRB can be configured as either a continuous "full closure" bridge or assembled and used for rafting operations. The IRB has a Military Load Capacity (MLC) of 105 wheeled / 85 tracked (normal) and MLC 110 wheeled / 90 tracked (caution) vehicles. This MLC will support the joint force commander's ability to employ and sustain forces throughout the world. The IRB is used to transport weapon systems, troops, and supplies over water when permanent bridges are not available. The M14 Improved Boat Cradle (IBC) and the M15 Bridge Adapter Pallet (BAP) are used to carry BEBs and IRB bays on the CBT.

SYSTEM INTERDEPENDENCIES

IRB operations rely and are interdependent upon fully mission-capable CBTs, BAPs, IBCs, and BEB assets within a fully MTOE-equipped MRBC.

PROGRAM STATUS

- **Current:** This system has been fielded since 2002

PROJECTED ACTIVITIES

- Fieldings are ongoing based on the Army Requirements Prioritization List

ACQUISITION PHASE

Technology Development | Engineering & Manufacturing Development | **Production & Deployment** | Operations & Support

Improved Ribbon Bridge (IRB)

FOREIGN MILITARY SALES
None

CONTRACTORS
General Dynamics European Land Systems–Germany (GDELS-G) (Kaiserslautern, Germany)
Logistic support:
AM General (AMG) (Livonia, MI)
CBT manufacturer:
Oshkosh Truck Corp. (Oshkosh, WI)
BEB manufacturer:
FBM Babcock Marine (Isle of Wight, United Kingdom)

Improved Target Acquisition System (ITAS)

INVESTMENT COMPONENT

Modernization

Recapitalization

Maintenance

MISSION

To provide long-range sensor and anti-armor/precision assault fire capabilities, enabling the Soldier to shape the battlefield by detecting and engaging targets at long range with tube-launched, optically tracked, wire-guided (TOW) missiles or directing the employment of other weapon systems to destroy those targets.

DESCRIPTION

The ITAS is multipurpose weapon system that is used as a reconnaissance, surveillance, and target acquisition sensor that also provides long range anti-armor/precision assault fires capabilities to the Army's Infantry and Stryker BCTs as well as Marines. ITAS is a major product upgrade that greatly reduces the number of components, minimizing logistics support and equipment requirements. Built-in diagnostics and improved interfaces enhance target engagement performance. ITAS's second-generation infrared sensors double the range of its predecessor, the M220 TOW system. It offers improved hit probability with aided target tracking, improved missile flight software algorithms, and an elevation brake to minimize launch transients. The ITAS includes an integrated target acquisition subsystem (day/night sight with laser rangefinder), a position attitude determination subsystem, a fire-control subsystem, a lithium-ion battery power source, and a modified traversing unit. Soldiers can also detect and engage long-range targets with TOW missiles or, using the ITAS far-target location (FTL) enhancement, direct other fires to destroy them. The FTL enhancement consists of a position attitude determination subsystem (PADS) that provides the gunner with his own GPS location and a 10-digit grid location to his target through the use of differential global positioning system. With the PAQ-4/PEQ-2 Laser Pointer, ITAS can designate .50 caliber or MK-19 grenade engagements. The ITAS can fire all versions of the TOW family of missiles.

The TOW 2B Aero and the TOW Bunker Buster have an extended maximum range to 4,500 meters. The TOW 2B Aero flies over the target (offset above the gunner's aim point) and uses a laser profilometer and magnetic sensor to detect and fire two downward-directed, explosively formed penetrator warheads into the target. However, the TOW Bunker Buster impacts the target. With its high-explosive blast-fragmentation warhead, the TOW Bunker Buster is optimized for performance against urban structures, earthen bunkers, field fortifications, and light-skinned Armor threats. ITAS operates from the High Mobility Multipurpose Wheeled Vehicle, the dismount tripod platform, and Stryker anti-tank guided missile vehicles (ATGMs).

SYSTEM INTERDEPENDENCIES

The ITAS system is integrated on the M1121/1167 HMMWV and the Stryker ATGM. The ITAS system is the guidance for the TOW missile.

PROGRAM STATUS

- **Current:** ITAS has been fielded to 18 active and six reserve component Infantry Brigade Combat Teams and seven Stryker Brigade Combat Teams.
- **Current:** The Marine Corps has begun fielding the ITAS to infantry and tank battalions to replace all Marine Corps M220A4 TOW 2 systems by 2012.

PROJECTED ACTIVITIES

- **2QFY10–2QFY12:** Fielding of 16 IBCTs and 12 separate battalions; ITAS production concludes; sustainment training for fielded units; pre-deployment training; anticipated continuation of border patrol activities; contractor logistics support

ACQUISITION PHASE

Technology Development | Engineering & Manufacturing Development | Production & Deployment | Operations & Support

Improved Target Acquisition System (ITAS)

FOREIGN MILITARY SALES
NATO Maintenance and Supply Agency, Canada

CONTRACTORS
Raytheon (McKinney, TX)
Training Devices:
Intercoastal Electronics (Mesa, AZ)

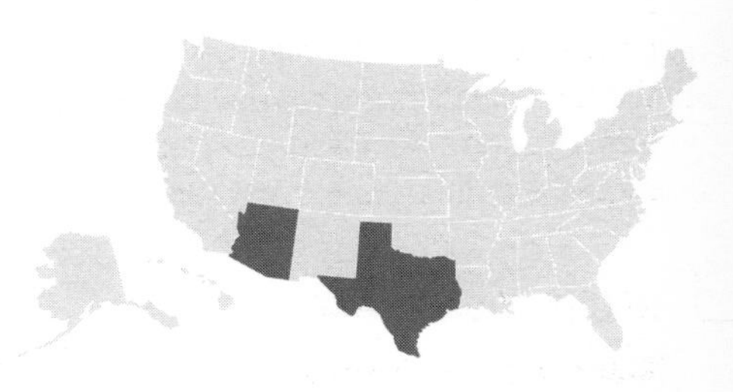

Installation Protection Program (IPP) Family of Systems

INVESTMENT COMPONENT

Modernization

Recapitalization

Maintenance

MISSION

To provide an effective chemical, biological, radiological, and nuclear (CBRN) protection, detection, identification, and warning system for military installations.

DESCRIPTION

The Installation Protection Program (IPP) will allow DoD installations to effectively protect personnel and critical operations against a chemical, biological, radiological, and nuclear (CBRN) event, to effectively respond with trained and equipped emergency personnel, and to ensure installations can continue critical operations during and after an attack.

IPP uses a tiered approach of government and commercial off-the-shelf capabilities optimized for an installation. The Baseline Tier provides a foundation for installations to maintain a standard level of preparedness for a CBRN incident. This tier consists of non-materiel solutions that address military-civilian interoperability, system architecture, policy, doctrine, training, and administration. It includes joint training products, planning templates, Mutual Aid Agreement templates, and exercise templates and scenarios. The IPP Portal (IP3) makes these solutions available through Joint Knowledge Online (JKO) and Army Knowledge Online (AKO) at https://www.us.army.mil/suite/page/449823 or through a link on the Joint Acquisition CBRN Knowledge System (JACKS) website at https://jacks.jpeocbd.osd.mil.

Tier 1 focuses on enhancing an installation's existing emergency responder capabilities and enables an installation to prepare, respond, and transfer the mission after a CBRN attack. Tier 1 installations are critical to the overall accomplishment of the national military strategy or installations that provide combat service support. Tier 1 includes all Baseline Tier capabilities and adds individual protective equipment for emergency responders and first receivers; portable radiological and chemical detection equipment; portable biological collectors with analysis and identification laboratory support; personal dosimeters; hazard marking and controlling equipment; medical countermeasures for first responders/receivers; mass casualty decontamination showers and tents; mass casualty litters and support equipment; mass notification systems; an incident management system; and new equipment training and field exercise support.

Tier 2 applies to installations hosting one-of-a-kind, critical strategic missions or capabilities. The objective of Tier 2 is to provide installations with the capability to prepare, react, and continue critical missions or capabilities without significant interruption. The Tier 2 capability package includes Baseline and Tier 1 capabilities plus fixed chemical detectors for warfare agents and toxic industrial materials/chemicals; fixed biological collectors with analysis and identification laboratory support; radiological monitoring equipment for entry controllers; collective protection for one of a kind strategic assets (up to 3,000 square feet); and a decision support system of software tools and networked sensors.

SYSTEM INTERDEPENDENCIES

None

PROGRAM STATUS

- **4QFY09:** Completed 31 additional installations

PROJECTED ACTIVITIES

- **4QFY10:** Complete 18 additional installations
- **4QFY11:** Complete 16 additional installations
- **4QFY12:** Complete 16 additional installations

ACQUISITION PHASE

Technology Development | Engineering & Manufacturing Development | Production & Deployment | Operations & Support

Installation Protection Program (IPP) Family of Systems

FOREIGN MILITARY SALES
None

CONTRACTORS
Science Applications International Corp. (SAIC) (Falls Church, VA)
AIE:
Computer Sciences Corp. (CSC) (Falls Church, VA)

Instrumentable–Multiple Integrated Laser Engagement System (I–MILES)

INVESTMENT COMPONENT

Modernization

Recapitalization

Maintenance

MISSION

To provide force-on-force and force-on-target capabilities to support live collective training at home station, and Combat Training Centers (CTC).

DESCRIPTION

The Instrumentable–Multiple Integrated Laser Engagement System (I–MILES) is composed of several component systems.

The I–MILES Combat Vehicle System (CVS) provides live training devices for armored vehicles with fire control systems including Bradley Fighting Vehicles and Abrams Tanks. It interfaces and communicates with CTC and home station instrumentation, providing casualty and battlefield damage assessments for after-action reporting (AAR).

The I–MILES Individual Weapons System (IWS) is a man-worn dismounted system, providing real-time casualty effects necessary for tactical engagement training in direct fire force-on-force and instrumented training scenarios. Event data can be downloaded for use in an AAR and training assessment. The IWS replaces Basic MILES Individual Weapon Systems (IWS) at home stations and Maneuver Combat Training Centers Army wide.

The Wireless Independent Target System (WITS) provides real-time casualty effects necessary for tactical engagement training in direct fire force-on-force training scenarios and instrumented training scenarios. It replaces the previously fielded Independent Target System (ITS) and other Basic MILES currently fielded on non-turret military vehicles. WITS designs include a tactical wheeled vehicle configuration and a separate configuration for tracked vehicles such as the M113.

The Shoulder Launched Munitions (SLM) provides real-time casualty effects necessary for tactical engagement training in direct fire force-on-force training scenarios and instrumented scenarios. It replaces Basic MILES currently fielded and provides better training fidelity for blue forces' weapons and a more realistic simulation of threat weapons using opposing force visual modifications.

The Universal/Micro Controller Devices (UCD/MCD) are low-cost, lightweight devices used by observer controllers and maintenance personnel to initialize, set up, troubleshoot, reload, reset, and manage participants during live force-on-force training exercises. These modular, self-contained devices interact and provide administrative control of all other MILES devices.

SYSTEM INTERDEPENDENCIES

None

PROGRAM STATUS

IWS:
- Fielded approximately 6,000 IWS kits to the National Training Center

SLM:
- Fielded approximately 2,400 SLM Kits to National Training Center, Joint Maneuver Training Center, and various home stations

UCD/MCD:
- Fielded approximately 2,700 MCD/UCD Kits to National Training Center, Joint Maneuver Training Center, and various home stations

WITS:
- Fielded approximately 2,000 WITS kits to various home stations

PROJECTED ACTIVITIES

IWS:
- **FY09:** Recompete contract

SLM:
- **FY09:** Complete basis of issue

UCD/MCD:
- **FY09:** Complete basis of issue

CVS:
- **FY09:** Continue fielding

WITS:
- **FY09:** Combine with Mine Resistant Ambush Protected vehicle system and recompete

ACQUISITION PHASE

Technology Development | Engineering & Manufacturing Development | **Production & Deployment** | Operations & Support

Instrumentable–Multiple Integrated Laser Engagement System (I–MILES)

FOREIGN MILITARY SALES
None

CONTRACTORS
I–MILES IWS:
Cubic Defense Systems (San Diego, CA)
I–MILES SLM, WITS, and UCD/MCD:
Universal Systems and Technology (Centreville, VA)
I–MILES CVS:
Science Applications International Corp. (SAIC) (San Diego, CA)

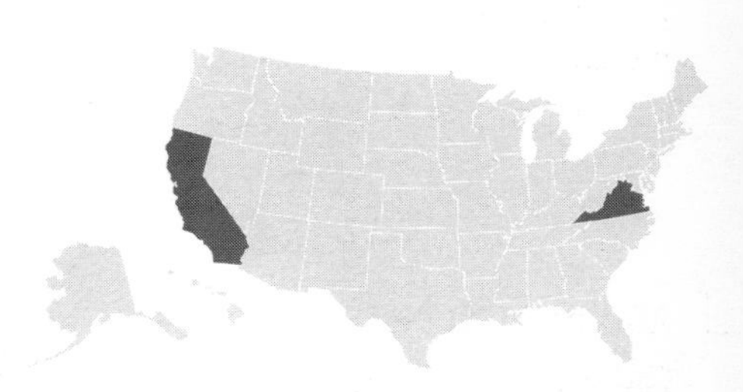

Integrated Air and Missile Defense (IAMD)

INVESTMENT COMPONENT
- Modernization
- Recapitalization
- Maintenance

MISSION

To provide the full combat potential of an Integrated Air and Missile Defense capability through a network-centric "plug and fight" architecture at the component level (e.g. launchers and sensors) and a common command and control system.

DESCRIPTION

Army Integrated Air and Missile Defense (IAMD) will enable the integration of modular components (current and future Air and Missile Defense [AMD] sensors, weapons, and command and control [C2]) with a common C2 capability in a networked and distributed "plug and fight" architecture. This common C2, called the IAMD Battle Command System (IBCS), will provide standard configurations and capabilities at each echelon. This allows joint, interagency, intergovernmental, and multinational (JIIM) AMD forces to organize based on mission, enemy, terrain and weather, troops and support available, time available, and civil considerations (METT–TC). Shelters and vehicles may be added to enable broader missions and a wider span of control executed at higher echelons. A network-enabled "plug and fight" architecture and common C2 system will enable dynamic defense design and task force reorganization, and provide the capability for interdependent, network-centric operations that link joint IAMD protection to the supported force scheme of operations and maneuver. This Army IAMD system of systems will enable extended range and non-line-of-sight engagements across the full spectrum of aerial threats, providing fire control quality data to the most appropriate weapon to successfully complete the mission. Furthermore, it will mitigate the coverage gaps and the single points of failure that have plagued AMD defense design in the past.

SYSTEM INTERDEPENDENCIES

Patriot, SLAMRAAM, Improved Sentinel, JLENS, THAAD, ABCS, E-IBCT, BMDS, JTAGS, MEADS, AEGIS, E-10, AWACS, CAC2S, BCS, E-2C, and DD(X)

PROGRAM STATUS

- **4QFY07:** Approval of two-contractor competition strategy
- **1QFY08:** Approval of acquisition strategy
- **2QFY08:** Request for proposal released
- **4QFY08:** IAMD Battle Command System Contract Award

PROJECTED ACTIVITIES

- **4QFY09:** IAMD Increment 2 Preliminary Design Review
- **4QFY09:** Milestone B approval to enter Engineering and Manufacturing Development (EMD)
- **3QFY11:** IAMD Increment 2 critical design review (CDR)

ACQUISITION PHASE
- Technology Development
- Engineering & Manufacturing Development
- Production & Deployment
- Operations & Support

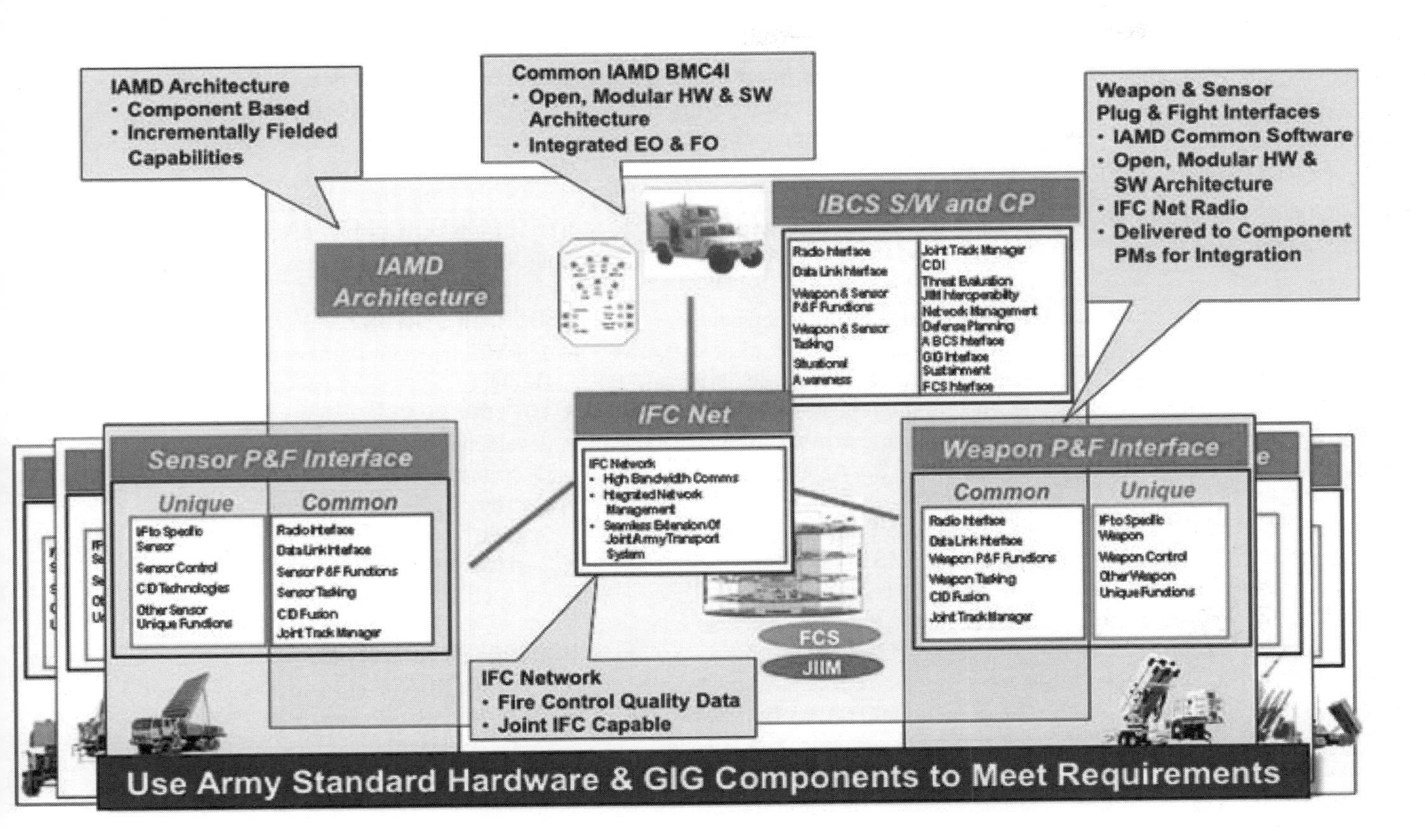

Integrated Air and Missile Defense (IAMD)

FOREIGN MILITARY SALES
None

CONTRACTORS

Concept and Technology Development Phase Competing Contractors:
Northrop Grumman (Huntsville, AL)
Raytheon (Huntsville, AL; Andover, MA)

System Engineering Technical Analysis (SETA) support:
Dynetics, Millennium, Davidson (DMD) (Huntsville, AL)

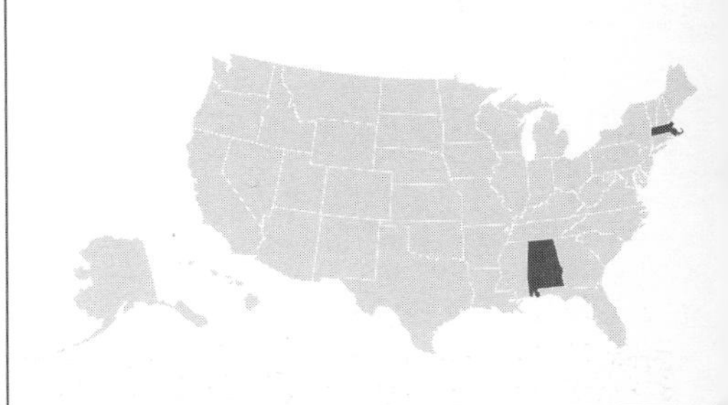

Integrated Family of Test Equipment (IFTE)

INVESTMENT COMPONENT

Modernization

Recapitalization

Maintenance

Maintenance Support Device (MSD-V2)

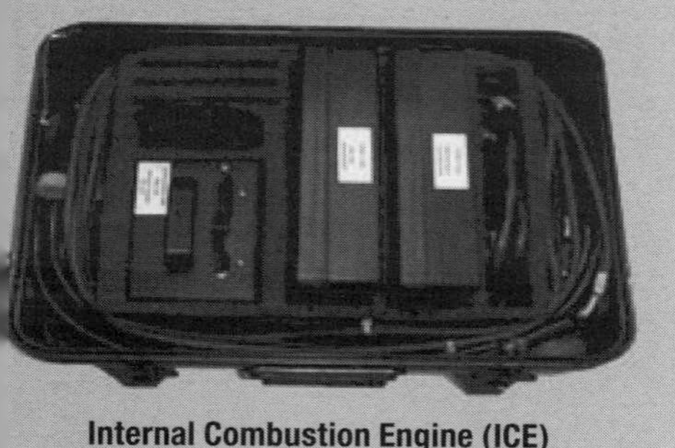

Internal Combustion Engine (ICE)

MISSION

To enable maintenance, verification, testing, and repair of Army weapon systems and their components; to diagnose, isolate, and repair faults through mobile, general purpose, and automatic test systems.

DESCRIPTION

The Integrated Family of Test Equipment (IFTE) consists of interrelated, integrated, mobile, tactical, and man-portable systems. These rugged, compact, lightweight, general-purpose systems enable verification of the operational status of weapon systems, as well as fault isolation to the line replaceable unit at all maintenance levels, both on and off the weapon system platform. IFTE is a Early Infantry Brigade Combat Team (E-IBCT) Associate Program.

Electro-Optics Test Facility–Version 5 (EOTF-V5)

The EOTF-V5 tests the full range of Army electro-optical systems, including laser transmitters, receivers, spot trackers, forward-looking infrared systems, and television systems. It is fully mobile with VXI instrumentation, touch-screen operator interface, and an optical disk system for test program software and electronic technical manuals.

Next Generation Automatic Test System (NGATS)

The NGATS is the follow-on reconfigurable, rapidly deployable, automatic test equipment that supports joint operations, reduces logistics footprint, and replaces/consolidates obsolete, unsupportable automatic test equipment in the Army's inventory.

Maintenance Support Device–Version 2 (MSD-V2)

The second-generation MSD is a lightweight, rugged, compact, man-portable, general-purpose automatic tester. It is used to verify the operational status of aviation, automotive, electronic, and missile weapon systems and to isolate faulty components for immediate repair or replacement. MSD-V2 hosts Interactive Electronic Technical Manuals, is used as a software uploader/verifier to provide or restore mission software to weapon systems, and supports testing and diagnostic requirements of current and Future Combat Systems. MSD-V2 supports more than 40 weapon systems and is used by more than 30 military occupational specialties.

SYSTEM INTERDEPENDENCIES

None

PROGRAM STATUS

- **1QFY09:** NGATS limited user test (LUT)
- **1QFY09:** NGATS system development and demonstration (Increment 2)
- **Current:** MSD-V2 production and fielding
- **Current:** EOTF-V5 operations and support

PROJECTED ACTIVITIES

- **3QFY09:** NGATS milestone C
- **4QFY09:** MSD-Version 3 (MSD-V3) contract award
- **4QFY10:** MSD-V3 production and fielding
- **2QFY12:** NGATS first-unit equipped (FUE)

ACQUISITION PHASE

Technology Development | Engineering & Manufacturing Development | Production & Deployment | Operations & Support

Integrated Family of Test Equipment (IFTE)

Automatic Test Equipment (ATE)

Next Generation Automatic Test System – Version 6 (NGATS-V6)

Electro-Optics Test Facility – Version 5 (EOTF-V5)

Maintenance Support Device – Version 2 (MSD-V2)

FOREIGN MILITARY SALES

MSD:

Australia, Afghanistan, Bahrain, Chile, Djibouti, Egypt, Ethiopia, Germany, Israel, Iraq, Jordan, Korea, Kuwait, Lithuania, Macedonia, Morocco, Netherlands, Oman, Poland, Portugal, Saudi Arabia, Taiwan, Turkey, United Arab Emirates, Uzbekistan, Yemen

CONTRACTORS

MSD-V2:

Science and Engineering Services, Inc. (SESI) (Huntsville, AL)

Vision Technology Miltope Corp. (Hope Hull, AL)

NGATS-V6:

Northrop Grumman (Rolling Meadows, IL)

DRS Technologies (Huntsville, AL)

EOTF-V5:

Northrop Grumman (Rolling Meadows, IL)

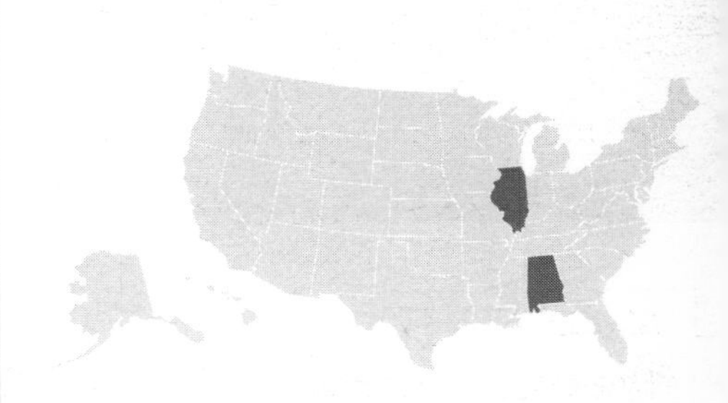

Interceptor Body Armor

INVESTMENT COMPONENT

- **Modernization**
- Recapitalization
- Maintenance

MISSION

To protect individual Soldiers from ballistic and fragmentation threats in a lightweight, modular body armor package.

DESCRIPTION

Interceptor Body Armor (IBA) is modular, multiple-threat body armor, consisting of an Improved Outer Tactical Vest (IOTV); Enhanced Small Arms Protective Inserts (ESAPI); Enhanced Side Ballistic Inserts (ESBI); and Deltoid and Auxiliary Protector (DAP).

Eleven sizes of IOTVs and five sizes of ESAPI plates are being fielded. The basic system weight (IOTV, ESAPI, ESBI, size medium) is 30 pounds and provides increased area coverage and greater protection. The medium IOTV, without plates, weighs 15.9 pounds and protects against fragmentation and 9mm rounds. The ESAPI plates provide additional protection and can withstand multiple small arms hits. IBA includes attachable throat, groin, and neck protectors. It also has webbing attachment loops on the front and back of the vest for attaching pouches for the Modular Lightweight Load-Carrying Equipment (MOLLE). DAP provides additional protection from fragmentary and 9mm projectiles to the upper arm and underarm areas. During Operation Iraqi Freedom combat operations, the side and underarm areas not covered by the ESAPI component of the IBA were identified by combat commanders and medical personnel as a vulnerability that needed to be addressed. To meet this threat and provide an increased level of protection, the ESBI was developed. Commanders have the flexibility to tailor the IBA to meet the specific mission needs or changing threat conditions.

SYSTEM INTERDEPENDENCIES

None

PROGRAM STATUS

- **Current:** In production and being fielded
- **4QFY09:** 294,309 IOTVs and 913,437 ESAPI sets fielded

PROJECTED ACTIVITIES

- **Continue:** Fielding

ACQUISITION PHASE

Technology Development | Engineering & Manufacturing Development | **Production & Deployment** | Operations & Support

Interceptor Body Armor

FOREIGN MILITARY SALES
None

CONTRACTORS
Armacel Armor (Camarillo, CA)
Ceradyne, Inc. (Costa Mesa, CA)
UNICOR Protective Materials Company (Miami Lakes, FL)
BAE Systems (Phoenix, AZ)

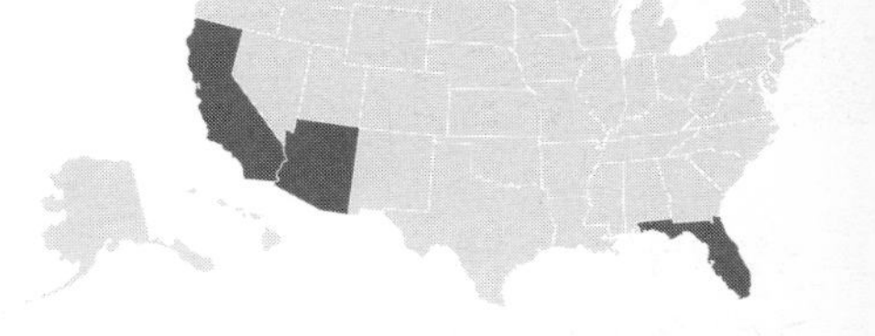

Javelin

INVESTMENT COMPONENT

Modernization

Recapitalization

Maintenance

MISSION

To provide a man-portable, fire-and-forget, medium-range missile with enhanced situational awareness and precision direct-fire effects to defeat armored vehicles, fortifications, and soft targets across the spectrum of operations.

DESCRIPTION

The Close Combat Missile System–Medium (CCMS–M) Javelin is highly effective against a variety of targets at extended ranges under day/night, battlefield obscurants, adverse weather, and multiple counter-measure conditions. The system's soft-launch feature permits firing from a fighting position or an enclosure. Javelin's modular design allows the system to evolve to meet changing threats and requirements via both software and hardware upgrades. The system consists of a reusable command launch unit (CLU) with a built-in-test (BIT), and a modular missile encased in a disposable launch tube assembly. The Javelin missile and command launch unit together weigh 48.8 pounds. The system also includes training devices for tactical training and classroom training. Javelin's fire-and-forget technology allows the gunner to fire and immediately take cover, to move to another fighting position, or to reload. The Javelin provides enhanced lethality through the use of a tandem warhead that will defeat all known armor threats. It is effective against both stationary and moving targets. This system also provides defensive capability against attacking/hovering helicopters. The performance improvements in current production Javelin Block I CLUs are: increased target identification range, increased surveillance time with new battery and software management of the "on" time, and external RS-170 interface for video output. The performance improvements in current production Javelin Block I missiles are: increased probability of hit/kill at 2,500 meters, improved warhead lethality, and reduced time of flight. In current conflicts the CLU is being used as a stand-alone surveillance and target acquisition asset. The Army is the lead for this joint program with the Marine Corps. Javelin is a Early Infantry Brigade Combat Team (E-IBCT)-complementary system as the lethality solution for the E-IBCT Armed Robotic Vehicle–Assault (Light).

SYSTEM INTERDEPENDENCIES

None

PROGRAM STATUS

- **3QFY07:** Received full material release on Block I CLU
- **4QFY08:** Received full material release on Block I missile
- **Current:** Missile and CLU production
- **Current:** CLU total package fielding
- **Current:** Javelin has been fielded to more than 95 percent of active duty units. Fielding is underway to the National Guard.

PROJECTED ACTIVITIES

- **FY09–10:** Final CLU procurements; achieves Army acquisition objective
- **Continue:** CLU production
- **Continue:** CLU total package fielding
- **Continue:** Missile production

ACQUISITION PHASE

Technology Development | Engineering & Manufacturing Development | **Production & Deployment** | Operations & Support

Javelin

FOREIGN MILITARY SALES
United Kingdom, Australia, Ireland, Jordan, Lithuania, Taiwan, Norway, New Zealand, Czech Republic, Oman, United Arab Emirates

CONTRACTORS
Javelin Joint Venture:
Raytheon (Tucson, AZ)
Lockheed Martin (Orlando, FL)

Joint Air-to-Ground Missile (JAGM)

INVESTMENT COMPONENT

- **Modernization**
- Recapitalization
- Maintenance

MISSION

To provide a single variant, precision-guided, air-to-ground weapon for use by joint service manned and unmanned aircraft to destroy stationary and moving high-value land and naval targets.

DESCRIPTION

The Joint Air-to-Ground Missile (JAGM) System will replace the Hellfire, Maverick families of missiles, and the aviation-launched TOW missile with a single-variant, multi-mode weapon. JAGM is a joint program with the Army, Navy, and Marine Corps and will be fired from helicopters, aircraft, and unmanned aircraft systems (UAS). JAGM will increase the warfighter's operational flexibility by effectively engaging a variety of stationary and mobile targets on the battlefield from longer ranges, including advanced heavy/light armored vehicles, bunkers, buildings, patrol craft, command and control vehicles, transporter/erector (e.g., SCUD) launchers, artillery systems, and radar/air defense systems. Its multi-mode seeker will provide robust capability in adverse weather, day or night, and in an obscured/countermeasured environment. The warhead is designed for high performance against both armored and non-armored targets. The JAGM System includes missile, trainers, containers, support equipment, and launchers. JAGM will be fielded to the Super Hornet (F/A-18E/F), Apache (AH-64D), and the Super Cobra (AH-1Z) in 2016. Follow-on fieldings of JAGM on the Seahawk (MH-60R) and the SkyWarrior UAS are planned for 2017.

Diameter: 7 inches
Weight: 108 pounds
Length: 70 inches

SYSTEM INTERDEPENDENCIES

Rotary-wing Launcher/Rack: M299, Fixed-wing Launcher Rack: Design to be determined

PROGRAM STATUS

- **4QFY08:** Competitive Technology Development contracts awarded to Lockheed Martin and Raytheon

PROJECTED ACTIVITIES

- **1QFY09:** Integrated baseline review (IBR)
- **4QFY09:** System requirements review (SRR)
- **3QFY10:** Preliminary design review (PDR)
- **1QFY11:** Milestone B

ACQUISITION PHASE

Technology Development | Engineering & Manufacturing Development | Production & Deployment | Operations & Support

Joint Air-to-Ground Missile (JAGM)

FOREIGN MILITARY SALES
None

CONTRACTORS
Lockheed Martin (Orlando, FL)
Raytheon (Tucson, AZ)
Boeing (St. Louis, MO)

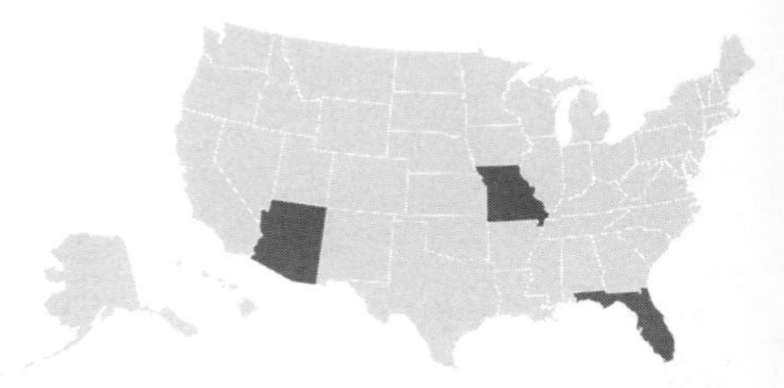

Joint Biological Point Detection System (JBPDS)

INVESTMENT COMPONENT

Modernization
Recapitalization
Maintenance

MISSION

To protect the Soldier by providing rapid and fully automated detection, identification, warning, and sample isolation of high-threat biological warfare agents.

DESCRIPTION

The Joint Biological Point Detection System (JBPDS) is the first joint biological warfare agent (BWA) detection system designed to meet the broad spectrum of operational requirements encountered by the services, across the entire spectrum of conflict.

It consists of a common biosuite that can be integrated onto Service platform, shipboard, or trailer mounted to provide biological detection and identification to all service personnel. The JBPDS is portable and can support bare-base or semi-fixed sites. JBPDS will presumptively identify 10 BWAs simultaneously. It will also collect a liquid sample for confirmatory analysis and identification. Planned product improvements will focus on identification sensitivity, life cycle costs, and system reliability.

JBPDS can operate from a local controller on the front of each system, remotely, or as part of a network of up to 26 systems. JBPDS meets all environmental, vibration, and shock requirements of its intended platforms, as well as requirements for reliability, availability, and maintainability.

The JBPDS includes both military and commercial global positioning, meteorological, and network modem capabilities. The system will interface with the Joint Warning and Reporting Network (JWARN).

SYSTEM INTERDEPENDENCIES

Joint Warning and Reporting Network (JWARN), Nuclear Biological Chemical Reconnaissance Vehicle (NBCRV), Biological Integrated Detection System (BIDS)

PROGRAM STATUS

- **2QFY08:** Extended low-rate initial production
- **4QFY09:** Full-rate production decision

PROJECTED ACTIVITIES

- **2QFY10:** Continue unit fieldings
- **2QFY10:** Follow-on operational test and evaluation

ACQUISITION PHASE

Technology Development | Engineering & Manufacturing Development | Production & Deployment | Operations & Support

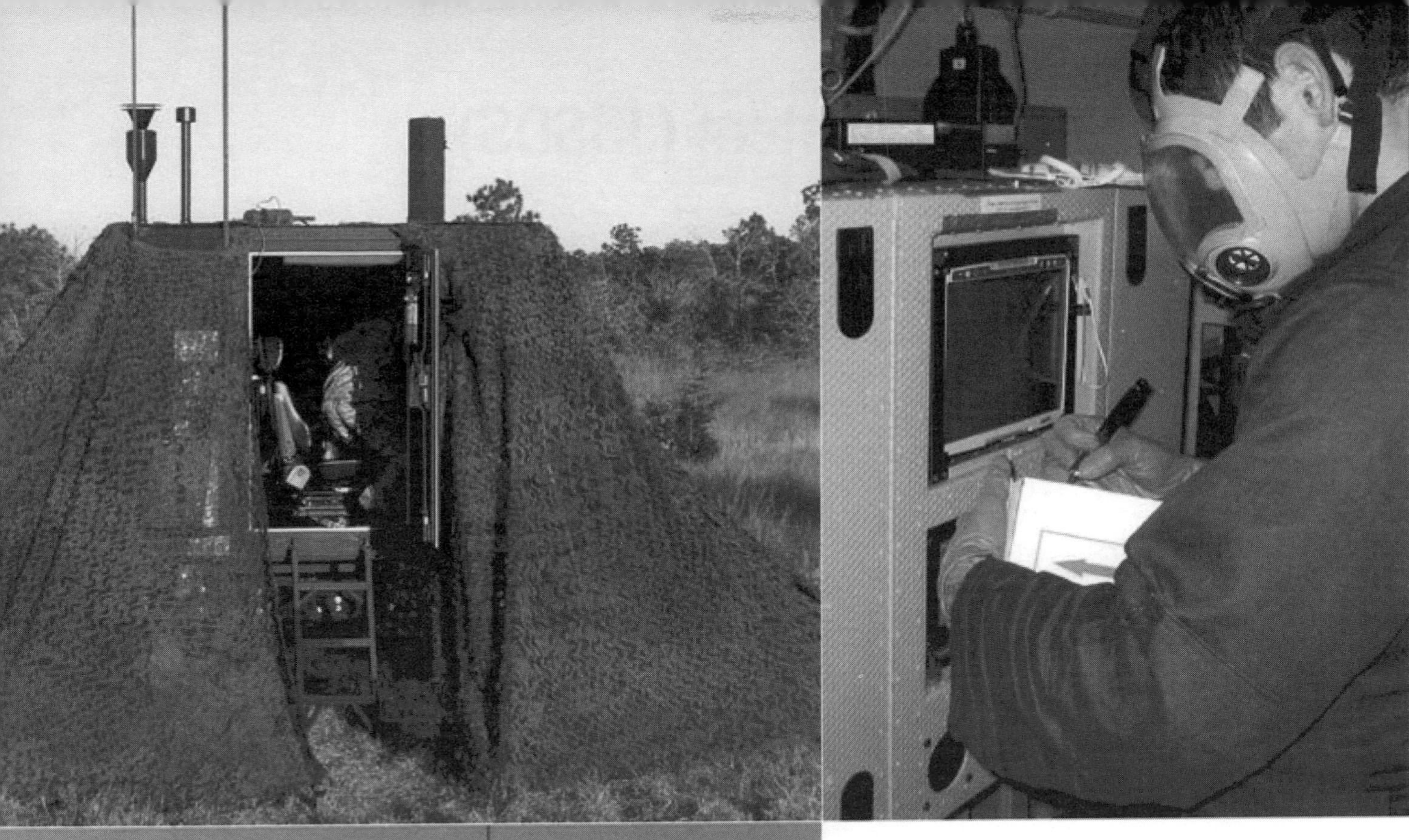

Joint Biological Point Detection System (JBPDS)

FOREIGN MILITARY SALES
Japan

CONTRACTORS
General Dynamics Armament and Technical Products (GDATP) (Charlotte, NC)

Joint Biological Standoff Detection System (JBSDS)

INVESTMENT COMPONENT

- **Modernization**
- Recapitalization
- Maintenance

MISSION

To protect U.S. forces by detecting, tracking, and discriminating aerosol clouds of biological warfare agents.

DESCRIPTION

The Joint Biological Standoff Detection System (JBSDS) is a detector, or network of detectors, that provides standoff detection, discrimination, and warning of Biological Warfare Agent (BWA) clouds. The JBSDS is the first biological defense detect-to-warn capability to protect individual warfighters. The JBSDS provides initial early warning capability against BWA attack by detecting aerosol clouds out to five kilometers with Infrared (IR) Light Detection and Ranging (LIDAR) and discriminating biological versus non-biological particles in clouds out to one kilometer with Ultra-Violet (UV) and IR LIDAR. JBSDS operates at fixed sites or in a stationary mode on mobile platforms.

JBSDS meets all environmental, vibration, and shock requirements of its intended platforms, as well as requirements for reliability, availability, and maintainability. JBSDS includes military global positioning capabilities.

SYSTEM INTERDEPENDENCIES

Single Channel Ground and Airborne Radio System (SINCGARS), Combat Service Support Automated Information Systems Interface (CAISI)

PROGRAM STATUS

- **1QFY09:** Completed product verification test
- **1QFY10:** Receive full materiel release and full-rate production decision
- **1QFY10:** First-unit equipped/initial operational capability Army

PROJECTED ACTIVITIES

- **2QFY12:** Full operational capability

ACQUISITION PHASE

Technology Development | **Engineering & Manufacturing Development** | Production & Deployment | Operations & Support

Joint Biological Standoff Detection System (JBSDS)

FOREIGN MILITARY SALES
None

CONTRACTORS
Science and Engineering Services, Inc. (SESI) (Columbia, MD)

Joint Cargo Aircraft (JCA)

INVESTMENT COMPONENT

Modernization

Recapitalization

Maintenance

MISSION

To transport time-sensitive/mission-critical supplies and key personnel over operational and tactical distances to forward-deployed forces in remote and austere locations. To provide routine and combat aerial sustainment to the Joint Force.

DESCRIPTION

The Joint Cargo Aircraft (JCA) is a fixed-wing, multipurpose cargo aircraft for today's diverse missions. It provides the Army and Air Force with a mid-range, multifunctional, and interoperable aircraft able to perform: logistical resupply, casualty evacuation, troop movement, airdrop operations, humanitarian assistance, and missions in support of Homeland Security.

Extremely maneuverable and versatile, the JCA has a high power-to-weight ratio, and the ability to perform fighter-aircraft-like 3.0g-force maneuvers, enabling it to make tight turns, and climb and descend quickly. JCA features the following performance: 2,300 nautical miles with over 13,227 pounds (6,000 kilograms) of payload, 3,200 nautical miles ferry flight, and 325 knots true airspeed maximum cruise speed. JCA has a state-of-the-art suite of standard off-the-shelf military equipment integrated with military standard digital data bus architecture. It is powered by two Rolls-Royce E 2100D2 engines capable of up to 4,637 shaft horsepower each, and equipped with full authority digital electronic control. The strong propulsion system allows for access to a wide range of airfields; enabling landings on short, unprepared strips, and in hot weather and high-altitude conditions, all while transporting heavy loads.

JCA has a full suite of navigation aids: a Night Vision Imaging System (NVIS) compatible cockpit, NVIS-compatible internal and external lighting, and can operate in all weather conditions, day and night. JCA is capable of varying the cargo floor height and continuously adjusting altitude, ensuring easy loading and unloading of large-volume high-density payloads without ground support equipment and easy drive in/out of vehicles, enabling their immediate operability.

SYSTEM INTERDEPENDENCIES

None

PROGRAM STATUS

- **1QFY09:** Delivery of JCA 1 and 2
- **1QFY09:** Began production qualification testing
- **2QFY09:** Seven aircraft ordered
- **3QFY09:** Resource Management Decision (RMD) 802 directed the transfer of mission and program to Air Force

PROJECTED ACTIVITIES

- **1QFY10:** Begin implementation of RMD 802
- **2QFY10:** Complete production qualification testing
- **2QFY10:** Multi-service operational test and evaluation
- **2QFY10:** Air Force order 8 aircraft
- **2QFY10:** Complete transfer of program to Air Force IAW RMD 802"
- **1QFY11:** Full-rate production decision

ACQUISITION PHASE

Technology Development | Engineering & Manufacturing Development | Production & Deployment | Operations & Support

Joint Cargo Aircraft

FOREIGN MILITARY SALES
None

CONTRACTORS
L-3 Communications Integrated Systems, L.P. (Greenville, TX)
Alenia Aeronautica (Rome, Italy)

Joint Chem/Bio Coverall for Combat Vehicle Crewman (JC3)

INVESTMENT COMPONENT

Modernization

Recapitalization

Maintenance

MISSION

To provide the Combat Vehicle Crewman (CVC) with flame resistant (FR), percutaneous protection against chemical and biological (CB) agents, radioactive particles, and toxic industrial materials (TIMs).

DESCRIPTION

The JC3 is a lightweight, one-piece, flame resistant, chemical, and biological protective coverall that resembles a standard CVC coverall. The JC3 is intended to be worn as a duty uniform; however, it may be worn as an overgarment. It will resist ignition and will provide thermal protection to allow emergency egress. The JC3 will not be degraded by exposure to petroleum, oils, and lubricants present in the operational environment. The JC3 will be compatible with current and developmental protective masks and mask accessories, headgear, gloves/mittens, footwear, and other CVC ancillary equipment (e.g. Spall vest).

SYSTEM INTERDEPENDENCIES

The JC3 interfaces with existing and co-developmental protective masks, appropriate mask accessories, protective headwear, hand-wear, footwear, and USA and Marine Corps armored vehicles.

PROGRAM STATUS

- **2QFY09:** First article testing
- **1QFY10:** First unit equipped

PROJECTED ACTIVITIES

- **4QFY10:** Continue production
- **3QFY11:** Complete production

ACQUISITION PHASE

Technology Development | Engineering & Manufacturing Development | Production & Deployment | Operations & Support

Joint Chem/Bio Coverall for Combat Vehicle Crewman (JC3)

FOREIGN MILITARY SALES
None

CONTRACTORS
Group Home Foundation, Inc. (Belfast, ME)

Joint Chemical Agent Detector (JCAD)

INVESTMENT COMPONENT

Modernization

Recapitalization

Maintenance

MISSION

To provide advanced detection, warning, identification of contamination on personnel and equipment, and monitoring for the presence of chemical warfare agent contamination.

DESCRIPTION

The Joint Chemical Agent Detector (JCAD) is a detector or an array of networked detectors capable of automatically detecting, identifying, and quantifying chemical agents, providing handheld monitoring capabilities, protecting the individual Soldier, Airman, and Marine through the use of pocket-sized detection and alarm.

The JCAD program will provide the services a handheld, combined-portable monitoring and small-point chemical agent point detector for ship, aircraft, and individual warfighter applications.

JCAD will automatically and simultaneously detect, identify, and quantify chemical agents in their vapor form. The detector will provide visual and audible indicators and display the chemical agent class and relative hazard level dosage. The services will deploy the system on mobile platforms to include vehicles, at fixed sites, and on individuals designated to operate in a chemical threat area (CTA). The system will operate in a general chemical warfare environment, and can undergo conventional decontamination procedures by the warfighter. JCAD is designed to interface and be compatible with current and future anti-chemical, nuclear, and biological software.

The JCAD acquisition program market survey found that commercially available detectors could satisfy revised JCAD requirements. JCAD's restructured acquisition strategy assessed commercially available products to provide the most capable, mature system, at the best life-cycle cost. This strategy provides opportunities to leverage commercial developments for fielding expanded capabilities. In 2009, the acquisition strategy incorporated a production cut-in of an Enhanced JCAD (M4E1 JCAD) to provide an additional capability to meet the objectives requested in the capability production document.

JCAD Increment 1 systems are being purchased to replace the Automatic Chemical Agent Detector and Alarm (ACADA or M22), M90, and M8A1 systems.

Specific capabilities include:

- Instant feedback of hazard (mask only or full Mission-Oriented Protective Posture)
- Real-time detection of nerve, blister, and blood agents
- Stores up to 72 hours of detection data
- Will be net-ready through implementation of the Common Chemical, Biological, Radiological, and Nuclear (CBRN) Standard Interface (Increment 2)

SYSTEM INTERDEPENDENCIES

Modular Lightweight Load-carrying Equipment (MOLLE), HMMWV, M113, M2 (Bradley)

PROGRAM STATUS

- **FY09:** Full-rate production and fielding of JCAD Increment 1

PROJECTED ACTIVITIES

- **FY10:** Customer testing of M4E1 JCAD
- **FY11:** Production cut-in decision

ACQUISITION PHASE

Technology Development | Engineering & Manufacturing Development | Production & Deployment | Operations & Support

Joint Chemical Agent Detector (JCAD)

FOREIGN MILITARY SALES
None

CONTRACTORS
Smiths Detection, Inc. (Edgewood, MD)

Joint Chemical Biological Radiological Agent Water Monitor (JCBRAWM)

INVESTMENT COMPONENT

- Modernization
- Recapitalization
- Maintenance

MISSION

To protect U.S. forces by detecting and identifying the presence of biological warfare agents, and radiological contaminant, in water supplies.

DESCRIPTION

The Joint Chemical Biological Radiological Agent Water Monitor (JCBRAWM) provides a waterborne biological and radiological agent detection capability. JCBRAWM provides the ability to: detect, identify, and quantify chemical (future), biological, and radiological (CBR) contamination during three water-monitoring missions; source site selection/reconnaissance, treatment verification, and quality assurance of stored and distributed product water.

JCBRAWM will provide the first biological and radiological detection capability in water. The system is designed to be one-man portable for use by the warfighter. JCBRAWM provides detection and identification capability for two biological agents and detection of alpha and beta radiological contaminants in water. The system performs biological detection and identification functions with an immunoassay ticket.

JCBRAWM leverages commercial technologies and fielded systems to the greatest extent possible. JCBRAWM will neither replace nor displace a current system, but rather supplements the currently fielded M272 Water Testing Kit (WTK).

SYSTEM INTERDEPENDENCIES

Reverse Osmosis Purification Units, Modernization Mission Oriented Protective Posture (MOPP), Arctic MOPP, Fixed Sites (Ports/Airfields/FOB)

PROGRAM STATUS

- **FY09:** Completed production verification testing
- **FY08:** Full-rate production, full materiel release and fielding of the system

PROJECTED ACTIVITIES

- **FY10:** Continuation of fielding

ACQUISITION PHASE

Technology Development | Engineering & Manufacturing Development | Production & Deployment | Operations & Support

Joint Chemical Biological Radiological Agent Water Monitor (JCBRAWM)

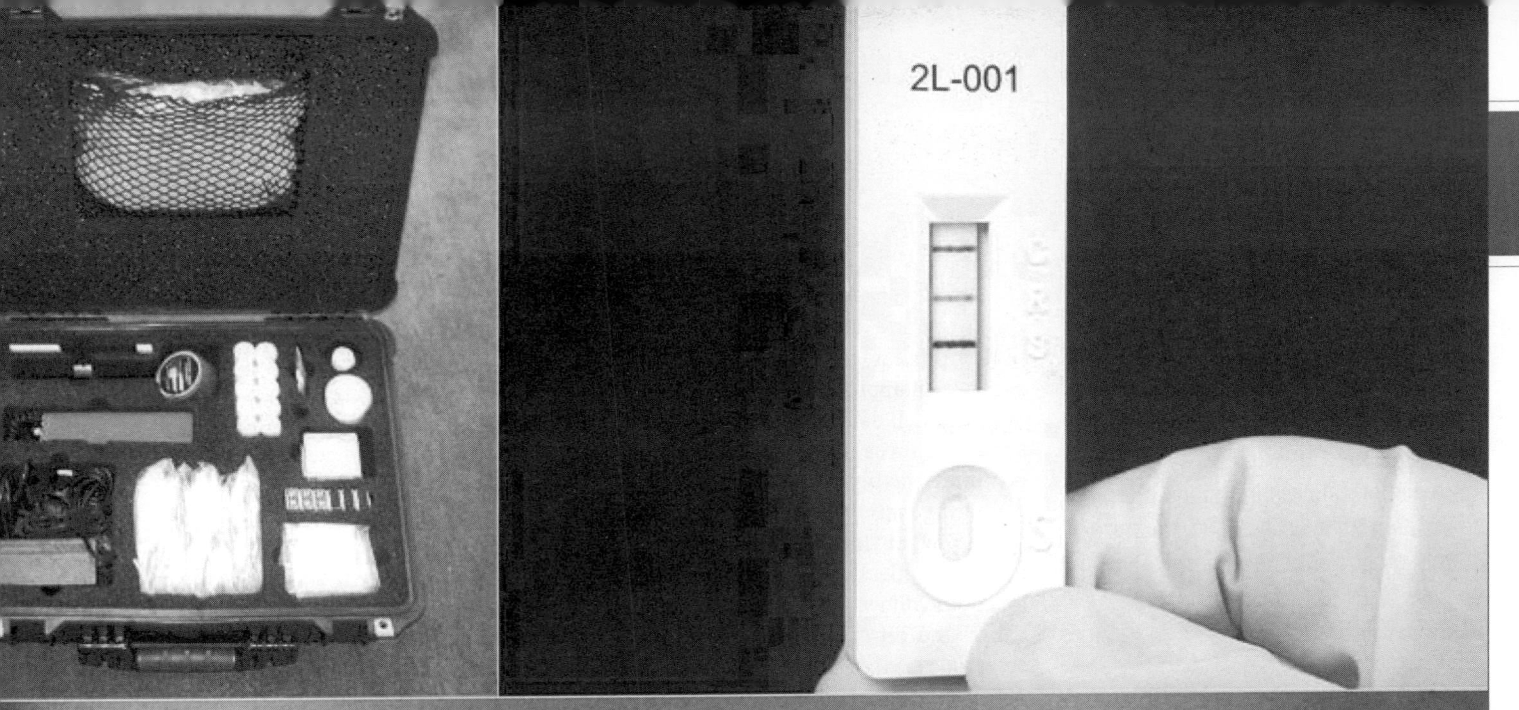

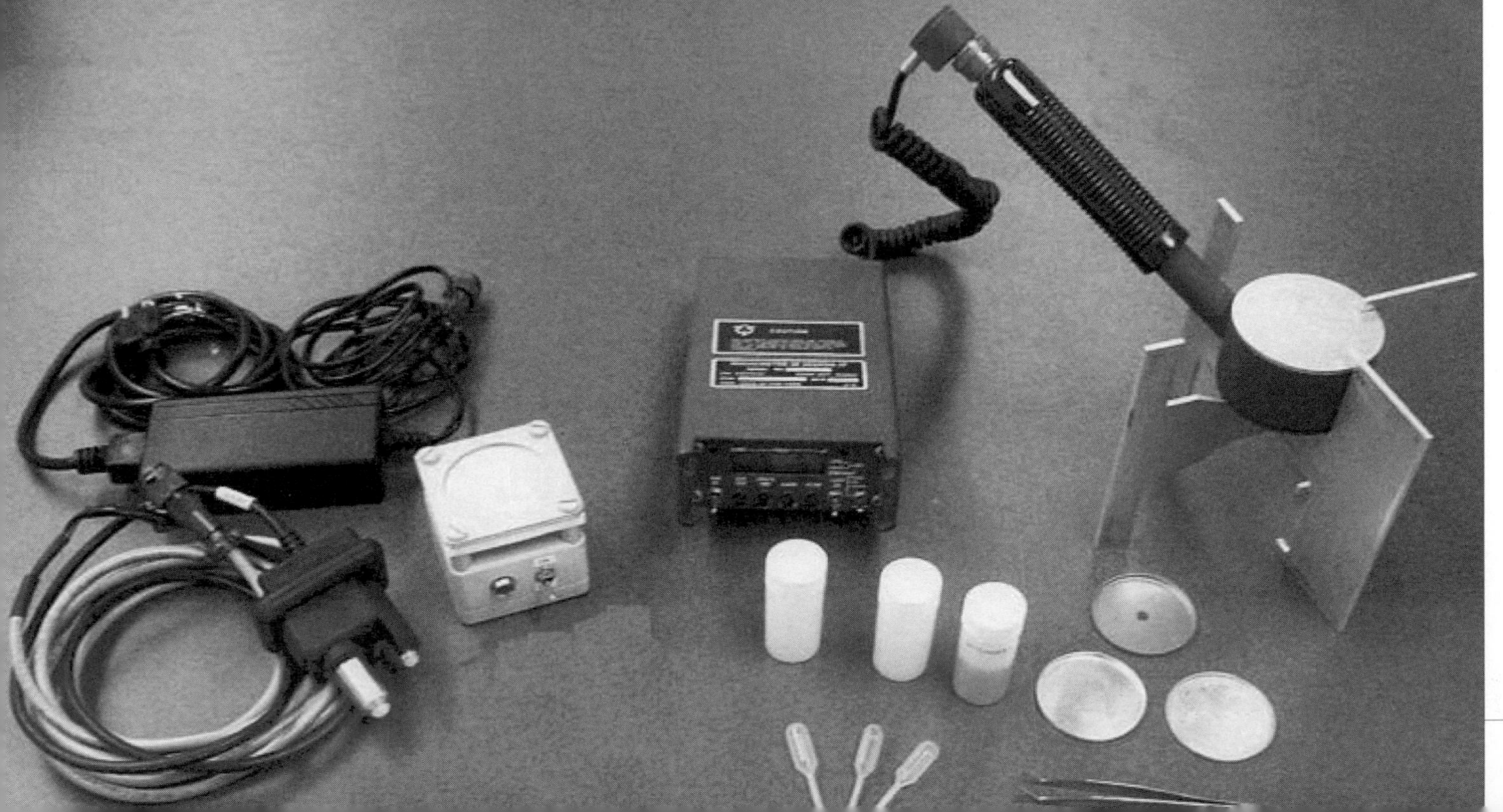

FOREIGN MILITARY SALES
None

CONTRACTORS
ANP Technologies (Newark, DE)

Joint Effects Model (JEM)

INVESTMENT COMPONENT

Modernization

Recapitalization

Maintenance

MISSION

To provide enhanced operational and tactical level situational awareness of the battlespace and to provide real-time hazard information during, and after, an incident, to influence and minimize effects on current operations.

DESCRIPTION

Joint Effects Model (JEM) is an Acquisition Category III software program. It is the only accredited DoD computer-based tactical and operational hazard prediction model capable of providing common representation of chemical, biological, radiological, nuclear (CBRN) and toxic industrial chemicals/toxic industrial material (TIC/TIM) hazard areas and effects. It may be used in two variants as either a standalone system or as a resident application on host command, control, communications, computers, and intelligence (C4I) systems. It is capable of modeling hazards in various scenarios, including: counterforce, passive defense, accidents, incidents, high-altitude releases, urban environments, building interiors, and human performance degradation.

JEM will follow an evolutionary acquisition approach. The JEM program will deliver a full-capability system in three increments, each retaining the functionality of the preceding increment(s). JEM Increment 1 will predict the probable hazard areas and effects for geographic locations following selected uses of CBRN and TIC/TIM by hostile forces; selected releases of CBRN materials resulting from offensive conventional strike missions performed by U.S. or allied forces on CBRN facilities; and selected accidental releases of TIC/TIM. Increment 2 and 3 will add additional capability and improve model performance. JEM will also support planning to mitigate the effects of weapons of mass destruction.

Chemical staff sections at the battalion, brigade, division, corps and echelons above corps levels, as well as Special Forces chemical recon detachments, will use JEM. Brigade, division, and corps-level CBRN staff planners will also have a reconnaissance version of JEM.

SYSTEM INTERDEPENDENCIES

Resides on and interfaces with C4I systems, which will use JEM to predict hazard areas and provide warning to U.S. Forces within those areas.

PROGRAM STATUS

- **2QFY09:** Increment 1 FRP decision

PROJECTED ACTIVITIES

- **FY10–11:** Continue Increment 2 developmental testing
- **FY10–11:** Continue Increment 2 software development

ACQUISITION PHASE

Technology Development | Engineering & Manufacturing Development | Production & Deployment | **Operations & Support**

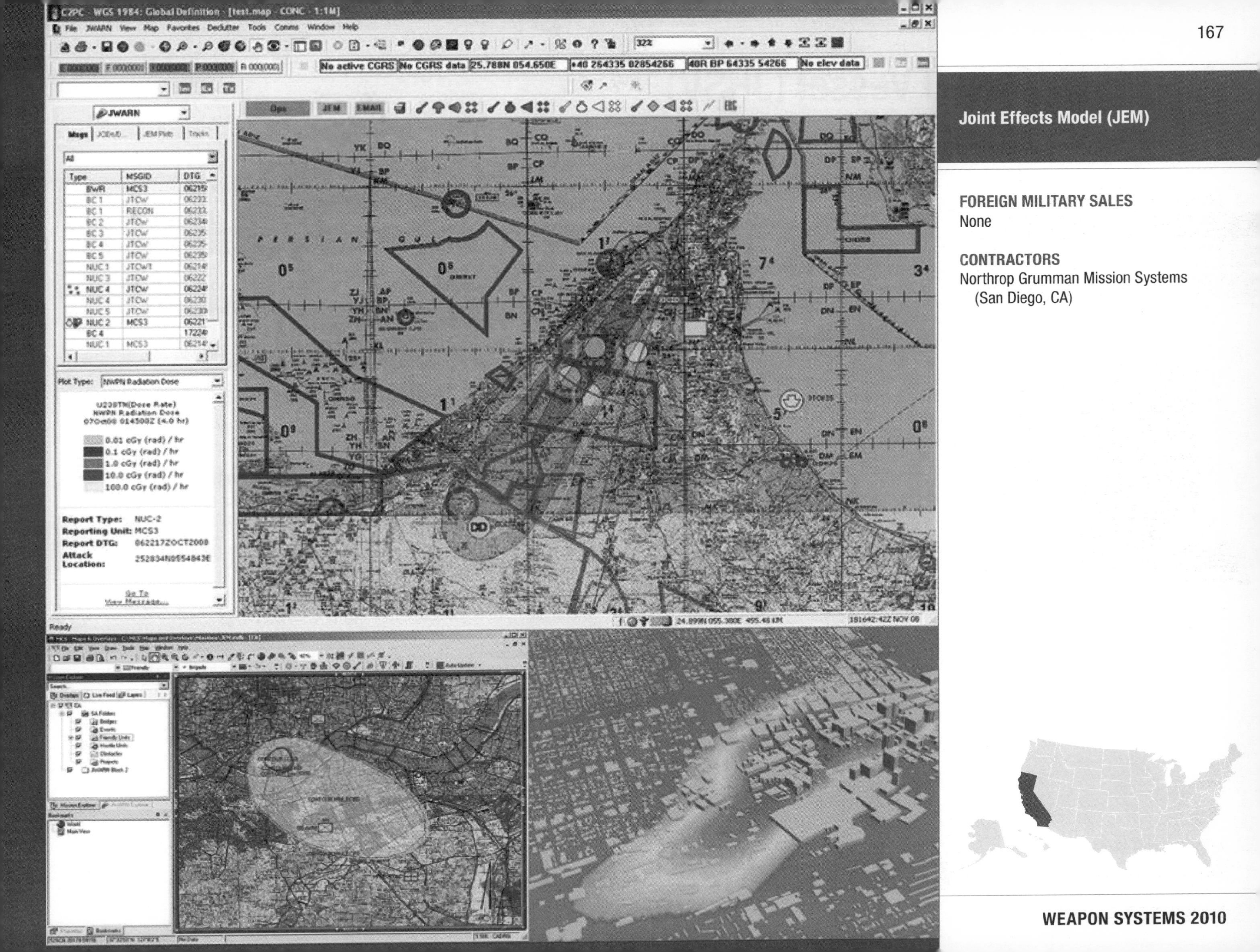

Joint Effects Model (JEM)

FOREIGN MILITARY SALES
None

CONTRACTORS
Northrop Grumman Mission Systems
(San Diego, CA)

Joint High Speed Vessel (JHSV)

INVESTMENT COMPONENT

Modernization

Recapitalization

Maintenance

MISSION

To provide high-speed intra-theater transport of Soldiers, military vehicles, equipment and cargo.

DESCRIPTION

The Joint High Speed Vessel (JHSV) is a 103meter (338 feet) high-speed, shallow-draft catamaran capable of transporting 600 short tons 1,200 nautical miles at an average speed of 35 knots. The JHSV is capable of interfacing with roll-on/roll-off discharge facilities, and on/off-loading a combat-loaded M1A2 Abrams Tank. JHSV has an aviation flight deck to support day and night air vehicle launch and recovery operations. The JHSV also has seating for more than 300 embarked Soldiers and fixed berthing for approximately 100 more. The JHSV represents the next-generation of Army watercraft to support the Army's doctrinal intra-theater lift mission. The JHSV provides flexibility and agility within a theater, enabling the Joint Force Commander to insert combat power and sustainment into austere ports worldwide. The JHSV bridges the gap between low-speed sealift and high-speed airlift. Supporting Army prepositioned stocks and joint logistics over-the-shore, the JHSV expands the reach and possibilities of prepositioning both on land and afloat. The JHSV provides the capability to conduct operational maneuver and repositioning of intact unit sets. This affords the combatant commander with increased throughput, increased survivability, increased responsiveness, and improved closure rates. This transport transformation-enabler helps achieve force deployment goals and full distribution-based logistics. The JHSV offers the Joint Force Commander a multi-modal and multipurpose platform to support joint operations that complements airlift capabilities, thereby minimizing the need for large-scale reception, staging, onward movement, and integration of Soldiers, vehicles, and equipment within the battlespace.

JHSV features:

- Flight deck
- Joint interoperable, command, control, communications, computers, intelligence, surveillance, and reconnaissance (C4ISR)
- Underway refueling
- Electronic navigation
- Anti-Terrorism/Force Protection capabilities

SYSTEM INTERDEPENDENCIES

None

PROGRAM STATUS

- **1QFY09:** Defense Acquisition Board Milestone B review
- **1QFY09:** Contract Award for Detail Design and Construction of one JHSV

PROJECTED ACTIVITIES

- **FY09–12:** Contract Option Awards for construction of four additional JHSVs

ACQUISITION PHASE

Technology Development | Engineering & Manufacturing Development | Production & Deployment | Operations & Support

Joint High Speed Vessel (JHSV)

FOREIGN MILITARY SALES
None

CONTRACTORS
Detail Design and Construction:
Austal USA (Mobile, AL)
Electronic Systems:
General Dynamics Advanced Information Systems (Fairfax, VA)

Joint Land Attack Cruise Missile Defense Elevated Netted Sensor System (JLENS)

INVESTMENT COMPONENT

Modernization

Recapitalization

Maintenance

MISSION

To provide over-the-horizon detection, tracking, classification, and engagement of cruise missiles and other air targets, enabling defensive engagement by air-directed, surface-to-air missiles or, air-to-air missile systems.

DESCRIPTION

The Joint Land Attack Cruise Missile Defense Elevated Netted Sensor System (JLENS) comprises two systems: a fire control radar system and a surveillance radar system. Each fire control radar system has a 74-meter tethered aerostat, a mobile mooring station, radar, communications payload, processing station, and associated ground support equipment. The JLENS mission is achieved by both the fire control radar and the surveillance radar systems operating as an "orbit"; however, each system can operate autonomously and contribute to the JLENS mission.

JLENS uses its advanced sensor and networking technologies to provide 360-degree wide-area surveillance and tracking of cruise missiles and other aircraft. Operating as an orbit, the surveillance radar generates information that enables the fire control radar to readily search for, detect, and track low-altitude cruise missiles and other aircraft. Once the fire control radar develops tracks, this information is provided to tactical data networks so other network participants can assess threat significance and assign systems to counter the threat. The fire control data supports extended engagement ranges by other network participants by providing high-quality track data on targets that may be terrain-masked from surface-based radar systems. JLENS information is distributed via the Link 16 Tactical Data Link and the Cooperative Engagement Capability (CEC) Network and adds to the single integrated air picture.

JLENS also performs as a multirole platform, enabling extended range C2 linkages, communications relay, and battlefield situational awareness. JLENS can stay aloft up to 30 days, providing 24-hour radar coverage of the assigned areas. The radar systems can be transported by aircraft, railway, ship, or roadway.

SYSTEM INTERDEPENDENCIES

The JLENS program is interdependent with PATRIOT Advance Capability–3, Surfaced Launched Advanced Medium Range Air-to-Air Missile (SLAMRAAM), and Navy Integrated Fire Control–Counter Air (NIFC–CA). The JLENS System is dependent on capabilities provided by CEC, Multifunctional Information Distribution System (MIDS), Integrated Broadcast System (IBS), and the Warfighter Information Network–Tactical (WIN–T)

PROGRAM STATUS

- **3QFY07:** Fire Control Radar critical design readiness review
- **2QFY08:** Orbit preliminary design review
- **4QFY08:** Surveillance radar critical design readiness review
- **1QFY09:** Orbit critical design review

PROJECTED ACTIVITIES

- **4QFY10:** Orbit 1 system integration begins
- **4QFY11:** Limited users test (LUT)

ACQUISITION PHASE

Technology Development | Engineering & Manufacturing Development | Production & Deployment | Operations & Support

Joint Land Attack Cruise Missile Defense Elevated Netted Sensor System (JLENS)

FOREIGN MILITARY SALES
None

CONTRACTORS
Raytheon (Andover, MA; El Segundo, CA; Dallas, TX)
TCOM (Columbia, MD)
CAS, Inc. (Huntsville, AL)

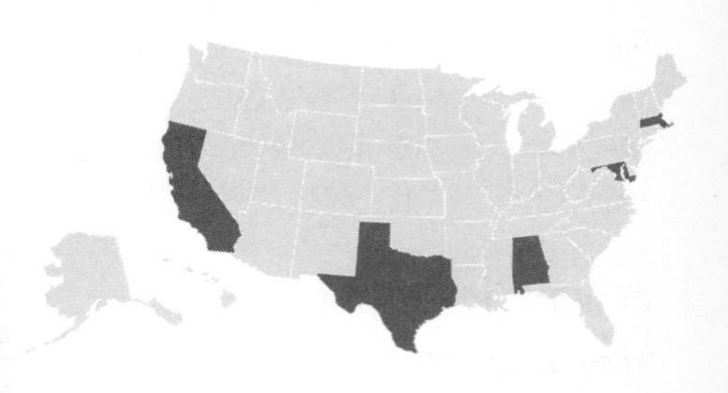

Joint Land Component Constructive Training Capability (JLCCTC)

INVESTMENT COMPONENT

Modernization

Recapitalization

Maintenance

MISSION

To provide tools to train unit commanders and their staffs from battalion through theater levels.

DESCRIPTION

The Joint Land Component Constructive Training Capability (JLCCTC) is a software modeling and simulation capability. It contributes to the joint training functional concept and the Army training mission area by providing the appropriate levels of model and simulation resolution. It also provided the fidelity needed to support both Army and joint training requirements. The JLCCTC is composed of two separate federations: JLCCTC–Multi-Resolution Federation (MRF), and JLCCTC–Entity-Resolution Federation (ERF). The MRF is a federated set of constructive simulation software supported by commercial software and commercial off-the-shelf (COTS) hardware that will support training of commanders and their staffs in maneuver, logistics, intelligence, air defense, and artillery. The federate models are connected by a combination of standard High-Level Architecture (HLA), Run-Time Infrastructure (RTI), Distributed Interactive Simulation (DIS), custom interfaces, Master Interface (MI), and Point-to-Point (PTP). The JLCCTC–MRF is a Command Post Exercise (CPX) driver designed to train Army commanders and their staffs at division through echelons-above-corps. It provides a simulated operational environment in which computer-generated forces simulate and respond to the C2 processes of the commanders and staffs. The JLCCTC models will provide full training functionality for leader and battle staff for the Army and the Joint, Intergovernmental, and Multinational (JIIM) spectrum. The JLCCTC provides an interface to Army Battle Command System (ABCS) equipment, allowing commanders and their staffs to train with their "go-to-war" systems. JLCCTC–ERF is a federation of simulations, simulation command, control, communications, computers and information (C4I) interfaces, data collection, and after action review (AAR) tools. It simulates the ABCS to facilitate battle staff collective training by requiring staff reaction to incoming digital information while executing the commander's tactical plan. The targeted training audience is composed of brigade and battalion battle staffs, functional command post (CP) training and full CP training. Battle staffs of higher echelons may also employ JLCCTC–ERF to achieve specific training objectives.

SYSTEM INTERDEPENDENCIES

None

PROGRAM STATUS

- **3QFY08:** JLCCTC MRF–W V5 verification event (VE) and operational readiness event (ORE)
- **4QFY08:** JLCCTC ERF V5 VE/ORE
- **4QFY08–1QFY09:** JLCCTC MRF–W Fielding to the National Simulation Center and Battle Command Training Program
- **1QFY09:** JLCCTC ERF V5 Fielding to Fort Bragg and Fort Indiantown Gap
- **1QFY09:** JLCCTC V5.5 integration and test events

PROJECTED ACTIVITIES

- **2QFY09–1QFY11:** JLCCTC fieldings
- **3QFY09:** JLCCTC V5.5 MRF–W VE/ORE
- **3QFY09:** JLCCTC ERF V5.2 software version release
- **4QFY09:** JLCCTC MRF–C software version release
- **4QFY09–4QFY10:** JLCCTC MRF–W V6 development, integration and test events

ACQUISITION PHASE

Technology Development | Engineering & Manufacturing Development | Production & Deployment | Operations & Support

Joint Land Component Constructive Training Capability (JLCCTC)

FOREIGN MILITARY SALES
None

CONTRACTORS
Lockheed Martin Information Systems (Orlando, FL)
Tapestry Solutions (San Diego, CA)

Joint Light Tactical Vehicle (JLTV)

INVESTMENT COMPONENT

Modernization

Recapitalization

Maintenance

MISSION

To provide a family of vehicles with companion trailers, capable of performing multiple mission roles that will be designed to provide protected, sustained, networked mobility for personnel and payloads across the full range of military operations.

DESCRIPTION

The Joint Light Tactical Vehicle (JLTV) Family of Vehicles (FoV) is a Joint Service and International program that will be capable of operating across a broad spectrum of terrain and weather conditions. The Joint Services require enhanced capabilities, greater than those provided by the existing High Mobility Multipurpose Wheeled Vehicle, to support the Joint Functional Concepts of Battlespace Awareness, Force Application, and Focused Logistics.

Payloads: Category A–3,500 pounds, Category B–4,500 pounds, Category C–5,100 pounds

Transportability: Internal–C-130, External–CH-47/53, Sea–Height-restricted decks

Protection: Scalable armor to provide mission flexibility while protecting the force.

Mobility: Maneuverability to enable operations across the spectrum of terrain, including urban areas.

Networking: Connectivity for improved Battlespace Awareness and responsive, well-integrated command and control for embarked forces.

Sustainability: Reliable, maintainable, maximum commonality across mission role variants, onboard and exportable power, and reduced fuel consumption.

The JLTV FoV balances the "Iron Triangle" of payload, protection, and performance.

SYSTEM INTERDEPENDENCIES

None

PROGRAM STATUS

- **1QFY09:** Awarded three Technology Development contracts
- **2QFY09:** Start of Work Meetings
- **4QFY09:** Preliminary design reviews
- **1QFY10:** Critical design reviews

PROJECTED ACTIVITIES

- **3QFY10:** Milestone Decision Authority Review
- **2QFY11:** Capability Development Document (CDD) approved
- **4QFY11:** Milestone B, Enter Engineering and Manufacturing Development (EMD
- **4QFY11:** Award two EMD contracts

ACQUISITION PHASE

Technology Development | Engineering & Manufacturing Development | Production & Deployment | Operations & Support

Solutions in Payload Categories A,B & C

Payload Category A

Payload: 3,500 lbs

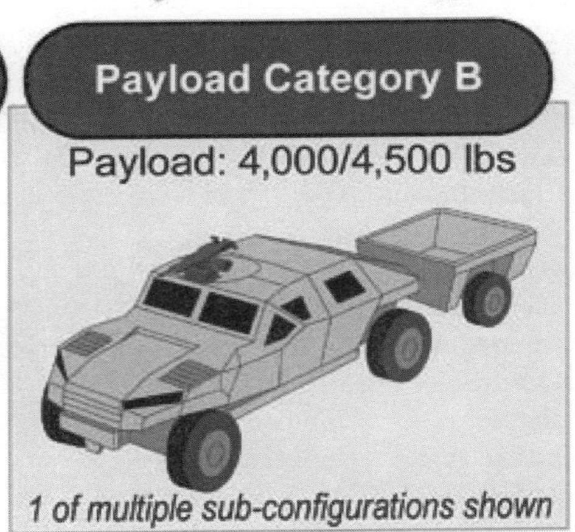

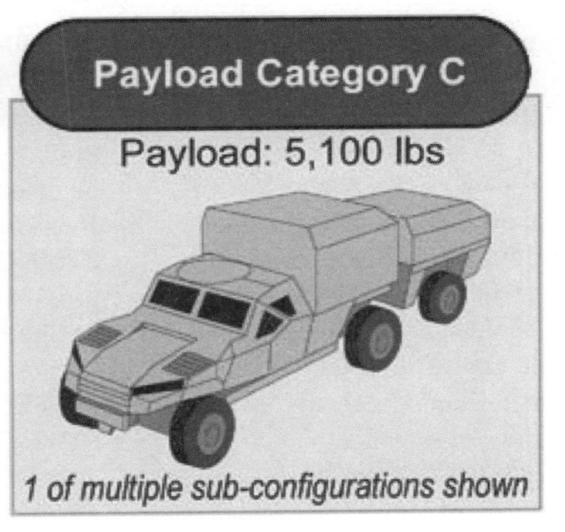

Joint Light Tactical Vehicle (JLTV)

FOREIGN MILITARY SALES
None

CONTRACTORS
BAE Systems Land & Armaments (Santa Clara, CA)
General Tactical Vehicle (Sterling Heights, MI)
Lockheed Martin (Owego, NY)

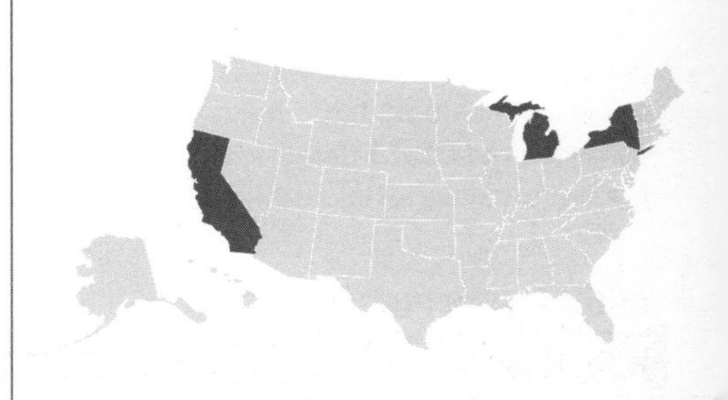

Joint Nuclear Biological Chemical Reconnaissance System (JNBCRS)

INVESTMENT COMPONENT

- **Modernization**
- Recapitalization
- Maintenance

MISSION

To provide Chemical, Biological, Radiological, and Nuclear (CBRN) reconnaissance in confined spaces and terrain that is inaccessible to CBRN reconnaissance vehicles, homeland defense consequence management operations, tactical force protection operations, and to characterize hazardous material events or accidents

DESCRIPTION

The Joint Nuclear Biological, Chemical Reconnaissance System (JNBCRS) is comprised of CBRN sets, kits, and outfits tailorable to mission requirements. Each system consists of both commercial and government off-the-shelf equipment to provide personnel protection from CBRN hazards, including toxic industrial materiel and CBRN detection, presumptive materiel and CBRN detection, identification, sample collection, decontamination, marking, and hazard reporting.

The Joint Urgent Operational Need Statement (JUONS) phase supports CBRN and Explosive Consequence Management Response Force, Central Command's Urgent Need for Toxic Industrial Detection Equipment, and Service Urgent Needs for similar equipment.

The Dismounted Reconnaissance Sets, Kits, and Outfits (DR-SKO) phase will provide a modular, scalable, mission tailorable equipment package. This will provide expeditionary CBRN capabilities to conduct reconnaissance missions and consequence management. It will provide conventional forces with the capability to confirm or deny the presence of Weapons of Mass Destruction (WMD) in support of WMD Eliminations (WMD-E), WMD Interdiction (WMD-I) and a capability to respond to a hazardous material event or accident.

The CBRN monitor and survey sets, kits, and outfits (MS-SKO) will provide the next increment of dismounted capability for the Joint Forces.

SYSTEM INTERDEPENDENCIES

None

PROGRAM STATUS

- **FY09:** Approval to field to an additional 8 JUONS Systems to provide a total of 27 systems to be fielded under Urgent Material Releases
- **FY09:** Continue development of program documentation, system design, and integrated logistics support activities for DR-SKO and MS-SKO phases
- **FY09:** Continue systems engineering support (government)
- **FY09:** Materiel Development Decision for the MS-SKO

PROJECTED ACTIVITIES

- **FY10:** DR-SKO Milestone C, Low Rate Initial Production Decision
- **FY10:** MS-SKO Milestone B Decision
- **FY11:** MS-SKO Production Qualification Testing
- **FY12:** MS-SKO Operational Assessment

ACQUISITION PHASE

- **Technology Development**
- Engineering & Manufacturing Development
- Production & Deployment
- **Operations & Support**

Joint Nuclear Biological Chemical Reconnaissance System (JNBCRS)

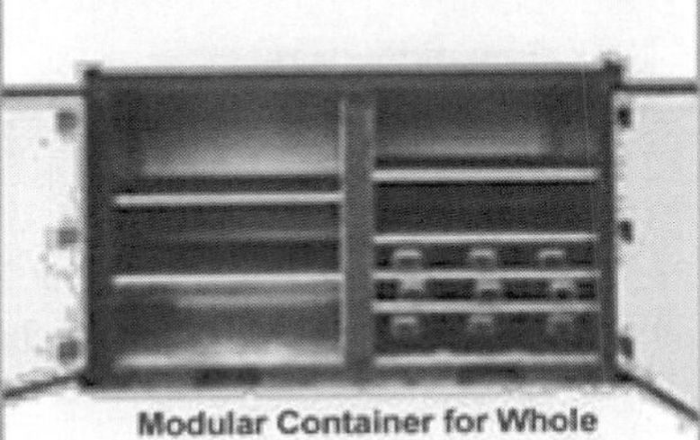

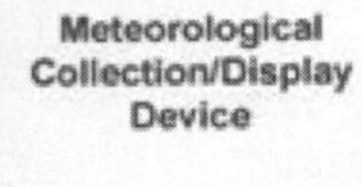

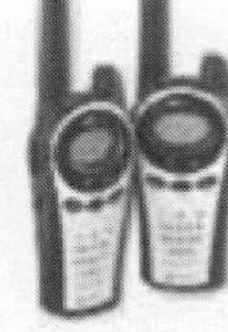

(Notional)

FOREIGN MILITARY SALES
None

CONTRACTORS
Prime Integrator:
ICx™ Technologies, Inc. (Glen Burnie, MD)

Joint Precision Airdrop System (JPADS)

INVESTMENT COMPONENT

Modernization

Recapitalization

Maintenance

MISSION

To provide the warfighter with precision airdrop ensuring accurate delivery of supplies to forward operating forces, reducing vehicular convoys, and allowing aircraft to drop cargo at safer altitudes and off-set distances.

DESCRIPTION

The Joint Precision Airdrop System (JPADS) integrates a parachute decelerator, an autonomous guidance unit, and a load container or pallet to create a system that can accurately deliver critical supplies with great precision. The system is being developed in two weight classes: 2,000 pounds and 10,000 pounds, with potential future requirements for 30,000 pounds, and an objective system of 60,000 pounds. The guidance system uses military global positioning satellite data for precise navigation and interfaces with a Mission Planning module on board the aircraft to receive real-time weather data and compute aerial release points. JPADS is being designed for aircraft to drop cargo from altitudes of up to 24,500 feet mean sea level. It will release cargo from a minimum off-set of 8 kilometers from the intended point of impact, with an objective capability of 25 kilometers off-set. This off-set allows aircraft to stay out of range of many anti-aircraft systems. It also enables aircraft to drop systems from a single aerial release point and deliver them to multiple or single locations, thus reducing aircraft exposure time. Once on the ground, the precise placement of the loads greatly reduces the time needed to recover the load. Exposure to ground forces is minimized as well.

SYSTEM INTERDEPENDENCIES

None

PROGRAM STATUS

- **3QFY07–4QFY08:** Testing for 2,000-pound variant completed
- **1QFY08:** Milestone B (permission to enter system development and demonstration phase) received for 10,000-pound variant
- **1QFY08:** Testing began for 10,000-pound variant
- **4QFY08–1QFY09:** Milestone C (full-rate production and fielding decision) preparation underway for 2,000-pound variant

PROJECTED ACTIVITIES

- **2QFY09:** Milestone C for the 2,000-pound variant subsequent, with production contract
- **4QFY09:** Fielding begins for 2,000-pound variant and will continue until FY12, assuming projected funding remains
- **1QFY11:** Complete testing of the 10,000-pound variant
- **2QFY11:** Milestone C (full-rate production and fielding decision) for 10,000-pound variant with subsequent award production contract
- **4QFY11:** Fielding begins for 10,000-pound variant

ACQUISITION PHASE

Technology Development | Engineering & Manufacturing Development | Production & Deployment | Operations & Support

Joint Precision Airdrop System (JPADS)

FOREIGN MILITARY SALES
None

CONTRACTORS
Airborne Systems North America (Pennsauken, NJ)

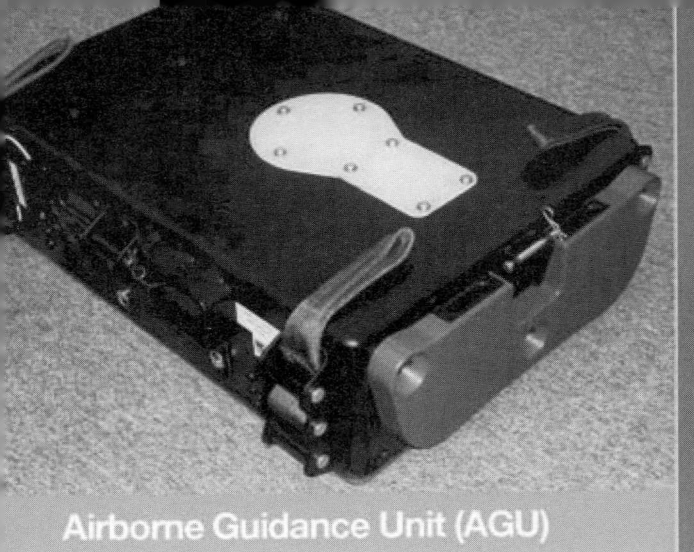

Airborne Guidance Unit (AGU)

Joint Service General Purpose Mask (JSGPM)

INVESTMENT COMPONENT

Modernization

Recapitalization

Maintenance

MISSION

To enable warfighters' survival in a chemical, biological, radiological, and nuclear (CBRN) environment by providing chemical, biological, toxin, radioactive particulate, and toxic industrial material protection.

DESCRIPTION

The Joint Service General Purpose Mask (JSGPM) is a lightweight protective mask system incorporating state-of-the-art technology to protect U.S. Joint Forces from actual or anticipated threats. The JSGPM will provide above-the-neck, head-eye-respiratory protection against CBRN threats, including toxic industrial chemicals (TIC). The mask component designs will be optimized to minimize their impact on the wearer's performance and to maximize its ability to interface with current and future Service equipment and protective clothing. The JSGPM mask system replaces the M40/M42 series of protective masks for the Army and Marine Corps ground and combat vehicle operations, as well as the MCU-2/P series of protective masks for Air Force and Navy shore-based and shipboard applications.

SYSTEM INTERDEPENDENCIES

The JSGPM will interface with Joint service vehicles, weapons, communication systems, individual clothing and protective equipment, and CBRN personal protective equipment.

PROGRAM STATUS

- **1QFY09:** In production and fielding.

PROJECTED ACTIVITIES

- **FY10–12:** Continue production and fielding

ACQUISITION PHASE

Technology Development | Engineering & Manufacturing Development | Production & Deployment | Operations & Support

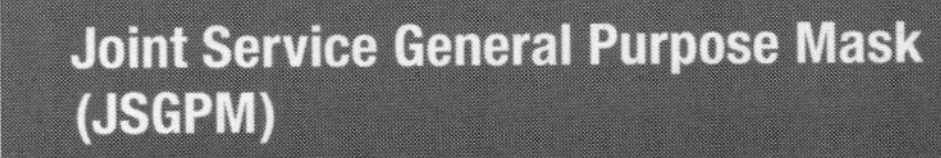

Joint Service General Purpose Mask (JSGPM)

FOREIGN MILITARY SALES
None

CONTRACTORS
Avon Protection Systems (Cadillac, MI)

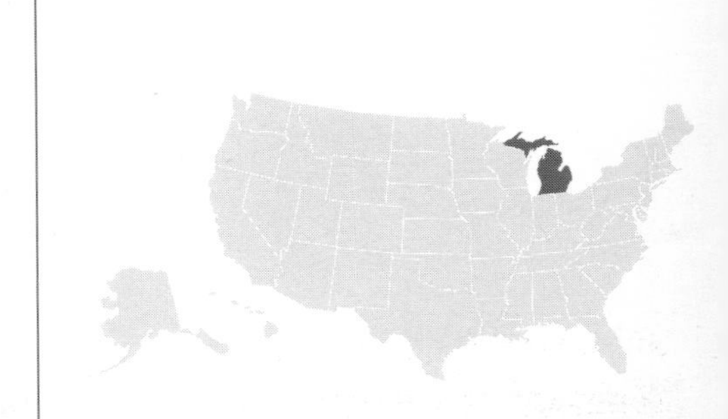

Joint Service Personnel/Skin Decontamination System (JSPDS)

INVESTMENT COMPONENT

Modernization

Recapitalization

Maintenance

MISSION

To provide the warfighter with skin decontamination capacity, after exposure to chemical/biological warfare agents, in support of immediate and thorough personnel decontamination operations.

DESCRIPTION

The Joint Service Personnel/Skin Decontamination System (JSPDS) Increment I, Reactive Skin Decontamination Lotion (RSDL), provides enhanced capabilities to immediately reduce chemical warfare agents and biological warfare agents from skin. RSDL provides the warfighter with improved capability over the existing M291 Skin Decontaminating Kit to reduce lethal and performance-degrading effects. Additionally it can be used to decontaminate individual equipment, weapons, and casualties on unbroken skin. RSDL is a commercially available product that has been approved for medical use by the U.S. Food and Drug Administration.

The JSPDS program supports an evolutionary acquisition strategy using incremental and spiral development. Increment I will provide hazard reduction efficacy capabilities greater than the M291 SDK for chemical warfare agents and equal capabilities for biological hazard reduction. The structure of the JSPDS program will allow the flexibility to accelerate fielding of capability enhancements through the use of commercial products. Increment II will provide increased hazard reduction efficacy capabilities on skin of nontraditional agents and toxic industrial chemicals, BW agents, and radiological and nuclear hazards. It can also be used for open-wound exposures.

SYSTEM INTERDEPENDENCIES

RSDL is compatible with all chemical, biological, radiological, and nuclear individual protective equipment, detectors, and decontaminants as well as small arm weapons.

PROGRAM STATUS

- **1QFY09:** Continue production

PROJECTED ACTIVITIES

- **2QFY10:** Initial operational capability

ACQUISITION PHASE

Technology Development | Engineering & Manufacturing Development | **Production & Deployment** | **Operations & Support**

Joint Service Personnel/Skin Decontamination System (JSPDS)

FOREIGN MILITARY SALES
None

CONTRACTORS
Bracco Diagnostics, Inc. (Montreal, Canada)

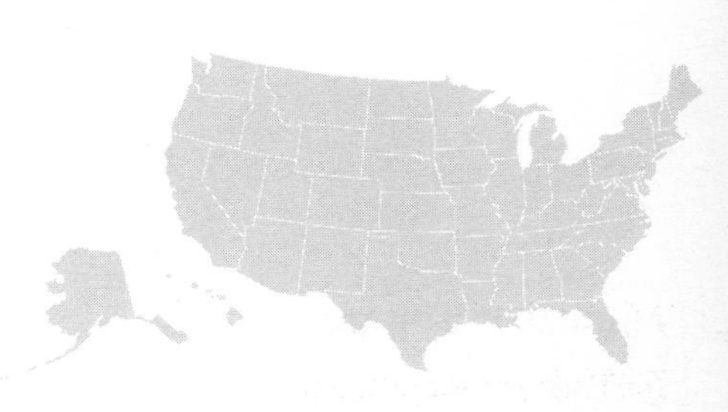

Joint Service Transportable Decontamination System (JSTDS)–Small Scale (SS)

INVESTMENT COMPONENT

- Modernization
- Recapitalization
- Maintenance

MISSION

To rapidly and effectively support operational and thorough decontamination of military equipment in accordance with Field Manual 3.11-5.

DESCRIPTION

The Joint Service Transportable Decontamination System–Small Scale (JSTDS–SS) will enable warfighters to conduct operational and support thorough decontamination of non-sensitive military materiel, limited facility decontamination at logistics bases, airfields (and critical airfield assets), naval ships, ports, key command and control centers, and other fixed facilities that have been exposed to chemical, biological, radiological, and nuclear (CBRN) warfare agents/contamination and toxic industrial materials. The system may also support other hazard abatement missions as necessary.

The JSTDS–SS is being developed using an incremental approach. Increment I provides improved capabilities over current systems to decontaminate tactical and non-tactical vehicles, ship exterior surfaces, aircraft, crew served weapons, and aircraft support equipment. Increment II will focus on improving overarching decontamination processes, efficacy, and system capabilities for operational and thorough decontamination of non-sensitive equipment. The JSTDS–SS is supported with one accessory kit and one water blivet per system.

SYSTEM INTERDEPENDENCIES

All individual protective equipment, decontaminants, and detectors.

PROGRAM STATUS

- **4QFY09:** Fielding and materiel release decision

PROJECTED ACTIVITIES

- **FY10–11:** Procure and field systems

ACQUISITION PHASE

Technology Development | Engineering & Manufacturing Development | Production & Deployment | Operations & Support

Joint Service Transportable Decontamination System (JSTDS)–Small Scale (SS)

FOREIGN MILITARY SALES
None

CONTRACTORS
DRS Technologies (Florence, KY)

Joint Tactical Ground Stations (JTAGS)

INVESTMENT COMPONENT

Modernization

Recapitalization

Maintenance

MISSION

To disseminate early warning, alerting, and cueing information of ballistic missile attack and other infrared events to theater combatant commanders by using real-time, direct down-linked satellite data.

DESCRIPTION

Joint Tactical Ground Stations (JTAGS) are forward-deployed, echelon-above-corps, transportable systems designed to receive, process and disseminate direct down-linked infrared data from space-based sensors. Ongoing product improvement efforts will integrate JTAGS with the next-generation Space Based Infrared System (SBIRS) satellites. SBIRS sensors will significantly improve theater missile warning parameters. Expected improvements include higher quality cueing of active defense systems, decreased missile launch search area, faster initial report times, and improved impact ellipse prediction.

JTAGS processes satellite data and disseminates ballistic missile warning or special event messages to warfighters in support of regional combatant commanders over multiple theater communication systems. Five JTAGS are deployed worldwide as part of the U.S. Strategic Command's Tactical Event System. The Army Space and Missile Defense Command Soldiers operate JTAGS, providing 24/7/365 support to theater operations.

SYSTEM INTERDEPENDENCIES

JTAGS improvements are dependent upon successful development, launch, test and certification of the U.S. Air Force's ACAT I, Space-Based Infrared System (SBIRS) satellite program. SBIRS will provide multiple new sensors on orbit, substantially enhancing the ability of JTAGS to improve all aspects of missile warning and battlespace awareness.

PROGRAM STATUS

- **1QFY09–4QFY09:** Worldwide fielding of JTAGS upgrades: Common Data Link Interface (CDLI), Joint Tactical Terminal (JTT), Multifunctional Information Distribution System (MIDS), and information assurance improvements; upgrades to all five JTAGS units and the JTAGS Development Lab

PROJECTED ACTIVITIES

- **4QFY09–4QFY11:** Field JTAGS block upgrades including: communication systems, information assurance, antennas, the addition of a new on-orbit SBIRS satellite sensor (highly elliptical orbit), and Initial SBIRS Geosynchronous Orbit (GEO) satellite capability. Software support, contractor logistics support, and depot operations continue
- **3QFY10:** New JTAGS contract awarded to develop and deliver full SBIRS GEO starer capability in a desheltered system integrated into operation centers

ACQUISITION PHASE

Technology Development | Engineering & Manufacturing Development | Production & Deployment | Operations & Support

Joint Tactical Ground Stations (JTAGS)

FOREIGN MILITARY SALES
None

CONTRACTORS
Develop, Deploy, Sustain (CLS):
Northrop Grumman Electronic Systems (Colorado Springs, CO)
SETA support:
BAE Systems (Huntsville, AL)

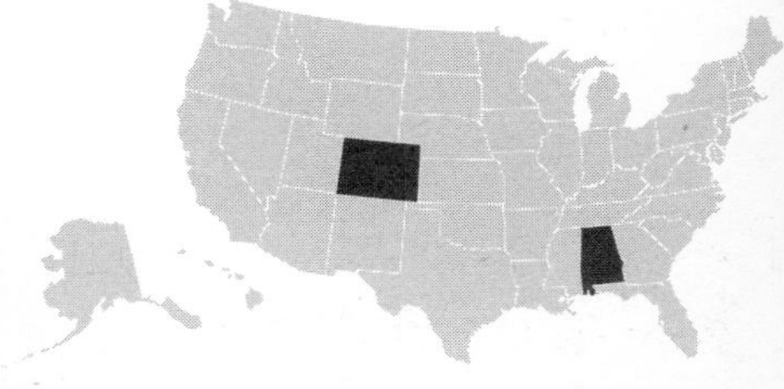

Joint Tactical Radio System Airborne, Maritime/Fixed Station (JTRS AMF)

INVESTMENT COMPONENT

Modernization

Recapitalization

Maintenance

MISSION

To provide scalable and modular networked radio frequency installed communication capability to meet Joint Service requirements through two Joint tactical radio (JTR) sets: Small Airborne (SA) and Maritime/Fixed Station (M/F) with common ancillary equipment for both radio form factors, aircraft such as the Air Force C-130, Army Rotary Wing, and Navy E-2, along with maritime and shore sites

DESCRIPTION

The Joint Tactical Radio System Airborne, Maritime/Fixed Station (JTRS AMF) will provide a four-channel, full duplex, software-defined radio integrated into airborne, shipboard, and fixed-station platforms, enabling maritime and airborne forces to communicate seamlessly and with greater efficiency through implementation of five initial waveforms (i.e., Ultra-High Frequency Satellite Communications, Mobile User Objective System, Wideband Network Waveform, Soldier Radio Waveform, and Link 16) providing data, voice, and networking capabilities.

SYSTEM INTERDEPENDENCIES

JTRS Network Enterprise Domain (NED) products and services; JTRS Ground Mobile Radio (GMR); JTRS Handheld Manpack and Small Form Fit (HMS); Multiple aircraft, maritime and fixed site platforms

PROGRAM STATUS

- **1QFY10:** Critical Design Review

PROJECTED ACTIVITIES

- **2QFY11:** Delivery begins for AMF SA engineering development models (EDM)
- **4QFY11:** Delivery begins for AMF M/F EDM

ACQUISITION PHASE

Technology Development | **Engineering & Manufacturing Development** | Production & Deployment | Operations & Support

Airborne and Maritime/Fixed Station

Increment 1 JTR Sets

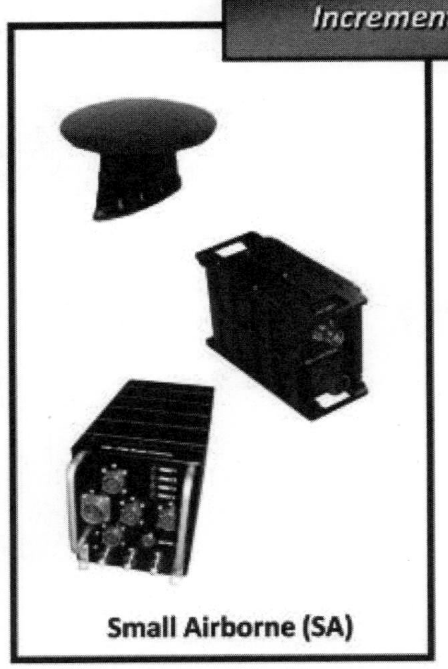

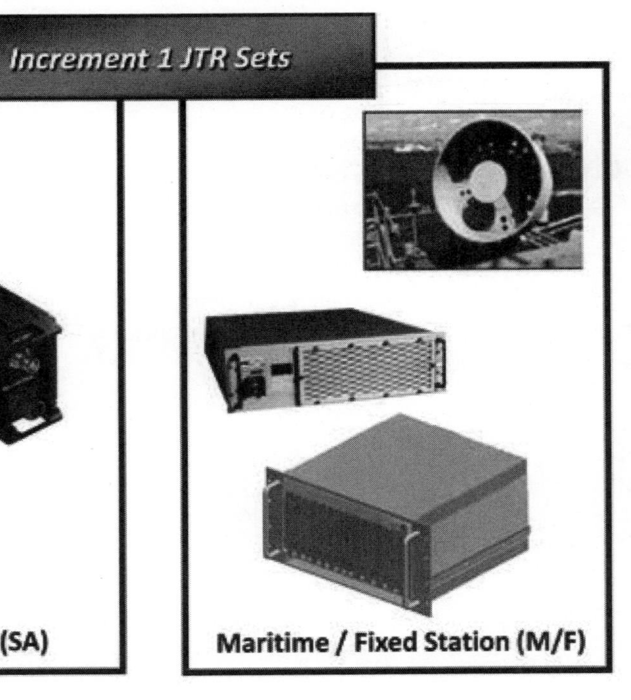

Joint Tactical Radio System Airborne, Maritime/Fixed Station (JTRS AMF)

FOREIGN MILITARY SALES
None

CONTRACTORS
Lockheed Martin (Chantilly, VA)
General Dynamics C4 Systems, Inc. (Scottsdale, AZ)
BAE Systems (Wayne, NJ)
Northrop Grumman (San Diego, CA)
Raytheon (Waltham, MA)

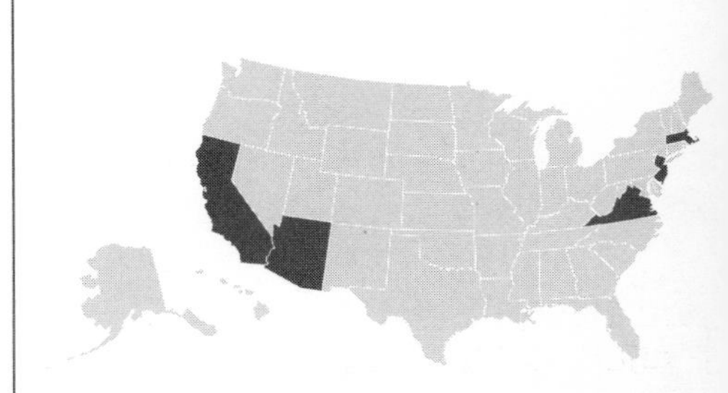

Joint Tactical Radio System Ground Mobile Radios (JTRS GMR)

INVESTMENT COMPONENT

Modernization

Recapitalization

Maintenance

MISSION

To provide mobile internet-like capability and interoperability with Current Force radios through simultaneous and secure voice, data, and video communications supporting battle command, sensor-to-shooter, sustainment, and survivability applications in full-spectrum operations on vehicular platforms.

DESCRIPTION

Through software reconfiguration, the Joint Tactical Radio System Ground Mobile Radios (JTRS GMR) can emulate Current Force radios and operate new internet protocol-based networking waveforms offering increased data throughput utilizing self-forming, self-healing, and managed communication networks. The GMR route and retransmit functionality links various waveforms in different frequency bands to form one inter-network. GMR can scale from one to four channels supporting multiple security levels and effectively use the frequency spectrum within the 2 megahertz to 26 hertz band.

SYSTEM INTERDEPENDENCIES

Army modernization efforts, Abrams, HMMWV, Bradley, Command Post System Carrier, JTRS Network Enterprise Domain (NED) products and services, WIN–T

PROGRAM STATUS

- **2QFY09:** EDM radio deliveries begins; Test readiness review
- **3QFY09:** EDM production and deliveries complete; production qualification test begins, security verification tests begin

PROJECTED ACTIVITIES

- **3QFY10:** System integration test begins; security verification test concludes; production qualification test concludes; NSA approval to enter limited user test
- **1QFY11:** Limited user test concludes
- **2QFY11:** Milestone C decision approving entry into the production and deployment phase

ACQUISITION PHASE

Technology Development | Engineering & Manufacturing Development | Production & Deployment | Operations & Support

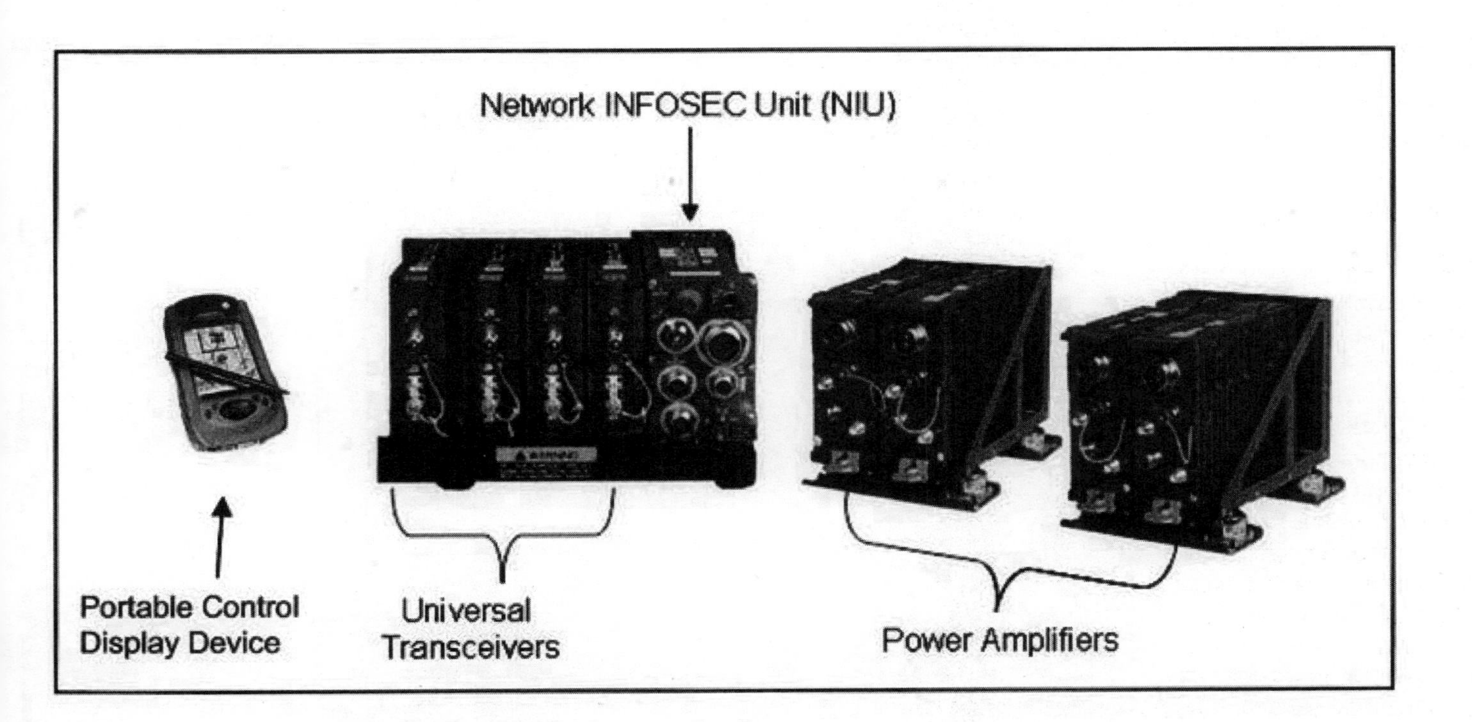

Joint Tactical Radio System Ground Mobile Radios (JTRS GMR)

FOREIGN MILITARY SALES
None

CONTRACTORS
Boeing (Huntington Beach, CA)
BAE Systems (Wayne, NJ)
Northrop Grumman (San Diego, CA)
Rockwell Collins (Cedar Rapids, IA)

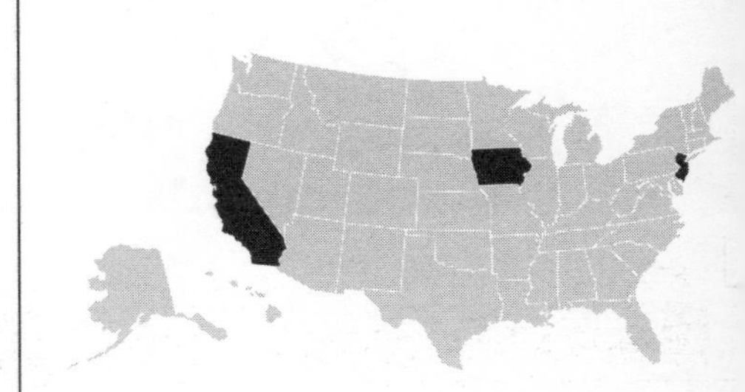

Joint Tactical Radio System Handheld, Manpack, and Small Form Fit (JTRS HMS)

INVESTMENT COMPONENT

Modernization

Recapitalization

Maintenance

MISSION

To provide a scalable and modular networked radio frequency communication capability to meet Joint handheld, manpack, and small form fit radio requirements at the tactical edge.

DESCRIPTION

Provides the warfighter with a software reprogrammable, networkable multi-mode system of systems capable of simultaneous voice, data, and/or video communications between 2 megahertz and 2.5 gigahertz. JTRS HMS satisfies joint service requirements for handheld, manpack, and small form fit applications, including support for Army modernization efforts and Ground Soldier System (GSS).

SYSTEM INTERDEPENDENCIES

UAV, SUGV, UGS, NLOS–LS, IMS, WIN–T, Army modernization efforts, GSS

PROGRAM STATUS

- **3QFY09:** Phase 1 limited user test
- **1QFY10:** Phase 1 Milestone C low rate initial production decision for SFF–C (v) 1

PROJECTED ACTIVITIES

- **3QFY10:** Delivery of Phase 2 engineering development models
- **4QFY10:** Phase 2 security verification test
- **1QFY11:** Phase 2 limited user testing
- **2QFY11:** Phase 2 Milestone C decision

ACQUISITION PHASE

Technology Development | **Engineering & Manufacturing Development** | Production & Deployment | Operations & Support

Joint Tactical Radio System Handheld, Manpack, and Small Form Fit (JTRS HMS)

FOREIGN MILITARY SALES
None

CONTRACTORS
General Dynamics C4 Systems, Inc. (Scottsdale, AZ)
BAE Systems (Wayne, NJ)
Rockwell Collins (Cedar Rapids, IA)
Thales Communications (Clarksburg, MD)

Joint Tactical Radio System Network Enterprise Domain (JTRS NED)

INVESTMENT COMPONENT

Modernization

Recapitalization

Maintenance

MISSION

To develop and deliver portable, interoperable, transformational networking waveforms (e.g., wide-band network waveforms, Soldier radio waveforms), as well as the software to manage the network services needed to fully enable JTRS' mobile, ad hoc networking capability. NED products will produce the networking capability that allows U.S. warfighters from all military branches to access and share relevant and timely information.

DESCRIPTION

The heart of the interoperable networking capability of JTRS, NED's product line consists of: 14 Legacy Waveforms (Bowman VHF, COBRA, EPLRS, Have Quick II, HF SSB/ALE, HF 5066, Link 16, SINCGARS, UHF DAMA SATCOM 181/182/183/184, UHF LOS, VHF LOS); three Mobile Ad-hoc Networking Waveforms (Wideband Networking Waveform [WNW], Soldier Radio Waveform [SRW], and Mobile User Objective System [MUOS]–Red Side Processing); and Network Enterprise Services (NES) including the JTRS WNW Network Manager (JWNM), JTRS Enterprise Network Manager (JENM), and Enterprise Network Services Phase 1 (ENS PH1).

SYSTEM INTERDEPENDENCIES

JTRS Ground Mobile Radio (GMR), JTRS Airborne and Maritime/Fixed Site (AMF), Multifunctional Information Distribution System (MIDS)–JTRS, JTRS Handheld, Manpack and Small Form Fit, NED provides SINCGARS and Link 16

PROGRAM STATUS

- **2QFY09:** SRW 1.0c final qualification test (FQT) delivered
- **3QFY09:** Link-16 FQT
- **1QFY10:** WNW v4.0; JWNM v4.0; UHF SATCOM; HF v4.0 FQT deliveries

PROJECTED ACTIVITIES

- **3QFY10:** SRWNM 1.0+ deliveries
- **4QFY10:** JENM Phase 1 FQT deliveries
- **1QFY11:** ENS Phase 1 FQT deliveries
- **2QFY11:** MUOS v3.1 FQT deliveries
- **4QFY11:** JENM Phase 2 FQT deliveries

ACQUISITION PHASE

Technology Development | **Engineering & Manufacturing Development** | Production & Deployment | Operations & Support

Joint Tactical Radio System Network Enterprise Domain (JTRS NED)

FOREIGN MILITARY SALES
None

CONTRACTORS
Boeing (Huntington Beach, CA)
Harris Corp. (Melbourne, FL)
ITT (Clifton, NJ)
Northrop Grumman (San Diego, CA)
Rockwell Collins (Cedar Rapids, IA)

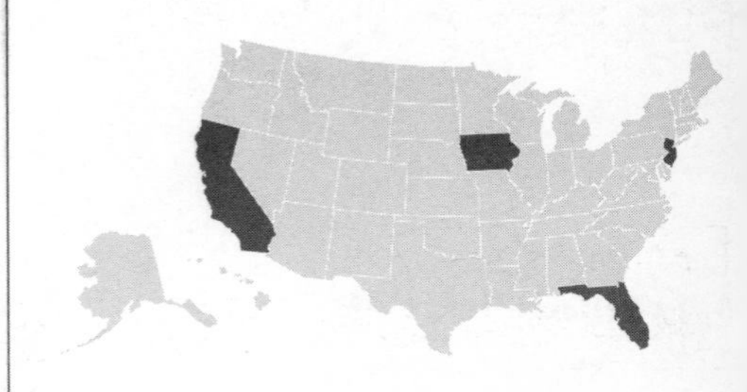

Joint Warning and Reporting Network (JWARN)

INVESTMENT COMPONENT

Modernization

Recapitalization

Maintenance

MISSION

To accelerate the warfighter's response to a nuclear, biological, or chemical attack by providing joint forces the capability to report, analyze, and disseminate detection, identification, location, and warning information.

DESCRIPTION

The Joint Warning and Reporting Network (JWARN) is a computer-based system designed to collect, analyze, identify, locate, and report information on nuclear, biological, or chemical (NBC) activity and threats from sensors in the field and to disseminate that information to decision-makers throughout the command. Located on mobile and fixed platforms, JWARN will be compatible and integrated with joint service command, control, and communications, computers, intelligence, and surveillance reconnaissance (C4ISR) systems. JWARN's component interface device connects to the sensors, which can detect various types of attack. The component device relays warnings to C4ISR systems via advanced wired or wireless networks. JWARN reduces the time from incident observation to warning to less than two minutes, enhances warfighters' situational awareness throughout the area of operations, and supports battle management tasks.

The JWARN full-capability system will be developed as a single increment. The development phase will be followed by a preplanned product improvement effort, which will include artificial intelligence modules for NBC operations, an upgrade to match future C4ISR systems, and standard interfaces for use with future detectors.

Block I

Initial acquisition and fielding of commercial off-the-shelf and government off-the-shelf software as standard for Armed Services.

Block II

Integration of NBC legacy and future detector modules, and NBC battlefield management modules

SYSTEM INTERDEPENDENCIES

Capable of two-way interface with current and planned individual service (C4ISR) hardware and software. JWARN will be compatible with Allied Technological Publication 45 (ATP-45).

PROGRAM STATUS

- **1QFY10:** JWARN full-rate production decision

PROJECTED ACTIVITIES

- **2QFY10:** Continue production and deployment

ACQUISITION PHASE

Technology Development | Engineering & Manufacturing Development | Production & Deployment | Operations & Support

Joint Warning and Reporting Network (JWARN)

FOREIGN MILITARY SALES
None

CONTRACTORS
Bruhn New-Tech (Ellicott City, MD)
Northrop Grumman Information Technology (NGIT) (Winter Park, FL)

Kiowa Warrior

INVESTMENT COMPONENT

Modernization

Recapitalization

Maintenance

MISSION

To support combat and contingency operations with a light, rapidly deployable helicopter capable of armed reconnaissance, security, target acquisition and designation, command and control, light attack, and defensive air combat missions.

DESCRIPTION

The Kiowa Warrior is a single-engine, two-man, lightly armed reconnaissance helicopter with advanced visionics, navigation, communication, weapons, and cockpit integration systems. Its mast-mounted sight houses a thermal imaging system, low-light television, and a laser rangefinder/designator permitting target acquisition and engagement at standoff ranges and in adverse weather. The navigation system can convey precise target locations to other aircraft or artillery via its advanced digital communications system. It provides anti-armor and anti-personnel capabilities at standoff ranges.

The Army is currently installing modifications to address safety, obsolescence and weight to keep the aircraft viable through its projected retirement date of FY20. Key among these modifications is the addition of an upgraded cockpit, a nose mounted sensor, and a dual channel full authority digital engine control. Additionally, the Army has started an aircraft replacement program to address Kiowa Warrior losses.

SYSTEM INTERDEPENDENCIES

Various communications, navigation, and weapons systems.

PROGRAM STATUS

- **1QFY09:** 4th Category B aircraft inducted for restoration to flyable status
- **2QFY09:** Awarded contract for Safety Enhancement Program (SEP) Lots 12 and 13 (27 aircraft/30 aircraft)
- **2QFY09:** First 10 aircraft equipped with Condition Based Maintenance units for operational test and calibration
- **2QFY09:** Completed fielding of new lightweight weapons rack
- **3QFY09:** Completed in-theatre fielding of M3P .50 Caliber Machine Gun; all squadrons in OIF/OEF equipped with M3P
- **3QFY09:** Completed SEP Lot 11 (27 aircraft)
- **3QFY09:** Kiowa Warrior Cockpit and Sensor Upgrade Program (CASUP) Acquisition Decision Memorandum released (ACAT II)
- **4QFY09:** 5th Category B aircraft inducted for restoration to flyable status
- **FY08-FY09:** Reset 121 aircraft re-deploying from OIF/OEF
- **FY08-FY09:** Installed 663 Modification Work Orders including lightweight weapons rack, personal computer data transfer system with video, APX-118 transponder, cockpit airbag system, and crash attenuating seats
- **4QFY09:** Successful flight demonstration of prototype level II UAV teaming capability
- **4QFY09:** Kiowa Warrior CASUP cockpit modification and prototyping initiated at prototyping and integration facility

PROJECTED ACTIVITIES

- **1QFY10:** Award contract for Control and Display Subsystem-5 upgrade as part of Kiowa Warrior CASUP modifications
- **1QFY10:** Award contract for dual channel engine full authority digital electronic fuel control system
- **1QFY10:** Induct 6th Category B aircraft for restoration to flyable status
- **1QFY10:** Award contract with Bell Helicopter for engineering services
- **3QFY10:** Complete delivery of lot 12 Safety Enhancement Program aircraft
- **3QFY10:** Award contract for OH-58A to D cabin conversion
- **3QFY10:** Fielding of lightweight, color, multi-function displays
- **3QFY10:** First production modifications of AN/AAR-57 Common Missile Warning System equipped Kiowa Warrior aircraft
- **3QFY10:** Kiowa Warrior CASUP Milestone B briefing

ACQUISITION PHASE

Technology Development | **Engineering & Manufacturing Development** | Production & Deployment | Operations & Support

Kiowa Warrior

FOREIGN MILITARY SALES
Taiwan

CONTRACTORS
Bell Helicopter, Textron (Fort Worth, TX)
DRS Optronics, Inc. (Palm Bay, FL)
Rolls Royce Corp. (Indianapolis, IN)
Honeywell (Albuquerque, NM)
Elbit Systems of America (Fort Worth, TX)

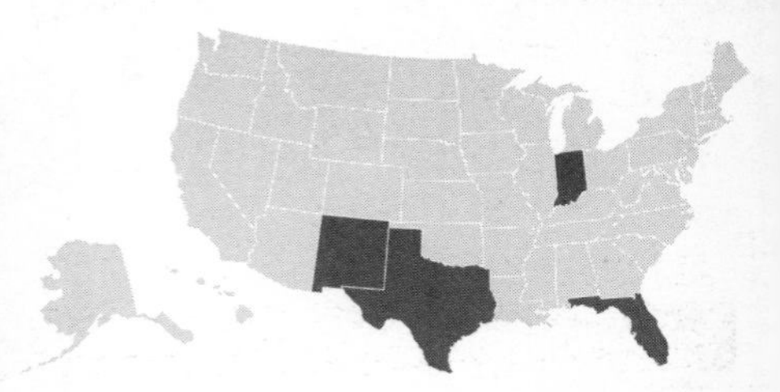

Light Tactical Trailer (LTT)

INVESTMENT COMPONENT

- Modernization
- Recapitalization
- Maintenance

MISSION

The Light Tactical Trailer (LTT) increases the cargo carrying capability of the High Mobility Multipurpose Wheeled Vehicle (HMMWV).

DESCRIPTION

The Light Tactical Trailer (LTT) is a general purpose cargo trailer that offers significant improvement over the M101 series trailers that it replaces. The LTT is produced in three configurations: the M1101 (Light), the M1102 (Heavy), and the LTT Chassis.

Gross vehicle weight:
M1101 is 3,400 pounds
M1102 is 4,200 pounds
Maximum payload weights:
M1101 is 1,940 pounds
M1102 is 2,740 pounds
LTT Chassis is 3,025 pounds

SYSTEM INTERDEPENDENCIES

The HMMWV is the prime mover for this trailer. The trailer is the mobile platform for various weapons and combat support systems.

PROGRAM STATUS

- **Current:** Continue fielding to Army, Marine Corps, Navy and Air Force customers

PROJECTED ACTIVITIES

- Increase production rate to meet demand

ACQUISITION PHASE

Technology Development | Engineering & Manufacturing Development | Production & Deployment | Operations & Support

Light Tactical Trailer (LTT)

FOREIGN MILITARY SALES
Kenya, Afghanistan

CONTRACTORS
Silver Eagle Manufacturing Company (SEMCO) (Portland, OR)
Schutt Industries (Clintonville, WI)

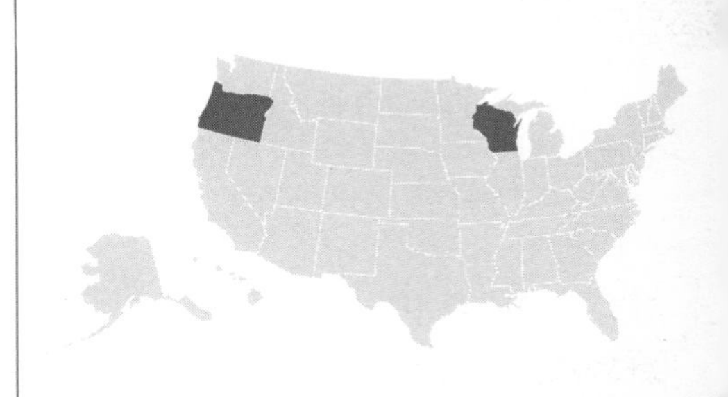

Light Utility Helicopter (LUH)/UH-72A Lakota

INVESTMENT COMPONENT

Modernization

Recapitalization

Maintenance

MISSION

To provide flexible response to homeland security requirements: to conduct civil search and rescue operations, support damage assessment, support test and training centers, perform generating force missions, reconnaissance and surveillance, and augment medical evacuation (MEDEVAC) missions.

DESCRIPTION

The UH-72A LAKOTA Light Utility Helicopter (LUH) will conduct general support utility helicopter missions and execute tasks as part of an integrated effort with other joint services, government agencies, and non-governmental organizations. The LUH is to be deployed only to non-combat, non-hostile environments. The UH-72A is a variant of the American Eurocopter U.S.-produced EC-145. The UH-72A is a twin-engine, single-main-rotor commercial utility helicopter of the 3–6 ton class. It has seating for two pilots and up to six passengers or two NATO standard litters, crew chief, and medical attendant. Two Turbomeca Arriel 1E2 engines, combined with an advanced four-blade rotor system, provide lift and speed in a wide range of operating conditions, including high-altitude and single-engine operation capability. Access to the aircraft is through sliding doors on each side of the cabin or through the wide rear clamshell doors.

Crew seating comprises two individual, longitudinally adjustable, energy-absorbing pilot and copilot seats with head rest and four-point safety belts with automatic locking system. The passenger seats have a four-point restraint harness. When equipped for medical evacuation (MEDEVAC) operations with two NATO standard litters, passenger seating is limited to a medical attendant and a crew chief.

The aircraft is equipped with modern communication and navigation avionics, which facilitate operation in civilian airspace systems. The cockpit is arranged and lit to be compatible with night vision devices. Included in the avionics are a radar altimeter, full autopilot, and a unique First Limit Indicator (FLI) that further simplifies engine monitoring and reduces pilot workload.

In addition to the MEDEVAC and hoist configuration, the UH-72A is also being fielded in a VIP, National Guard Homeland Security (HLS) and a Combined Training Center (CTC) configuration.

The United States Navy Test Pilot School (TPS) ordered five UH-72A aircraft in 2008.

SYSTEM INTERDEPENDENCIES

OH-58A/C, UH-1, ARC-231, UH-60 C-5 (RERP), C-17, GATM, Air Warrior, Sealift

PROGRAM STATUS

- **FY09:** 128 aircraft are on contract with 41 to be delivered.
- **2QFY09:** First fielding of MEDEVAC configuration to National Guard
- **3QFY09:** Field first aircraft overseas to National Guard units
- **4QFY09:** Initiate retrofit of ARC-231
- **1QFY10:** 4 Navy TPS aircraft delivered
- **Current:** Total of 72 aircraft delivered to units in Active Army and National Guard; receiving mission equipment packages, including vent kit, Environmental Control Unit (ECU), medical equipment storage kit, VIP kit, ARC-231 military radio, and equipment to support training operations/National Guard counter-drug mission.

PROJECTED ACTIVITIES

- **2QFY10:** 100 aircraft complete delivery
- **3QFY10:** Aircraft field to Europe; Aircraft field to Kwajalien
- **4QFY10:** 123 aircraft complete delivery, first three years of production
- **4QFY11:** 180 aircraft complete delivery

ACQUISITION PHASE

Technology Development | Engineering & Manufacturing Development | Production & Deployment | Operations & Support

Light Utility Helicopter (LUH)

FOREIGN MILITARY SALES
None

CONTRACTORS
EADS North America (Arlington, VA)
American Eurocopter (Columbus, MS; Grand Prairie, TX)
CAE USA (Tampa, FL)
Sikorsky Aircraft (Stratford, CT)

	UH-72A
Max Gross Weight:	7,903 pounds
Cruise Speed:	140 knots
Engines (2 each):	Turbomeca Arriel 1E2
External Load:	1,214 pounds
Internal Load:	1,214 troops/pounds
Crew:	Two pilots, one crew chief

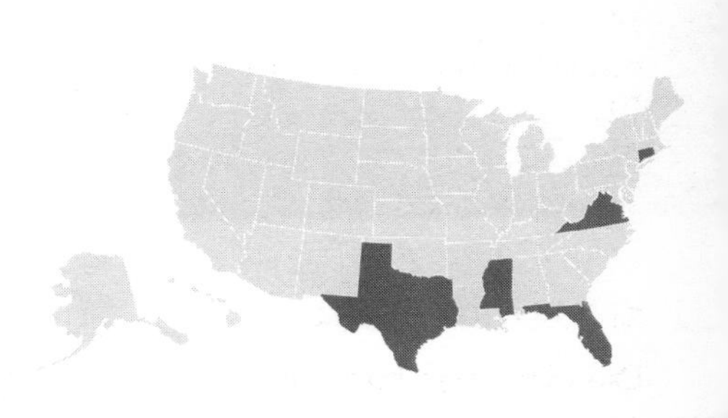

Lightweight 155mm Howitzer (LW155)

INVESTMENT COMPONENT

- **Modernization**
- Recapitalization
- Maintenance

MISSION

To provide direct, reinforcing, and general support fires to maneuver forces as well as direct support artillery for Stryker Brigade Combat Teams.

DESCRIPTION

The M777A2 Lightweight 155mm Howitzer (LW155) will replace all M198 155mm howitzers in operation with the Marine Corps and Army. The extensive use of titanium in all its major structures makes it 7,000 pounds lighter than its predecessor, the M198, with no sacrifice in range, stability, accuracy, or durability. Two M777A2s can be transported by a C-130, and it can be dropped by parachute. The M777A2's lighter weight, independent suspension, smaller footprint, and lower profile increase strategic deployability, and tactical mobility. The system uses a primer feed mechanism, projectile loader-assist, advanced digital fire control system, and other improvements to enhance reliability and accuracy, and significantly increase system survivability.

The M777A2 is jointly managed; the Marine Corps led the development of the howitzer and the Army led the development of Towed Artillery Digitization (TAD), the digital fire control system.

Software upgrades incorporating the Enhanced Portable Inductive Artillery Fuze Setter and the Excalibur Platform Integration Kit hardware gives the M777A2 the capability to program and fire the Excalibur precision-guided munition. Specifications for the M777A2 Excalibur-compatible howitzer are:

Weight: Less than 10,000 pounds
Emplace: Less than three minutes
Displace: Two to three minutes
Maximum range: 30 kilometer (rocket assisted round)
Rate-of-fire: Four to eight rounds per minute maximum; two rounds per minute sustained
Ground mobility: Family of Medium Tactical Vehicles (FMTV), Medium Tactical Vehicle Replacement, five-ton trucks
Air mobility: Two per C-130; six per C-17; 12 per C-5; CH-53D/E; CH-47D; MV-22
155mm compatibility: all fielded and developmental NATO munitions
Digital fire control: self-locating and pointing; digital and voice communications; self-contained power supply

SYSTEM INTERDEPENDENCIES

Army Software Blocking, FMTV Joint Light Tactical Vehicle Prime Movers, Excalibur and Precision Guidance Kit munitions, Single Channel Ground to Air Radio Station/Defense Advanced Global Positioning System Receiver/ Advanced Field Artillery Tactical Data System

PROGRAM STATUS

- **Current:** The M777A2 is in full-rate production with 572 systems having been delivered to the Army and Marine Corps (354/218 respectively)
- **Current:** Seven Stryker Brigades have been fielded and two Fires Brigades
- **Current:** Both the Army and Marine Corps have deployed the weapon to Iraq and Afghanistan where it is currently engaged in combat operations.

PROJECTED ACTIVITIES

- **FY10:** Award Performance Based Logistics Contract

ACQUISITION PHASE

Technology Development | Engineering & Manufacturing Development | **Production & Deployment** | Operations & Support

Lightweight 155mm Howitzer (LW155)

FOREIGN MILITARY SALES
Canada and Australia

CONTRACTORS
BAE Systems (United Kingdom; Hattiesburg, MS)
Castings:
Precision Castparts Corp. (Portland, OR)
Howmet Castings (Whitehall, MI)
Cannon Assembly:
Watervliet Arsenal (Watervliet, NY)
Howitzer body:
Triumph Systems Los Angeles (Chatsworth, CA)

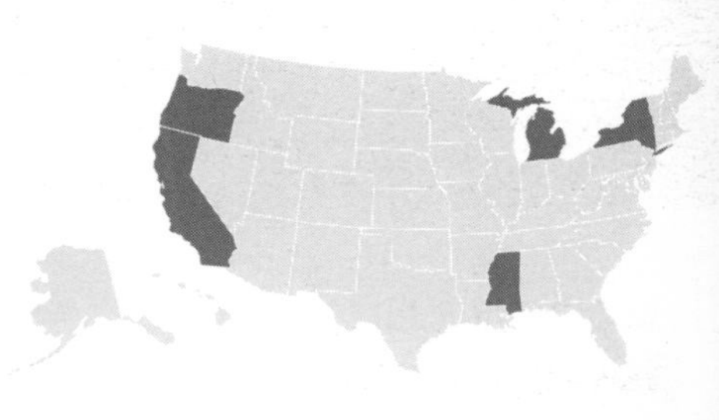

Lightweight .50 cal Machine Gun

INVESTMENT COMPONENT

Modernization

Recapitalization

Maintenance

MISSION

To provide a lighter, more accurate .50 caliber machine gun to reduce warfighter load, provide for more rapid emplacement and displacement, increase dismounted portability, and reduce the strain on vehicle mounts.

DESCRIPTION

The Lightweight .50 Caliber Machine Gun (LW50MG) is intended to provide .50 caliber lethality in a lightweight, two-man portable system. It will fire all standard inventory .50 caliber M9 linked ammunition. The LW50MG incorporates a manual safety and a quick-change barrel that eliminates headspace and timing adjustments. The LW50MG weapon will weigh 40 pounds with recoil loads less than 400 pounds. The lower recoil allows the use of a magnified optic, which enables the warfighter to rapidly acquire targets. Additionally the low recoil allows the warfighter to maintain control of the weapon and to minimize ammunition use.

SYSTEM INTERDEPENDENCIES

None

PROGRAM STATUS

- **3QFY08:** System development and demonstration contract awarded

PROJECTED ACTIVITIES

- **FY08–10:** Conduct system development and demonstration activities
- **3QFY09:** Award phase 2 contract
- **4QFY10:** Milestone C decision
- **FY11:** Low-rate initial production

ACQUISITION PHASE

Technology Development | Engineering & Manufacturing Development | Production & Deployment | Operations & Support

Lightweight .50 cal Machine Gun

FOREIGN MILITARY SALES
None

CONTRACTORS
General Dynamics Armament and Technical Products (GDATP) (Charlotte, NC)

Lightweight Laser Designator Range Finder (LLDR)

INVESTMENT COMPONENT

Modernization

Recapitalization

Maintenance

MISSION

To provide fire support teams and forward observers with a man-portable capability to observe and accurately locate targets, digitally transmit target location data to the tactical network, and laser-designate high-priority targets for destruction.

DESCRIPTION

The Lightweight Laser Designator Rangefinder (LLDR) is a man-portable, modular, target location, and laser designation system. The two primary components are the target locator module (TLM) and the laser designator module (LDM). The TLM can be used as a standalone device or in conjunction with the LDM. Total system weight to conduct a 24-hour mission is less than 30 pounds.

The TLM incorporates a thermal imager, day camera, electronic display, eye-safe laser rangefinder, digital magnetic compass, global positioning system electronics, and digital export capability. The TLM has the capability of seeing the laser designator spot, allowing the operator to more precisely aim the laser designator. At night and in obscured battlefield conditions, the operator can recognize vehicle-sized targets at greater than 3 kilometers. During day operations, the operator can recognize targets at a distance of greater than seven kilometers. At a range of 10 kilometers, the operator can locate targets to less than 40 meters. The LDM emits coded laser pulses compatible with DoD and NATO laser-guided munitions. Users can designate targets at ranges greater than five kilometers.

SYSTEM INTERDEPENDENCIES

None

PROGRAM STATUS

- **Current:** In full-rate production

PROJECTED ACTIVITIES

- **4QFY09:** Follow-on full-rate competitive contract award

ACQUISITION PHASE

Technology Development | Engineering & Manufacturing Development | Production & Deployment | Operations & Support

Lightweight Laser Designator Range Finder (LLDR)

FOREIGN MILITARY SALES
None

CONTRACTORS
Northrop Grumman Guidance and Electronics Company, Inc., Laser Systems (Apopka, FL)
Thermal Imager:
L-3 Communications Cincinnati Electronics (Mason, OH)
FLIR Systems, Inc. (Santa Barbara, CA)

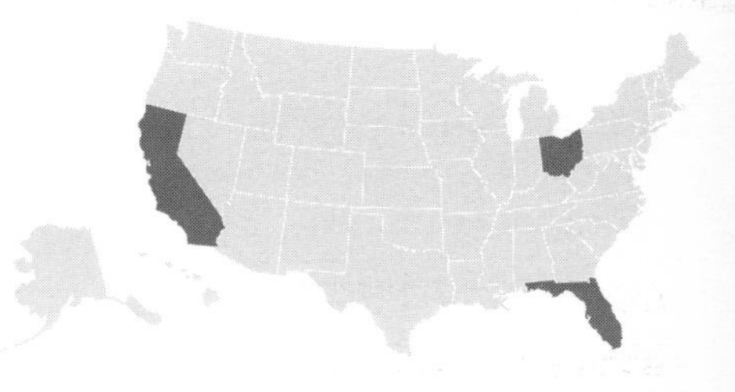

Line Haul Tractor

INVESTMENT COMPONENT

Modernization

Recapitalization

Maintenance

MISSION

To support combat service and support units with transportation of bulk petroleum products, containerized cargo, general cargo, and bulk water.

DESCRIPTION

The M915A3 Line Haul Tractor is the Army's key line haul distribution platform. It is a 6x4 tractor with a 2-inch kingpin and 105,000-pound gross combination weight capacity. The vehicle is transportable by highway, rail, marine, and air modes worldwide.

Gross vehicle weight: 52,000 pounds
Fifth-wheel capacity: 2-inch, 30,000 pounds
Diagnosis: Electronic
Brake system: Anti-lock
Towing speed: 65 miles per hour with full payload
Engine: Detroit Diesel S60 (430 horse power, 1450 pound-foot torque, DDEC IV engine controller)
Transmission: Allison HD5460P (six-speed automatic) with power take off

The M916A3 Light Equipment Transport (LET) is a 6x6 tractor with 68,000-pound gross vehicle weight tractor with 3-1/2-inch, 40,000-pound capacity, 45,000-pound winch for recovery and transport of engineering systems, and compensator fifth wheel. It has an electronic diesel engine, automatic electronic transmission, anti-lock brakes, and is capable of operating at speeds up to 60 miles per hour on flat terrain. This Non-Developmental Item (NDI) vehicle is used primarily to transport the M870 40-ton low-bed semi-trailer.

The M917A2 and M917A2 Truck Chassis, 75,000 gross vehicle weight rating, 8x6 (for 20-ton dump truck), 12-cubic yard dump truck vehicles are authorized in Corps units, primarily the construction and combat support companies and the combat heavy battalions. Freightliner produced the M917A1 and M917A2 vehicles. It has an electronic diesel engine, automatic electronic transmission, anti-lock brakes, and is capable of operating at speeds up to 55 miles per hour when on flat terrain.

The M915A5 Truck Tractor is a 6x4 semi-tractor used to perform the Line Haul mission. The truck is equipped with a two-passenger cab and powered by a 500 horse power diesel engine with an Allison Model 4500 SP electronically controlled automatic six-speed transmission. The M915A5 has a front and rear suspension system rated for Gross Combined Vehicle Weight Rating (GCVWR) of 120,000 pounds. The front axle is weight-rated at 20,000 pounds, the rear axle weight ratings are a combined 46,000 pounds. The electrical system is a 24-volt open-architecture starting system. The M915A5 has an updated power distribution module (PDM), upgraded wiring harnesses, a Roll Stability Control system (RSC), all around light emitting diode (LED) lighting. Auxiliary power connections have been added to supply emerging systems and added command, control, communications, computer and intelligence (C4I) communication systems. A pair of 60-gallon fuel tanks increase fuel capacity by 20 gallons to extend driving range. The cab is 10 inches wider and extends 34 inches behind the driver and passenger seats. The brakes have an improved anti-lock brake system (ABS), and an updated Collision Warning System (CWS) has been installed.

SYSTEM INTERDEPENDENCIES

Joint Land Attack Cruise Missile Defense Elevated Netted Sensor (JLENS); M872, 34-ton flatbed semi-trailer; M1062A1, 7,500-gallon semi-trailer; M967/M969, 5,000-gallon semi-trailer

PROGRAM STATUS

- **FY09:** Full production continues in support of Army operations in the United States and abroad.
- **FY09:** Completion of M915A5 development and operational testing; Full logistics demonstration.
- **4QFY09:** M915A5 contract award

PROJECTED ACTIVITIES

- **FY09:** Production verification testing of M915A5 block upgrade to M915 Series
- **2QFY10:** M915A5 type classification/materiel release; Production cut-in, ramp-up, and first-unit equipping.

ACQUISITION PHASE

Technology Development | Engineering & Manufacturing Development | **Production & Deployment** | Operations & Support

Line Haul Tractor

FOREIGN MILITARY SALES
Afghanistan

CONTRACTORS
Meritor (Troy, MI)
Holland Hitch (Holland, MI)
Pierce Manufacturing (Bradenton, FL)
Detroit Diesel (Detroit, MI)
Truck:
Daimler Truck, North America/Freightliner (Portland, OR)
Dump body:
Casteel Manufacturing (San Antonio, TX)

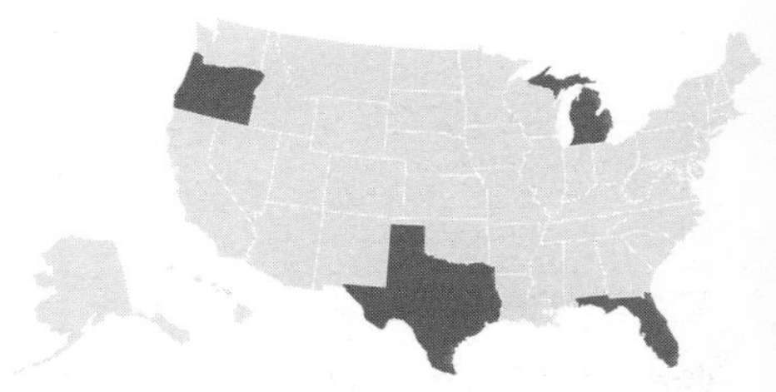

Load Handling System Compatible Water Tank Rack (Hippo)

INVESTMENT COMPONENT

- Modernization
- Recapitalization
- Maintenance

MISSION

To provide a mobile, hard-wall water tanker for bulk distribution of potable water to the division and brigade areas.

DESCRIPTION

The Load Handling System Compatible Water Tank Rack (Hippo) represents the latest in bulk water distribution systems technology. It replaces the 3K and 5K Semi-trailer Mounted Fabric Tanks (SMFTs). The Hippo consists of a 2,000-gallon potable water tank in an ISO frame with an integrated pump, engine, alternator, filling stand, and 70-foot hose reel with bulk suction and discharge hoses. It has the capacity to pump 125 gallons of water per minute. The Hippo is fully functional mounted or dismounted and is transportable when full, partially full, or empty. It is designed to operate in cold weather environments and can prevent water from freezing at -25 degrees Fahrenheit. The Hippo can be moved, set up, and established rapidly using minimal assets and personnel.

SYSTEM INTERDEPENDENCIES

HEMTT–LHS, PLS, and PLS Trailer

PROGRAM STATUS

- **2QFY07:** Full material release
- **FY08:** Production and fielding
- **2QFY08:** Hippo discoloration issue
- **4QFY08:** Placed additional quantities to current contract

PROJECTED ACTIVITIES

- **FY09:** Continue production and fielding
- **FY09:** New competitive production contract
- **FY09:** Update integrated electronic technical manuals
- **FY10:** Continue production and fielding
- **FY11:** Continue production and fielding

ACQUISITION PHASE

Technology Development | Engineering & Manufacturing Development | Production & Deployment | Operations & Support

Load Handling System Compatible Water Tank Rack (Hippo)

FOREIGN MILITARY SALES
None

CONTRACTORS
Mil-Mar Century, Inc. (Miamisburg, OH)

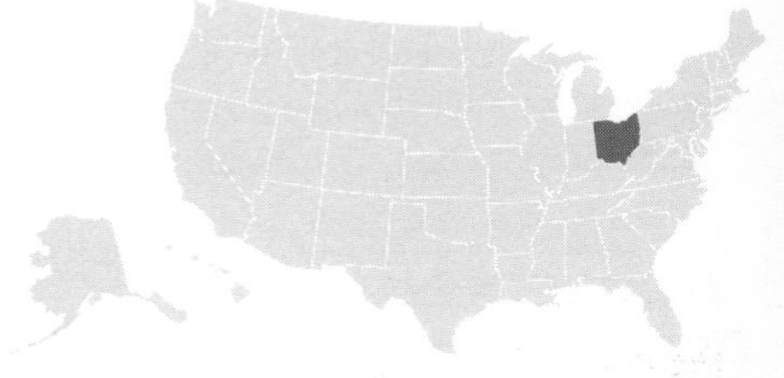

Longbow Apache

INVESTMENT COMPONENT

Modernization

Recapitalization

Maintenance

MISSION

To conduct close combat attack, deep precision strikes, and armed reconnaissance and security in day, night, and adverse weather conditions.

DESCRIPTION

The AH-64D Longbow Apache is the Army's only attack helicopter for both the Current and Future Force. It is capable of destroying armor, personnel, and materiel targets in obscured battlefield conditions. The Apache fleet includes the A model Apache and D model Longbow. The Longbow remanufacturing effort uses the A model and incorporates a millimeter wave fire control radar (FCR), radar frequency interferometer (RFI), fire-and-forget radar-guided Hellfire missiles, and other cockpit management and digitization enhancements. Both A and D models are undergoing recapitalization modifications such as upgraded forward-looking infrared (FLIR) technology with the Arrowhead Modernized Target Acquisition Designation Sight/ Pilot Night Vision Sensor (MTADS/ PNVS), non-line-of-sight communications, video transmission/reception, and maintenance cost reductions.

The Army's goal is to convert its remaining A models to the Longbow Apache configuration. The Longbow program began with two multi-year contracts: the first delivered 232 Longbows from FY96–FY01; the second delivered an additional 269 aircraft from FY02–FY06; 120 A to D conversions will occur between FY07–FY10. In addition, 52 Wartime Replacement Aircraft (WRA) are planned with deliveries complete CY2010. Apache is fielded to Active Army, National Guard (NG) and Army Reserve (AR) attack battalions, armed reconnaissance battalions, and cavalry units as defined in the Army Modernization Plan. Fielding of Longbow Apache began in 1QFY99 and will conclude in FY11. As of the end of FY08, 535 AH-64D Block I/ IIs have been fielded to 17 Longbow Battalions and Fort Rucker, AL. In addition, 219 (10 Battalions) out of a 634 planned deliveries of the MTADS/ PNVS sub-system have been completed and fielding will be complete in FY2010.

The Longbow Block III program is the next evolution of the Apache. Block III is the Army's only attack helicopter solution capable of interoperability with the Future Combat Force and will add significant combat capability while addressing obsolescence issues to ensure the aircraft remains a viable combat multiplier beyond 2030.

The Block III modernized Longbows will be designed and equipped with an open systems architecture to incorporate the latest communications, navigation, sensor, and weapon systems.

Combat mission speed: 167 miles per hour
Combat range: 300 miles
Combat endurance: 2.5 hours
Max. gross weight: 20,260 pounds
Armament: Hellfire missiles, 2.75-inch rockets, and 30mm chain gun
Crew: Two (pilot and copilot gunner)

SYSTEM INTERDEPENDENCIES

E-IBCT, Longbow Apache maintains digital interoperability with multiple battlefield systems through adherence of the Army's Software Blocking Policy.

PROGRAM STATUS

- **1QFY09:** Block III system development and demonstration contract currently 50% complete
- **4QFY08:** Block III first prototype flight conducted
- **Current:** Upgrade Block I and II Longbow to Block III configuration with eventual acquisition objective of 634 total airframes

PROJECTED ACTIVITIES

- **1QFY10:** Block III limited user test
- **3QFY10:** Block III milestone C and low-rate initial production award
- **FY10:** Block I inductions into Block III remanufacturing assembly line
- **2QFY11:** Initial Block III deliveries
- **3QFY12:** First unit equipped (FUE)
- **2QFY13:** Initial operating capacity (IOC)
- **FY25:** End of production

ACQUISITION PHASE

Technology Development | Engineering & Manufacturing Development | Production & Deployment | Operations & Support

Longbow Apache

FOREIGN MILITARY SALES
Egypt, Greece, Israel, Kuwait, Netherlands, Saudi Arabia, Singapore, United Arab Emirates
Direct commercial sales: Japan, Greece, United Kingdom

CONTRACTORS
Airframe/fuselage:
Boeing (Mesa, AZ)
Fire Control Radar:
Northrop Grumman (Linthicum, MD)
Lockheed Martin (Owego, NY; Orlando, FL)
MTADS/PNVS:
Lockheed Martin (Orlando, FL)
Boeing (Mesa, AZ)
Rotor blades:
Ducommun AeroStructures (Monrovia, CA)

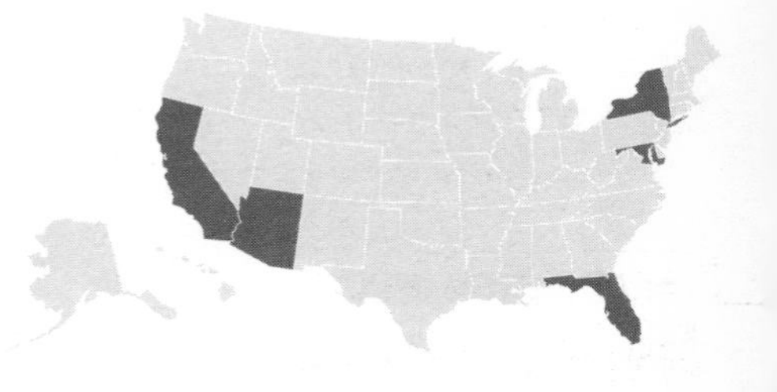

Maneuver Control System (MCS)

INVESTMENT COMPONENT

Modernization

Recapitalization

Maintenance

MISSION

To provide the tactical battle command core environment and common services baseline for executive decision-making capabilities, maneuver functional and battle staff tools, and enterprise services.

DESCRIPTION

Tactical Battle Command (TBC)/ Maneuver Control System (MCS) provides a suite of products and services that include the MCS, Command Post of the Future (CPOF), Joint Convergence effort with the Marine Corps, Battle Command Common Services (BCCS), and SharePoint web portal.

MCS 6.4 serves as a mission critical command and control (C2) system that allows commanders and staffs to visualize the battlespace and synchronize the elements of combat power for successful execution of tactical operations. MCS includes battle staff tools and maneuver functional capabilities. It also integrates Joint Program Management Information Systems (JPMIS) capabilities including chemical, biological, radiological, and nuclear (CBRN) tools and engineering capability for combat and construction engineers.

CPOF serves as a mission critical C2 system that provides collaborative and situational awareness tools to support decision-making, planning, rehearsal, and execution management. Near-real-time display on both 2D and 3D mapping and Voice Over Internet Protocol (VOIP) ensures geographical dispersed systems can collaborate seamlessly on data in a shared repository.

BCCS provides the network-enabling infrastructure for tactical battle command in support of Army battle command migration and DoD migration to Net Enabled Command Capability (NECC) and Net-Centric Enterprise Services (NCES) environment. The Battle Command Server (BC Server) provides interoperability services including the Publish and Subscribe Service (PASS) that allows for ABCS interoperability and Data Dissemination Service (DDS). It also provides tactical messaging, address capability, and MS Office SharePoint 2007 for both operational and business processes. The server also supports Joint Convergence with the Marine Corps by providing a data exchange gateway that allows the direct exchange of common operational picture data between the joint services.

SYSTEM INTERDEPENDENCIES

Army Systems: GCCS–A, IMETS, ASAS, AFATDS, FBCB2, FBCB2 L-Band Upgrade, DCGS–A, JBCP, AMDWS, BCS3, TAIS, AMPS
Joint Systems: TBMCS, JWARN, JEM, JOEF, JTCW/C2PC, GCCS, NECC, NCES
Other Systems: WIN–T, JTRS GMR, E-IBCT

PROGRAM STATUS

- **3QFY08:** Joint Requirements Oversight Council approves MCS 6.4 Capabilities Production Document
- **4QFY08:** BCCS v3 1st official fielding

PROJECTED ACTIVITIES

- **4QFY09:** Field test
- **1QFY10:** Limited user test
- **3QFY10:** Fielding decision (MCS/CPOF)

ACQUISITION PHASE

Technology Development | Engineering & Manufacturing Development | Production & Deployment | Operations & Support

Maneuver Control System (MCS)

FOREIGN MILITARY SALES
None

CONTRACTORS
General Dynamics (Taunton, MA; Scottsdale, AZ)
CECOM Software Engineering Center (Fort Monmouth, NJ)
GTSI (Chantilly, VA)
Lockheed Martin (Tinton Falls, NJ)
Viecore (Tinton Falls, NJ)

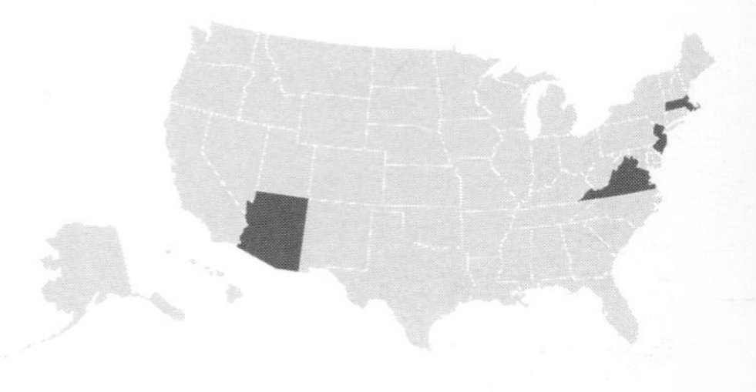

Medical Communications for Combat Casualty Care (MC4)

INVESTMENT COMPONENT

Modernization

Recapitalization

Maintenance

MISSION

To integrate, field, and support a medical information management/ technology system for Army tactical forces, enabling a lifelong electronic medical record for service members, enhancing medical situational awareness for operational commanders, and expanding medical logistics capabilities.

DESCRIPTION

The Medical Communications for Combat Casualty Care (MC4) system is a joint theater-level, automated combat health support system for the tactical medical forces. It serves three distinct user communities: warfighter commanders, healthcare providers, and medical staffs in theater. The system enhances medical situational awareness for the operational commander, enabling a comprehensive, lifelong electronic medical record for all service members. Using the Theater Medical Information Program (TMIP)–Joint software, MC4 receives, stores, processes, transmits, and reports medical command and control, medical surveillance, casualty movement and tracking, medical treatment, medical situational awareness, and medical logistics data across all levels of care.

The MC4 system provides the Army's solution to the Title 10 requirement for a medical tracking system for all deployed service members. The MC4 system is a fully operational standard Army system that operates on commercial off-the-shelf hardware. It supports commanders with a streamlined personnel deployment system using digital medical information.

The MC4 system comprises seven Army-approved line items that can be configured to support Army levels 1–4 and DoD roles 1-3 of the health care continuum. Future MC4 enhancements will be accomplished through minor system upgrades and major planned upgrades. The MC4 program completed a successful full-rate production decision review on July 21, 2005.

SYSTEM INTERDEPENDENCIES

MC4 relies on TMIP to provide global software databases to capture data generated by the MC4 system, in order to provide medical situational awareness for operational commanders.

PROGRAM STATUS

- **3QFY08–1QFY09:** Fielding per the Dynamic Army Resourcing Priority List

PROJECTED ACTIVITIES

- **3QFY09:** First planned upgrade (TMIP Block 2 Release 1)

ACQUISITION PHASE

Technology Development | Engineering & Manufacturing Development | Production & Deployment | Operations & Support

Medical Communications for Combat Casualty Care (MC4)

Joint, Gov't Software From TMIP-J (GOTS)

- AHLTA-T
- JMeWS
- TC2
- TMDS
- AHLTA-M
- DCAM

AND MORE

Commercial Software (COTS)

- Microsoft Office
- Micromedex
- Credant
- Adobe
- Active Client

AND MORE

Product Integration

Army Unique Software

- T-DBSS
- PDHA
- JTTR

U.S. ARMY

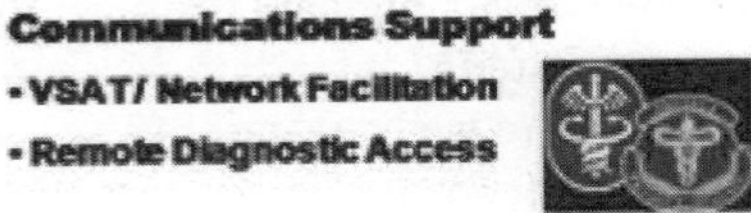

Future Systems

- TEWLS
- DMLSS
- FCS

Comprehensive Customer Services

Fielding

Training

Support

FOREIGN MILITARY SALES
None

CONTRACTORS
Hardware:
GTSI (Chantilly, VA)
CDW–G (Chicago, IL)
System engineering support:
Johns Hopkins University Applied Physics Laboratory (Laurel, MD)
System integration support:
L-3 Communications (Titan Group) (Reston, VA)
Fielding, training, and system administration support:
General Dynamics (Fairfax, VA)

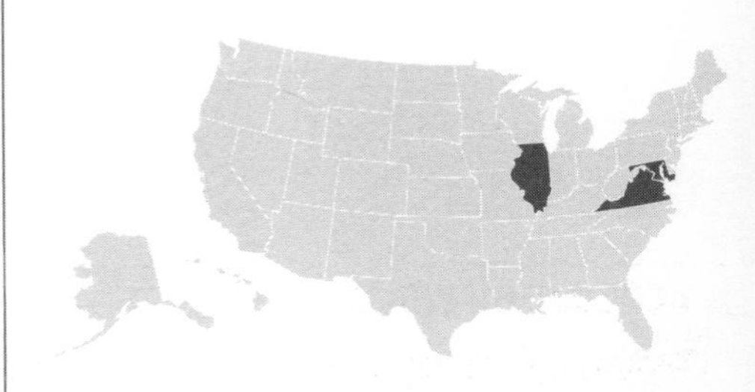

Medical Simulation Training Center (MSTC)

INVESTMENT COMPONENT
Modernization
Recapitalization
Maintenance

MISSION
To conduct standardized combat medical training for medical and non-medical personnel.

DESCRIPTION
The Medical Simulation Training Center (MSTC) systems are an Army training asset, with a regional training requirement, located at installations, delivering effective medical training with a standardized training platform for both classroom and simulated battlefield conditions. The goal is to better prepare warfighters for application of medical interventions under combat conditions. The MSTC is a standardized family of supporting component systems with the Virtual Patient System (VPS), Instruction Support System (ISS), Medical Training Command and Control (MT-C2) System, and the Medical Training Evaluation and Review System (MeTER), providing frameworks fitted with reconfigurable enabling technology and supporting training devices.

Enabling technology includes audio-visual enhancements, camera surveillance capability, computer labs, computerized control rooms, with a remotely managed training platform. Supporting training components include a computerized bleed-breathe mannequin that is weighted and airway equipped, partial task trainers, and associated equipment.

SYSTEM INTERDEPENDENCIES
None

PROGRAM STATUS
- **4QFY08:** All 18 initial systems fielded

PROJECTED ACTIVITIES
- Further development and procurement of a tetherless mannequin (TLM) training capability, the MeTER System, and the MT-C2 System

ACQUISITION PHASE
Technology Development | Engineering & Manufacturing Development | Production & Deployment | Operations & Support

Medical Simulation Training Center (MSTC)

FOREIGN MILITARY SALES
None

CONTRACTORS
Medical Education Technologies (Sarasota, FL)
Computer Sciences Corp. (CSC) (Orlando, FL)

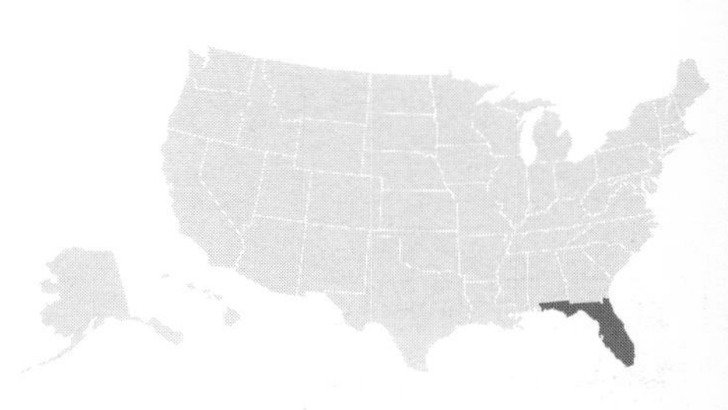

Medium Caliber Ammunition

INVESTMENT COMPONENT

- **Modernization**
- Recapitalization
- Maintenance

MISSION

To provide warfighters with overwhelming lethality overmatch in medium caliber ammunition for Current and Future Force systems.

DESCRIPTION

Medium caliber ammunition includes 20mm, 25mm, 30mm, and 40mm armor-piercing, high-explosive, smoke, illumination, tear gas, training, and antipersonnel cartridges with the capability to defeat light armor, materiel, and personnel targets. The 20mm cartridge is used in the Counter Rocket, Artillery, and Mortar (C-RAM) weapon system. The 25mm cartridges are fired from the M242 Bushmaster gun on the Bradley Fighting Vehicle. The 30mm cartridges are used in the Apache helicopter's M230 Chain Gun. A variety of 40mm cartridges are designed for use in the MK19 Grenade Machine Gun and the M203 Grenade Launcher.

SYSTEM INTERDEPENDENCIES

Medium caliber ammunition is dependent upon the weapons platforms currently in use.

PROGRAM STATUS

- **Current:** In production

PROJECTED ACTIVITIES

- **FY10:** Multiple year family buys for 25mm, 30mm, and 40mm ammunition

ACQUISITION PHASE

- Technology Development
- Engineering & Manufacturing Development
- **Production & Deployment**
- Operations & Support

Medium Caliber Ammunition

FOREIGN MILITARY SALES

25mm:
Israel, Philippines

30mm:
Egypt, Israel, Japan, Kuwait, Netherlands, Serbia, Taiwan, and UAE

40mm:
Afghanistan, Canada, Greece, Israel, Japan, Kenya, Philippines, and Tunisia

CONTRACTORS

General Dynamics Ordnance and Tactical Systems (Marion, IL; Red Lion, PA)
Alliant Techsystems (Radford, VA; Rocket City, WV)
AMTEC Corp. (Janesville, WI)
DSE (Balimoy) Corp. (Tampa, FL)

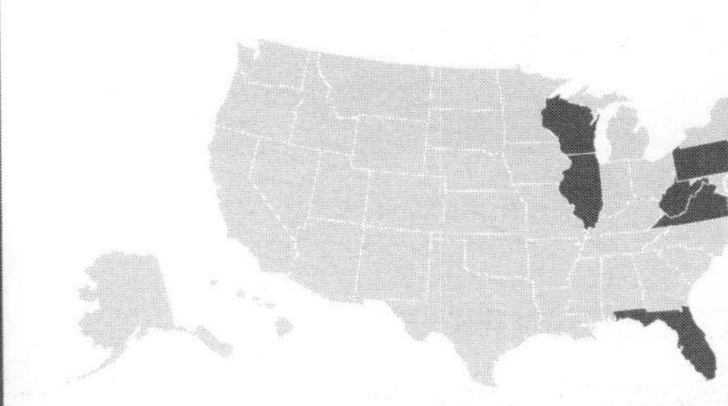

Medium Extended Air Defense System (MEADS)

INVESTMENT COMPONENT

- Modernization
- Recapitalization
- Maintenance

MISSION

To provide low- to medium-altitude air and missile defense to maneuver forces and other land component commanders' designated critical assets for all phases of tactical operations.

DESCRIPTION

The Medium Extended Air Defense System (MEADS) provides a robust, 360-degree defense using the PATRIOT PAC-3 hit-to-kill missile segment enhancement (MSE) against the full spectrum of theater ballistic missiles, anti-radiation missiles, cruise missiles, unmanned aerial vehicles, tactical air-to-surface missiles, and rotary and fixed wing threats. MEADS will also provide defense against multiple and simultaneous attacks by short-range ballistic missiles, low-radar cross-section cruise missiles, and other air-breathing threats. MEADS can be immediately deployed by air for early entry operations. MEADS also has the mobility to displace rapidly and protect maneuver forces assets during offensive operations. Netted, distributed, open architecture, and modular components are utilized in the MEADS to increase survivability and flexibility of use in a number of operational configurations. A significant increase in firepower with the PAC-3 MSE is also employed in the MEADS, with greatly reduced requirements for manpower, maintenance, and logistics.

The MEADS weapon system will use its netted and distributed architecture to ensure joint and allied interoperability, and to enable a seamless interface to the next generation of battle management command, control, communications, computers, and intelligence (BMC4I). The system's improved sensor components and its ability to link other airborne and ground-based sensors facilitate the employment of its battle elements.

The MEADS weapon system's objective battle management tactical operations center (TOC) will provide the basis for the future common air and missile defense (AMD) TOC, leveraging modular battle elements and a distributed and open architecture to facilitate continuous exchange of information to support a more effective AMD system of systems.

SYSTEM INTERDEPENDENCIES

PATRIOT, Terminal High Altitude Air Defense (THAAD), Ballistic Missile Defense System (BMDS), Integrated Air and Missile Defense (IAMD), E-2C, AWACS, Rivet Joint, DSP

PROGRAM STATUS

- **4QFY08–4QFY10:** Incremental critical design review (CDR) phase

PROJECTED ACTIVITIES

- **4QFY10:** System CDR

ACQUISITION PHASE

Technology Development | Engineering & Manufacturing Development | Production & Deployment | Operations & Support

Medium Extended Air Defense System (MEADS)

FOREIGN MILITARY SALES
None

CONTRACTORS
MEADS International (Orlando, FL)

Meteorological Measuring Set–Profiler (MMS–P)

INVESTMENT COMPONENT

Modernization

Recapitalization

Maintenance

MISSION

To provide real-time meteorological data for field artillery on demand over an extended battlespace.

DESCRIPTION

The AN/TMQ-52 Meteorological Measuring Set–Profiler (MMS–P) uses a suite of meteorological sensors, meteorological data from communications satellites, and an advanced weather model to provide highly accurate meteorological data for indirect fire, field artillery systems. The system uses common hardware, software, and operating systems and is housed in a command post platform shelter and transported on an M1152A High Mobility Multipurpose Wheeled Vehicle (HMMWV). Profiler measures and transmits meteorological conditions to indirect fire direction centers, such as wind speed, wind direction, temperature, pressure and humidity, rate of precipitation, visibility, cloud height, and cloud ceiling, all of which are necessary for precise targeting and terminal guidance of various munitions. Profiler uses this information to create a four-dimensional meteorological model (height, width, depth, and time) that includes terrain effects. This new capability increases the lethality of all field artillery platforms such as the Multiple Launch Rocket System (MLRS), Paladin, and self-propelled or towed howitzers by increasing the probability of first-round hit, resulting in significant ammunition cost savings for the Army. The current Profiler provides meteorological coverage throughout a 60 kilometers radius, while the follow-on Block II variant extends coverage to 500 kilometers. For the first time, Army field artillery systems can apply meteorological data along the trajectory from the firing platform to the target area.

SYSTEM INTERDEPENDENCIES

Navy Operational Global Atmospheric Prediction System, Global Broadcast System

PROGRAM STATUS

- **2QFY07–Present:** Continued full-rate production and fielded the system to 41 Interim Brigade Combat Teams (IBCTs) and four fires brigades.

PROJECTED ACTIVITIES:

- **2QFY09–2QFY11:** Continue full-rate production of the current system and complete fielding to remaining Brigade Combat Teams and fires brigades; begin development of Profiler Block II configuration to reduce the system footprint and leverage technology and software advancements to achieve improvements in accuracy to eliminate the need for balloons and radiosondes.

ACQUISITION PHASE

Technology Development | Engineering & Manufacturing Development | Production & Deployment | Operations & Support

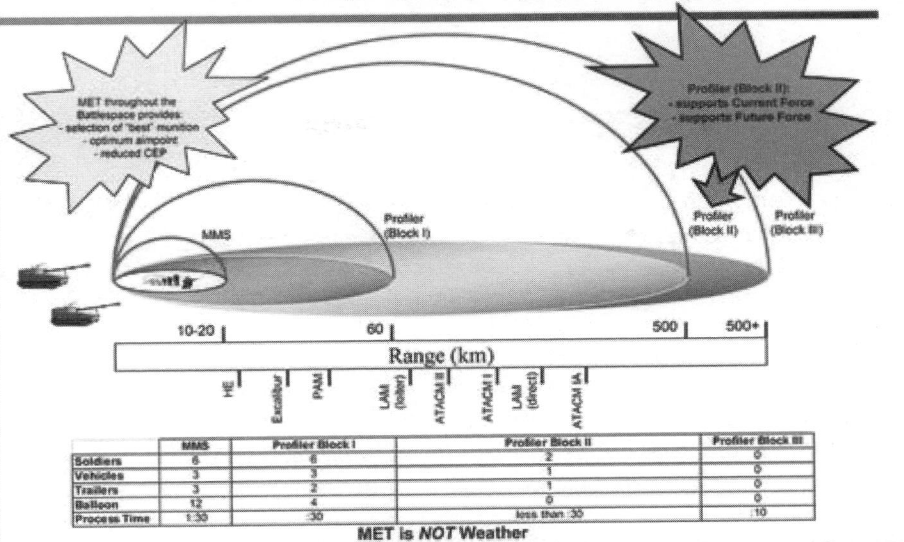

	MMS	Profiler Block I	Profiler Block II	Profiler Block III
Soldiers	6	6	2	0
Vehicles	3	3	1	0
Trailers	3	2	1	0
Balloon	12	4	0	0
Process Time	1:30	:30	less than :30	:10

Meteorological Measuring Set–Profiler (MMS–P)

FOREIGN MILITARY SALES

None

CONTRACTORS

Smiths Detection, Inc. (Edgewood, MD)
Pennsylvania State University (University Park, PA)

Mine Protection Vehicle Family (MPVF)

INVESTMENT COMPONENT
Modernization
Recapitalization
Maintenance

MISSION

To provide forces with blast-protected platforms capable of locating, interrogating, and classifying suspected explosive hazards, including improvised explosive devices (IEDs).

DESCRIPTION

The Mine Protection Vehicle Family (MPVF) consists of the Medium Mine Protected Vehicle (MMPV), the Vehicle Mounted Mine Detection (VMMD) system, and the Mine Protected Clearance Vehicle (MPCV). Each of the systems in the MPVF has a blast-deflecting, V-shaped hull, and each conducts specific missions.

The MMPV system is a blast-protected command and control vehicle platform that operates in explosive hazardous environments and is adaptable to a wide range of security and force protection activities. It will support Future Engineer Force (FEF) clearance companies in route and area clearance operations, explosive hazards teams in explosive hazards reconnaissance operations, and explosive ordinance disposal (EOD) companies in support operations. The MMPV will also support Chemical Biological Response Teams and Prophet signals intelligence (SIGINT) systems.

The VMMD is a blast-protected, vehicle-mounted mine-detection and lane-proofing system capable of finding and marking metallic explosive hazards, including metallic-encased IEDs and anti-tank mines on unimproved roads. It consists of two towing/mine detection "Husky" vehicles, and a set of three mine detonation trailers (MDTs). The Husky detection platform detects, locates, and marks suspected metallic explosive hazards over a three meters wide path. The Husky provides protection against mine blasts under the wheels and under the centerline, in addition to ballistic protection of the operator cab. The system is designed to be repairable in the field after a mine blast.

The MPCV provides deployed forces with an effective and reliable blast-protected vehicle capable of interrogating and classifying suspected explosive hazards, including IEDs. The MPCV has an articulating arm with a digging/lifting attachment and camera to remotely interrogate a suspected explosive hazard and allow the crew to confirm, deny, and/or classify the explosive hazard. It provides a blast-protected platform to transport Soldiers and allow them to dismount to mark and/or neutralize explosive hazards.

SYSTEM INTERDEPENDENCIES

None

PROGRAM STATUS

- **3QFY07:** Milestone C, type classification-generic, and low-rate initial production for VMMD
- **1QFY08:** Milestone C, type classification-generic, and low-rate initial production for MMPV
- **1QFY08:** Milestone C, type classification-generic, and low-rate initial production for MPCV

PROJECTED ACTIVITIES

MMPV:

- **4QFY10:** Full materiel release and full-rate production decision
- **4QFY10:** First unit equipped

MPCV:

- **4QFY10:** Full materiel release and full-rate production decision
- **4QFY10:** First unit equipped

VMMD:

- **2QFY10:** Full materiel release and full-rate production decision
- **2QFY10:** Type classification-standard
- **2QFY10:** First unit equipped

ACQUISITION PHASE
Technology Development | Engineering & Manufacturing Development | Production & Deployment | Operations & Support

Mine Protection Vehicle Family (MPVF)

FOREIGN MILITARY SALES

MPCV:
United Kingdom

VMMD:
Canada

CONTRACTORS

MMPV:
BAE Systems (York, PA)

MPCV:
Force Protection Industries, Inc. (Ladson, SC)

VMMD:
Critical Solutions International, Inc. (Dallas, TX)

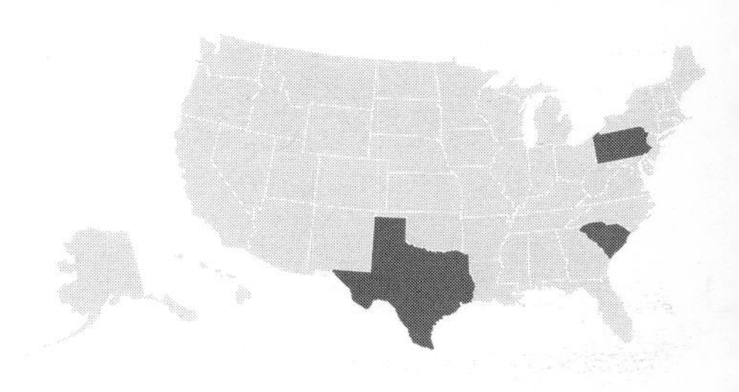

Mine Resistant Ambush Protected Vehicles (MRAP)

INVESTMENT COMPONENT

Modernization

Recapitalization

Maintenance

MISSION

To provide tactical mobility for warfighters with multimission platforms capable of mitigating the effects of improvised explosive devices (IEDs), underbody mines, and small arms fire threats.

DESCRIPTION

The Joint Mine Resistant Ambush Protected (MRAP) Vehicle Program (JMVP) is a multi-service program currently supporting the Army, Navy, Marine Corps, Air Force, and the U.S. Special Operations Command. The program procures, tests, integrates, fields, and supports highly survivable vehicles that provide protection from IEDs and other threats. These four- to six-wheeled vehicles are configured with government furnished equipment (GFE) to meet unique warfighting requirements. Vehicle combat weights (fully loaded without add-on armor) range from approximately 34,000 to 60,000 pounds, with payloads ranging from 1,000 to 18,000 pounds. Key components (e.g., transmissions, engines) vary between vehicles and manufacturers, but generally consist of common commercial and military parts.

Three categories of vehicles support the following missions:

- **Category (CAT) I:** Carries four to six passengers. Designed to provide increased mobility and reliability in rough terrain.
- **CAT II:** Multimission operations (such as convoy lead, troop transport, and ambulance). Carries 10 passengers.
- **CAT III:** Mine/IED clearance operations and explosive ordnance disposal (EOD). Carries six passengers, plus specialized equipment to support EOD operations. The Force Protection Industries. Buffalo is the only CAT III variant. This is the largest MRAP vehicle.

SYSTEM INTERDEPENDENCIES

MRAP vehicles are equipped with multiple GFE items, including communications equipment, mine and IED counter-measure equipment, in addition to weapons and crew protection systems.

PROGRAM STATUS

- **3QFY07–3QFY10:** Produce and field MRAP vehicles to Army, Marine Corps, Air Force, Navy, U.S. Special Operations and foreign military sales customers
- **1QFY08–2QFY09:** Upgrading vehicles to meet emerging threats, enhance survivability, vehicle mobility, and improve automotive performance by incorporating engineering changes in current production, planned orders and fielded vehicles
- **2QFY08:** Initiated capabilities insertion (CI) program to provide enhanced rocket-propelled grenade protection, integration of remote weapon system, increased vehicle power and enhanced Command, Control, Communications, Computers, and Intelligence (C4I) capability.

PROJECTED ACTIVITIES

- **4QFY09:** Begin fielding CI vehicles.

ACQUISITION PHASE

Technology Development | Engineering & Manufacturing Development | Production & Deployment | Operations & Support

Mine Resistant Ambush Protected Vehicles (MRAP)

FOREIGN MILITARY SALES

Canada, France, United Kingdom (UK), Italy

CONTRACTORS

BAE Systems Land & Armaments, Ground Systems Division (York, PA)
BAE-TVS (Sealy, TX)
Force Protection Industries, Inc. (Ladson, SC)
General Dynamics Land Systems, Canada (Ontario, Canada)
Navistar Defense (Warrenville, IL)
Oshkosh Corp. (Oshkosh, WI)

Mobile Maintenance Equipment Systems (MMES)

INVESTMENT COMPONENT

Modernization

Recapitalization

Maintenance

MISSION

To repair battle-damaged combat systems on site and up through the direct support level in the forward battle area.

DESCRIPTION

Mobile Maintenance Equipment Systems (MMES) employs a system-of-systems approach to provide two-level maintenance capability to the warfighter. These systems reduce common tool redundancy, provide tool standardization, minimize transport requirements, and are backed by the Product Manager Sets, Kits, Outfits, and Tools (PM–SKOT) Warranty/Replacement Program. MMES includes the following: Shop Equipment Contact Maintenance, Forward Repair System, Standard Automotive Tool Set, and Shop Equipment Welding.

The Shop Equipment Contact Maintenance (SECM) is a first responder to battle/IED-damaged tracked, wheeled, ground support, and aviation equipment and provides immediate field-level maintenance. Because the SECM's mobility, agility, and maintenance capability is a combat maintenance multiplier, it gets equipment back into the fight as far forward as possible. The SECM supports modularity and Army transformation. The SECM is a fabricated enclosure mounted on a separately authorized associated support item of equipment (ASIOE) High Mobility Multipurpose Wheeled Vehicle (HMMWV). It integrates commercial off-the shelf tools and components for engineer and ordnance maintenance units.

The Forward Repair System (FRS) is a high-mobility, forward maintenance/repair module system. Mounted to a flat rack, it is transported by Palletized Load System (PLS) trucks in Heavy Brigades, or by the Heavy Expanded Mobility Tactical Truck Load Handling System (HEMTT–LHS) in Stryker Brigade Combat Teams (SBCTs). Capabilities of the FRS include: crane capacity up to 10,000 pounds, 35 kilowatt generator, air compressor, welding and cutting equipment, and industrial grade hand/pneumatic/power tools.

The Standard Automotive Tool Set (SATS) provides the warfighter a common tool set that is capable of performing field-level maintenance of military vehicles and ground support equipment at all levels of materiel system repairs. SATS increases tactical independence and enables the unit to fight autonomously via self-maintaining capabilities for all organic systems. SATS features a modular containerized shop set that is deployable, mobile, and mission capable, with a tool load that supports two-level maintenance.

The Shop Equipment Welder (SEW) provides a full spectrum of welding capabilities throughout the battlefield in all weather, climatic, and light conditions. SEW provides heavy-duty, on-site welding capability, supporting two-level maintenance utilizing qualified Army welders. The SEW contains provisions for safe oxygen acetylene braze welding, straight-stick electric arc, metal inert gas, air-carbon arc cutting and flux-cored arc welding of ferrous and non-ferrous metals.

SYSTEM INTERDEPENDENCIES

None

PROGRAM STATUS

- **Current:** Production and fielding

PROJECTED ACTIVITIES

- **Continue:** Production and fielding

ACQUISITION PHASE

Technology Development | Engineering & Manufacturing Development | Production & Deployment | Operations & Support

SECM
Shop Equipment Contact Maintenance

FRS
Forward Repair System

SATS
Standard Automotive Tool Set

SEW
Shop Equipment Welding

Mobile Maintenance Equipment Systems (MMES)

FOREIGN MILITARY SALES

SECM:
Iraq, Kuwait, Afghanistan, Canada, Egypt

SEW:
Egypt, Greece, Saudi Arabia, Armenia, Afghanistan

CONTRACTORS

FRS and SECM:
Rock Island Arsenal (Rock Island, IL)
Snap-on Industrial (Crystal Lake, IL)

SATS:
Kipper Tool Company (Gainesville, GA)
AAR Mobility Systems (Cadillac, MI)

SEW:
Power Manufacturing, Inc. (Covington, TN)

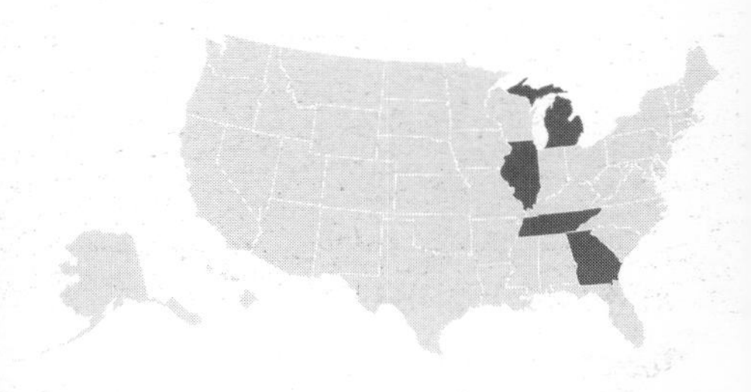

Modular Fuel System (MFS)

INVESTMENT COMPONENT

- Modernization
- Recapitalization
- Maintenance

MISSION

To provide the capability to receive, store, transport, distribute, issue, surge, and redistribute fuel between brigades, refuel on-the-move operations, and deploy without construction support.

DESCRIPTION

The Modular Fuel System (MFS), formerly known as the Load Handling System Modular Fuel Farm (LMFF), is transported by the Heavy Expanded Mobility Tactical Truck Load Handling System (HEMTT–LHS) and the Palletized Load System. It is composed of 14 2,500-gallon capacity tank rack modules (TRM) and two each pump and filtration modules. Each tank rack module has a baffled 2,500-gallon fuel storage tank and onboard storage compartments for hoses, nozzles, fire extinguishers, and grounding rods. The pump filtration module includes a 600-gallon-per-minute (GPM) diesel engine-driven centrifugal pump, filter separator, valves, fittings, hoses, refueling nozzles, and a manual hand pump for gravity discharge operations. Each pump filtration module has onboard storage for hoses, ground rods, water cans, and fire extinguishers. The pump module has an evacuation capability that allows the hoses in the system to be purged of fuel prior to recovery. The MFS's configuration can vary in size (total capacity) based on the type of force supported.

SYSTEM INTERDEPENDENCIES

MFS TRM is interdependent with HEMTT Palletized Load System (PLS) and LHS for transportation.

PROGRAM STATUS

- **3QFY08:** PEO CS&CSS approved MFS HEMTT/TRM interface
- **1QFY09:** ASA(ALT) signed un-termination letter restoring the MFS as an active program

PROJECTED ACTIVITIES

- **2QFY09:** Complete MFS–TRM interface hardware
- **3QFY09:** Complete MFS–TRM interface logistics
- **4QFY09:** MFS–TRM initial operational test
- **1QFY10:** MFS–TRM full materiel release, full-rate production, type classification-standard
- **2QFY10:** MFS contract award

ACQUISITION PHASE

Technology Development | Engineering & Manufacturing Development | **Production & Deployment** | Operations & Support

Modular Fuel System (MFS)

FOREIGN MILITARY SALES
None

CONTRACTORS
DRS Sustainment Systems, Inc.
(St. Louis, MO)
E.D. Etnyre and Co. (IL)

Mortar Systems

INVESTMENT COMPONENT

Modernization

Recapitalization

Maintenance

MISSION

To enhance mission effectiveness of the maneuver unit commander by providing organic indirect fire support.

DESCRIPTION

The Army uses three variants of 120mm mortar systems. All are smooth-bore, muzzle-loaded weapons in mounted or dismounted configurations. The M120 120mm Towed Mortar System mounts on the M1101 trailer and is emplaced and displaced using the M326 "quick stow" system. The mounted variants are the M121 120mm mortar, used on the M1064A3 Mortar Carrier (M113 variant), and the 120mm Recoiling Mortar System, used on the M1129 Stryker Mortar Carrier.

Lightweight variants of the M252 81mm Mortar System and M224 60mm Mortar System have been qualified and are in production. Both systems provide high-rate-of-fire capability and are man-portable. On the M224 mortar, cartridges can be drop-fired using the standard M7 baseplate or hand-held and trigger-fired using a smaller assault M8 baseplate.

The M95/M96 Mortar Fire Control System–Mounted (MFCS–M), used on the M1064A3 and M1129, and the M150/M151 Mortar Fire Control System–Dismounted (MFCS–D), used with the M120, combine a fire control computer with an inertial navigation and pointing system, allowing crews to fire in under a minute, greatly improving mortar lethality.

The M32 Lightweight Handheld Mortar Ballistic Computer (LHMBC) has a tactical modem and embedded global positioning system, allowing mortar crews to send and receive digital call-for-fire messages, calculate ballistic solutions, and navigate.

SYSTEM INTERDEPENDENCIES

M95/M96 MFCS–M and M150/M151 MFCS–D: Army Field Artillery Tactical Data System

PROGRAM STATUS

- **1QFY09–1QFY10:** MFCS fielded to two heavy Brigade Combat Teams (BCTs) and nine HBCTs reset
- **1QFY09–1QFY10:** LHMBC fielded to seven Infantry BCTs, four Special Forces groups, and 16 IBCTs reset
- **1QFY09–1QFY10:** Mortar weapons fielded to numerous IBCT, HBCT, SBCT and Special Forces groups
- **3QFY09:** Full materiel release MCFS-D

PROJECTED ACTIVITIES

- **4QFY09:** Full materiel release of M326 "quick stow" system
- **1QFY09–1QFY10:** Continue production and fielding of 60mm, 81mm, and 120mm mortar systems
- **2QFY09–1QFY10:** Production and initial fielding of M150/M151 MFCS–D
- **2QFY09–4QFY11:** Complete initial fielding of MFCS–M
- **1QFY10:** Initiate fielding of M326
- **2QFY12:** Complete production and fielding of M32 LHMBC

ACQUISITION PHASE

Technology Development | Engineering & Manufacturing Development | Production & Deployment | Operations & Support

Mortar Systems

Mortar Fire Control System (MFCS-Mounted)

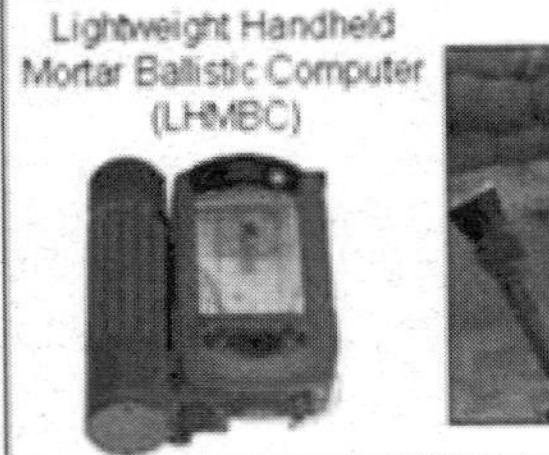

Motar Fire Control System (MFCS-Dismounted)

M1064A3 Motar Carrier

Lightweight Handheld Mortar Ballistic Computer (LHMBC)

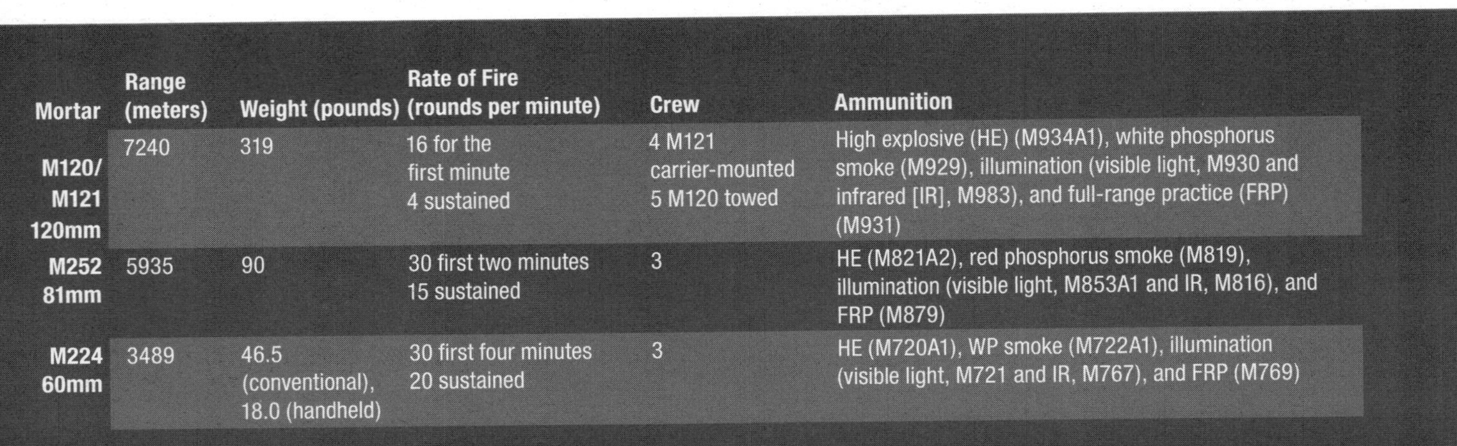

Mortar	Range (meters)	Weight (pounds)	Rate of Fire (rounds per minute)	Crew	Ammunition
M120/ M121 120mm	7240	319	16 for the first minute 4 sustained	4 M121 carrier-mounted 5 M120 towed	High explosive (HE) (M934A1), white phosphorus smoke (M929), illumination (visible light, M930 and infrared [IR], M983), and full-range practice (FRP) (M931)
M252 81mm	5935	90	30 first two minutes 15 sustained	3	HE (M821A2), red phosphorus smoke (M819), illumination (visible light, M853A1 and IR, M816), and FRP (M879)
M224 60mm	3489	46.5 (conventional), 18.0 (handheld)	30 first four minutes 20 sustained	3	HE (M720A1), WP smoke (M722A1), illumination (visible light, M721 and IR, M767), and FRP (M769)

FOREIGN MILITARY SALES
Afghanistan, Australia

CONTRACTORS

60mm and 81mm Mortar Bipod Production:
MaTech (Salisbury, MD)

60mm and 81mm Baseplate Production:
AMT (Fairfield, NJ)

MFCS-D and MFCS-M production, fielding, and installation:
Elbit Systems of America (Ft. Worth, TX)

M32 LHMBC (R-PDA):
General Dynamics C4 Systems (Taunton, MA)

120mm, 81mm, and 60mm cannons, 120mm baseplates:
Watervliet Arsenal (Watervliet, NY)

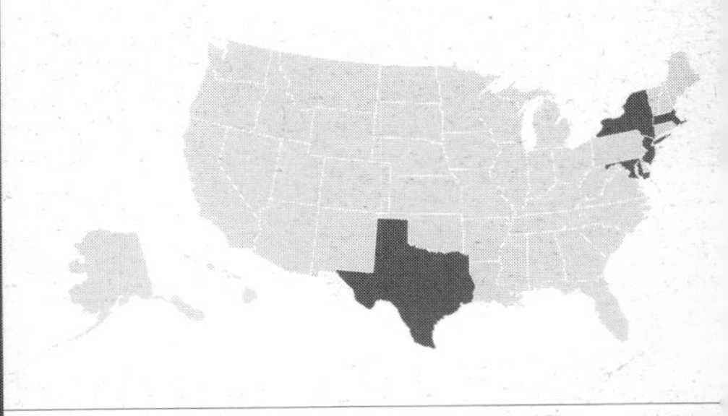

Mounted Soldier

INVESTMENT COMPONENT

- **Modernization**
- Recapitalization
- Maintenance

MISSION

To improve the tactical awareness, lethality, survivability, and to reduce heat stress for the combat vehicle crewmen.

DESCRIPTION

Mounted Soldier (MS) is an integrated system of systems designed for the combat vehicle crewmen. MS combines a cordless communications system, personal display system, and cooling system with Soldier mission equipment, and outfits all crew members (including vehicle commanders, drivers, and gunners) who operate ground platforms.

The system leverages capabilities developed in other Warrior programs such as Land Warrior (LW) and Air Warrior (AW). The system interfaces with other Army communications and command and control systems. MS includes lightweight, integrated, modular, mission-tailorable equipment carried or used by crewmen when conducting tactical operations with their assigned combat vehicles.

Team Soldier equipped the 4th Battalion, 9th Infantry, 4th Stryker Brigade Combat Team, 2nd Infantry Division at Fort Lewis, WA, with LW and the predecessor to MS, called Mounted Warrior (MW), to conduct a comprehensive assessment.

It covered the areas of doctrine, organization, training, materiel, leadership; education, personnel, and facilities; and tactics, techniques, and procedures. Results confirmed the operational need for this capability. An approval milestone decision allowed the MS program to enter engineering and manufacturing development.

SYSTEM INTERDEPENDENCIES

Vehicle for power and C2

PROGRAM STATUS

- **2QFY07:** Capabilities Development Document approved by Joint Requirements Oversight Council
- **1QFY08:** Milestone B
- **2QFY09:** MSS contracts awarded

PROJECTED ACTIVITIES

- **4QFY10-2QFY11:** Development Test, Limited User Test
- **3QFY11:** Milestone C
- **4QFY12** MSS Full-Rate Production Begins

ACQUISITION PHASE

- Technology Development
- **Engineering & Manufacturing Development**
- Production & Deployment
- Operations & Support

Mounted Soldier

FOREIGN MILITARY SALES
None

CONTRACTORS
General Dynamics (Falls Church, VA)

Movement Tracking System (MTS)

INVESTMENT COMPONENT

Modernization

Recapitalization

Maintenance

MISSION

To provide the Logistics Command with the technology necessary to track and communicate with its mobile assets in near real time through the full spectrum of military operations from peacetime to war.

DESCRIPTION

The Movement Tracking System (MTS) is a low-cost solution designed for the Army and its vehicle operators for tracking vehicles and communicating while on and off the road. MTS is a mobile satellite two-way messaging system that is totally wireless, from the MTS-equipped vehicles to the control station. The mobile configuration of the system is mounted on a unit's vehicles, and the control station configuration, in a fixed location, monitors vehicle locations. Both configurations use the same basic communications software and hardware, although the control station uses a computer with a larger display and faster processor. Communication between the two is provided by a commercial satellite vendor that enables units to send and receive traffic over the horizon, anytime, anywhere.

MTS technology allows the transportation coordinator to communicate with the driver of any truck, regardless of location, without having to put up antennas or involve more Soldiers. MTS has been adapted to incorporate radio frequency technology and an upgraded military global positioning system capability. In the future, it will incorporate an automatic reporting of vehicle diagnostics and other features that support in-transit visibility.

MTS will provide vehicles and watercraft visibility wherever they may be deployed throughout the world. Objectively, all common user logistic transport vehicles, selected combat support and combat service support Tactical Wheeled Vehicles, and watercraft will be fitted with MTS Mobile Units. In the future, a portable MTS unit may be made available to host nation or foreign national forces contributing to a combined operation for use in leased, contracted, or other vehicles that may be utilized in the distribution role, but that would not normally be equipped with MTS. MTS will provide watercraft transportation unit commanders with the capability to track and communicate with assets employed across the spectrum of operations such as established ports, logistics over-the-shore, intra-coastal, inland waterways, and amphibious operations.

When employed within the distribution system, MTS will improve the effectiveness and efficiency of limited-distribution assets, provide the ability to reroute supplies to higher priority needs, enable the avoidance of identified hazards, and inform operators of unit location changes. In addition, planned enhancements of MTS (referred to as Block II MTS enhancements) such as MTS's interface with embedded equipment diagnostic and prognostic systems, will provide accurate data that will aid fleet maintenance and improve availability and overall service life.

SYSTEM INTERDEPENDENCIES

None

PROGRAM STATUS

- **4QFY08:** Continued software development for MTS–Enhanced Software (ES)
- **1QFY09:** Conducted testing of new software upgrades

PROJECTED ACTIVITIES

- **2QFY09:** Develop MTS follow-on procurement strategies
- **2QFY09:** Begin testing MTS–ES software
- **3QFY09:** Continue testing MTS–ES software
- **4QFY09:** Continue testing MTS–ES software
- **2QFY10:** Field MTS–ES software
- **3QFY10:** MTS follow-on procurement

ACQUISITION PHASE

Technology Development | Engineering & Manufacturing Development | Production & Deployment | Operations & Support

Movement Tracking System (MTS)

FOREIGN MILITARY SALES
None

CONTRACTORS
System integrator:
COMTECH Mobile Datacom (Germantown, MD)
Software development:
Northrop Grumman (Redondo Beach, CA) via Force XXI Battle Command Brigade and Below (FBCB2) contract (Fort Monmouth, NJ)

Multifunctional Information Distribution System (MIDS)–Joint Tactical Radio System (JTRS)

INVESTMENT COMPONENT

Modernization

Recapitalization

Maintenance

MISSION

To provide real-time information and situational awareness to the warfighter in fast mover platforms (e.g., Navy F/A/18) via secure wireless, jam-resistant digital and voice communications.

DESCRIPTION

The Multifunctional Information Distribution System (MIDS) is a wireless, jam-resistant, and secure information system providing TACAN and Link-16 to Air, Land and Sea warfighting platforms. It provides real-time and low-cost information and situational awareness via digital and voice, communications. The MIDS program includes MIDS–Low Volume Terminal (LVT) full rate production and JTRS evolutionary development. MIDS–JTRS is a "form fit function" replacement for MIDS–LVT and adds three additional channels for JTRS waveforms as requested by platforms.

The MIDS–LVT program is a multinational cooperative development program with joint service participation. DoD established the program to design, develop, and deliver low-volume, lightweight tactical information system terminals for U.S. and allied aircraft, helicopters, ships, and ground sites. MIDS–LVT provides interoperability with NATO users, significantly increasing force effectiveness and minimizing hostile actions and friend-on-friend engagements. Three principal configurations of the terminal are in production and use an open-system, modular architecture. MIDS–LVT(1) includes voice, Tactical Air Navigation (TACAN) and variable-power transmission with maximum power of 200 watts. MIDS–LVT(1) also provides a Link 16 capability to Navy and Air Force platforms, which were previously unable to use Joint Tactical Information Distribution System (JTIDS) due to space and weight limitations. MIDS–LVT(2) is an Army variant of MIDS that is a functional replacement for the JTIDS Class 2M terminal. MIDS–LVT(3), also referred to as MIDS Fighter Data Link, is a reduced-function terminal for the Air Force (no voice, no TACAN, maximum power of 40 watts).

As the MIDS–LVT migrates to JTRS compliance, the system will maintain its Link 16 and TACAN functionality with Navy and Air Force platforms that use MIDS–LVT but also accommodate future technologies and capabilities. MIDS–JTRS improvements include enhanced Link 16 throughput, Link 16 frequency remapping, and programmable crypto. MIDS–JTRS will provide and additional three 2-megahertz or 2-gigahertz programmable channels to accommodate incremental delivery of the advanced JTRS waveforms through MIDS–JTRS Platform Capability Packages, such as the JAN–TE capability. Total program requirements include terminal development, F/A-18 Level 0 integration, software hosting (Operating Environment/JTRS Waveforms), implementation of National Security Agency guidelines and production transition.

SYSTEM INTERDEPENDENCIES

Link 16, TACAN, and JAN–TE Waveforms

PROGRAM STATUS

- **3QFY09:** PTT ongoing; Production verification terminals started
- **4QFY09:** Milestone C decision approving entry into production and deployment

PROJECTED ACTIVITIES

- **2QFY10:** Low-rate initial production (LRIP) award
- **2QFY10:** PTT concludes
- **3QFY10:** F/A-18 Initial Operational Capability
- **1QFY11:** LRIP concludes

ACQUISITION PHASE

Technology Development | Engineering & Manufacturing Development | Production & Deployment | Operations & Support

Multifunctional Information Distribution System (MIDS)–Joint Tactical Radio System (JTRS)

MIDS Low Volume Terminal Variants

MIDS-LVT (1) Family (includes LVT (4), (6), & (7))

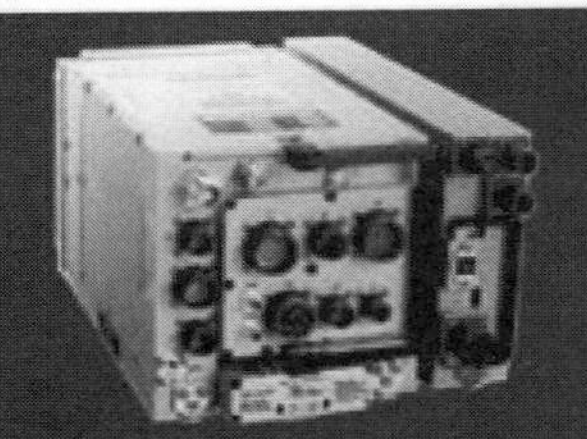

- Open and Modular Design
- 1553/Ethernet/X.25 Interfaces
- Optional J Voice and TACAN
- 200 Watt Transmitter
- Dual Antenna
- Primarily Airborne Applications

MIDS-LVT (2)/(11)

- MIDS-LVT Commonality
- ADDSI X.25, Ethernet Interface
- Optional J Voice
- No TACAN
- 200 Watt Transmitter
- Single Antenna
- Unique Blower and Power Supply
- Primarily Ground Applications

MIDS-LVT (3) (Fighter Data Link)

- MIDS-LVT Commonality
- MIL-STD 1553 Interface
- No Voice or TACAN
- 50 Watt Transmitter

FOREIGN MILITARY SALES

MIDS–LVT:
1,881 terminals (internationally)

JTRS MIDS:
None

CONTRACTORS

ViaSat (Carlsbad, CA)
Data Link Solutions (Cedar Rapids, IA)

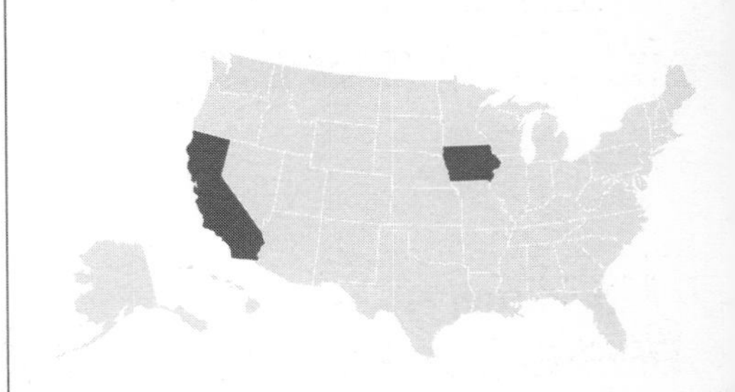

Multiple Launch Rocket System (MLRS) M270A1

INVESTMENT COMPONENT
- Modernization
- Recapitalization
- Maintenance

MISSION

To provide coalition ground forces with highly lethal, responsive and precise long-range rocket and missile fires which defeat point and area targets in both urban/complex and open terrain, with minimal collateral damage, via a highly mobile responsive multiple launch system.

DESCRIPTION

The combat-proven Multiple Launch Rocket System (MLRS) M270A1 is a mechanized artillery weapon system that provides the ground commander with round-the-clock, all-weather, lethal, close, and long-range precision rocket and missile fire support for joint forces, early-entry expeditionary forces, contingency forces, and modular fire brigades supporting Brigade Combat Teams.

The Multiple Launch Rocket System (MLRS) M270A1 is an upgraded version of the M270 launcher. The program entailed the concurrent incorporation of the Improved Fire Control System (IFCS) and the Improved Launcher Mechanical System (ILMS) on a rebuilt M993 Carrier (derivative of the Bradley Fighting Vehicle). With the IFCS, the M270A1 can fire future munitions and the ILMS reduces system load and reload times. The M270A1 provides responsive, highly accurate and extremely lethal, surface-to-surface, close- to long-range rocket and missile fires from 15 kilometers to a depth of 300 kilometers. It carries and fires either two launch pods containing six MLRS rockets each or two Army Tactical Missiles and is capable of firing all current and future MLRS family of rockets and missiles. It operates with the same MLRS command, control, and communications structure and has the same size crew as the M142 HIMARS. MLRS units are organic or assigned to modular fire brigades in support of Brigade Combat Teams.

SYSTEM INTERDEPENDENCIES

M993 Bradley derivative chassis, Advanced Field Artillery Tactical Data System (AFATDS)

PROGRAM STATUS

- **4QFY09:** Continue fleet modernization with Auxiliary Power Unit/Environmental Control Unit upgrades
- **4QFY09:** Initiate M270A1 launcher overhaul pilot program

PROJECTED ACTIVITIES

- Provide support to fielded units/ units in combat
- Provide sustainment and support activities for MLRS Strategic Partners and FMS customers
- Continue M270A1 launcher overhaul pilot program

ACQUISITION PHASE

Technology Development	Engineering & Manufacturing Development	Production & Deployment	Operations & Support

Multiple Launch Rocket System (MLRS) M270A1

FOREIGN MILITARY SALES
Bahrain, Denmark, Egypt, Finland, France, Germany, Greece, Israel, Italy, Japan, Korea (M270 & M270A1), Norway, Turkey, United Kingdom (M270 & M270B1)

CONTRACTORS
Prime and launcher:
Lockheed Martin (Dallas, TX; Camden, AR)
Chassis:
BAE Systems (Sealy, TX)
Improved Weapons Interface Unit:
Harris Corp. (Melbourne, FL)
Position Navigation Unit:
L-3 Communications Space & Navigation (Budd Lake, NJ)

NAVSTAR Global Positioning System (GPS)

INVESTMENT COMPONENT

Modernization

Recapitalization

Maintenance

MISSION

To provide real-time position, velocity, and timing data to tactical and strategic organizations.

DESCRIPTION

The Global Positioning System (GPS) is a space-based joint-service navigation program, led by the Air Force, which distributes position, velocity, and timing (PVT) data. The GPS has three segments: a space segment (nominally 24 satellites), a ground control segment, and a user equipment segment. User equipment consists of receivers configured for handheld, ground, aircraft, and watercraft applications. Military GPS receivers use the Precise Positioning Service (PPS) signal to gain enhanced accuracy and signal protection not available to commercial equipment. GPS receivers in the Army today are: the Precision Lightweight GPS Receiver (PLGR) with more than 100,000 in handheld, installed, and integrated applications; and the Defense Advanced GPS Receiver (DAGR) with more than 92,200 as handheld receivers and 62,000 distributed for platform installations to date for a total of 154,200 DAGRs fielded. In addition, GPS user equipment includes a Ground-Based GPS Receiver Applications Module (GB–GRAM). Over 78,000 GB–GRAMs have been procured and provide an embedded PPS capability to a variety of weapon systems. The Army represents more than 80 percent of the requirement for user equipment.

DAGR

Size: 6.37 x 3.4 x 1.56 inches
Weight: One pound; fits in a two-clip carrying case that attaches to Load-Bearing Equipment
Frequency: Dual (L1/L2)
Battery Life: 19 hours (4 AA batteries)
Security: Selective availability anti-spoofing module
Satellites: All-in-view

GB–GRAM

Size: 0.6 x 2.45 x 3.4 inches
Weight: 3.5 ounces
Frequency: Dual (L1/L2)
Security: Selective availability anti-spoofing module
Satellites: All-in-view

SYSTEM INTERDEPENDENCIES

Blue Force Tracking, PATRIOT, Excalibur, Paladin, mobile ballistic computers, laser rangefinders, movement tracking systems, and several unmanned aerial vehicle systems.

PROGRAM STATUS

- **3QFY08–1QFY09:** Continue DAGR fieldings

PROJECTED ACTIVITIES

- **2QFY09–1QFY11:** Continue DAGR fieldings

ACQUISITION PHASE

Technology Development | Engineering & Manufacturing Development | Production & Deployment | Operations & Support

NAVSTAR Global Positioning System (GPS)

FOREIGN MILITARY SALES

PPS-capable GPS receivers have been sold to 41 authorized countries.

CONTRACTORS

DAGR/GB–GRAM acquisition and PLGR support:

Rockwell Collins (Cedar Rapids, IA)

Non-Intrusive Inspection (NII) Systems

INVESTMENT COMPONENT

Modernization

Recapitalization

Maintenance

MISSION

To protect U.S. forces and critical warfighting materiel by inspecting cars, trucks, or cargo containers for the presence of explosives, weapons, drugs, or other contraband with nuclear (gamma) and X-ray technology.

DESCRIPTION

The Non-Intrusive Inspection (NII) systems consist of commercial off-the-shelf (COTS) products that are employed within a layered force protection system that includes security personnel trained to maintain situational awareness aided by a range of other products including military working dogs, under-vehicle scanning mirrors, and handheld or desktop trace explosive detectors. The NII systems produce a graphic image from which a trained operator can "look into" places such as false compartments that other systems cannot see.

NII systems currently include a variety of products with differing characteristics that are added to the Army commander's "tool box". They include mobile, rail-mounted but re-locatable, and fixed site characteristics. The primary systems employed are as follows:

The **Mobile Vehicle and Cargo Inspection System (MVACIS)** is a truck-mounted system that utilizes a nuclear source that can penetrate approximately 6.5 inches of steel. It can be employed in static locations or moved rapidly between access control points to provide protection where it is most needed.

The **Re-locatable Vehicle and Cargo Inspection System (RVACIS)** is a rail-mounted system that utilizes the same nuclear source as the MVACIS. It operates on rails and is employed in static locations or moved within 24 hours to locations where prepared use of the rail system eliminates the requirement to maintain a truck platform and the presence of an overhead articulated arm that can be struck and damaged by vehicles being scanned.

The **Militarized Mobile VACIS (MMVACIS)** uses the same gamma source as the other VACIS products. It is mounted, however, on a High Mobility Multipurpose Wheeled Vehicle (HMMWV). MMVACIS provides a capability that other NII systems do not: off installation external vehicle checkpoints in remote locations.

The **Z-Backscatter Van (ZBV)**, manufactured by American Science & Engineering (AS&E), is a van-mounted, system that utilizes backscatter X-ray technology. While it can penetrate only approximately ¼ inch of steel it can be employed in static locations where room is limited and scanning of cars, larger vehicles, or containers with smaller, less complex loads is expected.

SYSTEM INTERDEPENDENCIES

None

PROGRAM STATUS

- **FY08:** MMVACIS/RVACIS procurement and fielding
- **FY09:** ZBV procurement and fielding

PROJECTED ACTIVITIES

- **1QFY10:** MMVACIS Fielding

ACQUISITION PHASE

Technology Development | Engineering & Manufacturing Development | Production & Deployment | Operations & Support

Non-Intrusive Inspection (NII) Systems

FOREIGN MILITARY SALES
None

CONTRACTORS
American Science & Engineering, Inc. (Billerica, MA)
Rapiscan Systems (Torrance, CA)
Science Applications International Corp. (SAIC) (San Diego, CA)

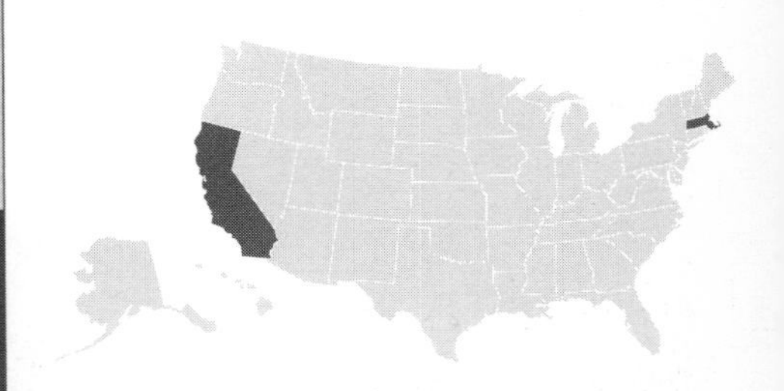

Non Line of Sight–Launch System (NLOS–LS)

INVESTMENT COMPONENT

Modernization

Recapitalization

Maintenance

MISSION

To enhance combat effectiveness and survivability by providing precise, highly deployable, non-line-of-sight lethal fires for the Early Infantry Brigade Combat Team (E-IBCT).

DESCRIPTION

The Non Line of Sight–Launch System (NLOS–LS) is a core system within the Brigade Combat Team (BCT), and provides unmatched lethality and "leap ahead" missile capability for U.S. forces. NLOS–LS consists of precision guided missiles loaded onto a highly deployable, platform-independent container launch unit (CLU) with self-contained technical fire control, electronics, and software to enable remote and unmanned fire support operations.

The precision guided munition being developed is the Precision Attack Missile (PAM). The NLOS–LS CLU will contain 15 missiles and one Missile Computer and Communications System (MCCS). The PAM, which launches vertically from the CLU, will be used primarily to defeat hard, soft, moving, or stationary target elements when fire mission orders are received by Advanced Field Artillery Tactical Data System (AFATDS). It will be able to receive in-flight target updates via its onboard network radio, and will have limited automatic target recognition capability. PAM will have a multi-functional warhead to effectively engage hard (armor) and soft targets. NLOS–LS CLUs were fielded to the Army Evaluation Task Force (AETF) in FY08 for integration into Current Forces as part of the E-IBCT spin-out strategy. NLOS–LS also supports the Navy's Littoral Combat Ship against small boat threats. Future missile variants may include air defense and non-lethal capabilities. Key NLOS–LS advantages include the following:

- Remote fire control
- Remote emplacement
- Extended-range target engagements and battle damage assessment
- Jam-resistant Global Positioning System
- Ability to engage moving targets

Weight: CLU with 15 missiles, approximately 3150 pounds
Width: 45 inches
Length: 45 inches
Height: 69 inches
Range: Approximately 40 kilometers

SYSTEM INTERDEPENDENCIES

AFATDS, Soldier Radio Waveform, Joint Tactical Radio System (JTRS) for Future Forces

PROGRAM STATUS

- **1QFY09:** Guided test vehicle (GTV) #1, #2 and #3
- **2QFY09:** GTV #4, #8
- **3QFY09:** GTV #9, #10
- **4QFY09:** GTV #11

PROJECTED ACTIVITIES

- **4QFY09–1QFY10:** Continue GTV testing
- **2QFY10:** Flight limited user test

ACQUISITION PHASE

Technology Development | Engineering & Manufacturing Development | Production & Deployment | Operations & Support

Non Line of Sight–Launch System (NLOS–LS)

FOREIGN MILITARY SALES
None

CONTRACTORS
Raytheon (Tucson, AZ; Fuller, CA)
Lockheed Martin (Baltimore, MD; Dallas, TX)
L-3/IAC (Anaheim, CA)

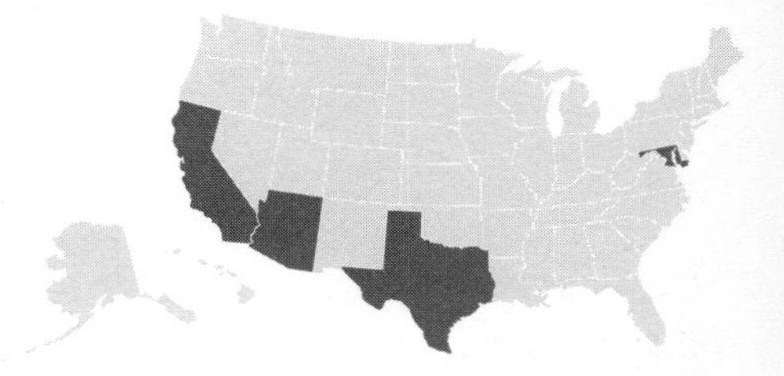

Nuclear Biological Chemical Reconnaissance Vehicle (NBCRV)–Stryker

INVESTMENT COMPONENT

Modernization

Recapitalization

Maintenance

MISSION

To perform nuclear, biological, and chemical (NBC) reconnaissance and to locate, identify, mark, sample, and report NBC contamination on the battlefield.

DESCRIPTION

The Nuclear Biological Chemical Reconnaissance Vehicle (NBCRV)–Stryker is the chemical, biological, radiological, and nuclear (CBRN) reconnaissance configuration of the infantry carrier vehicle in the Stryker Brigade Combat Team (SBCT).

The NBCRV–Stryker sensor suite is a dedicated system of nuclear and chemical detection and warning equipment, and biological sampling equipment. Its sensor suite provides outstanding capability on a common platform by use of a single, integrated reconnaissance and surveillance system. The NBCRV will be able to detect and collect chemical and biological contamination in its immediate environment, on the move, through point detection Chemical Biological Mass Spectrometer (CBMS) and Joint Biological Point Detection System (JBPDS), and at a distance through the use of the Joint Service Lightweight Standoff Chemical Agent Detector (JSLSCAD). It automatically integrates contamination information from detectors with input from on-board navigation and meteorological systems and automatically transmits digital NBC warning messages through the vehicle's command and control equipment to warn follow-on forces. NBCRV may replace the need for separate M93A1 Fox NBC reconnaissance systems and biological integrated detection systems.

SYSTEM INTERDEPENDENCIES

Joint Service Lightweight Standoff Chemical Agent Detector (JSLSCAD), Automatic Chemical Agent Detector Alarm (ACADA), Chemical Biological Mass Spectrometer (CBMS), Joint Biological Point Detection System (JBPDS), AN/UDR-13 Radiac Detector, Chemical Vapor Sampler System (CVSS), Nuclear Biological Chemical Sensor Processing Group (NBCSPG), and Double Wheel Sampler System (DWSS)

PROGRAM STATUS

- **FY09:** Continuing additional low-rate initial production

PROJECTED ACTIVITIES

- **FY10:** Initial operational test and evaluation
- **FY11:** Full-rate production

ACQUISITION PHASE

Technology Development | Engineering & Manufacturing Development | Production & Deployment | Operations & Support

Nuclear Biological Chemical Reconnaissance Vehicle (NBCRV)–Stryker

FOREIGN MILITARY SALES
None

CONTRACTORS
Prime vehicle:
General Dynamics Land Systems (Sterling Heights, MI)
Sensor software integrator:
CACI Technologies (Manassas, VA)

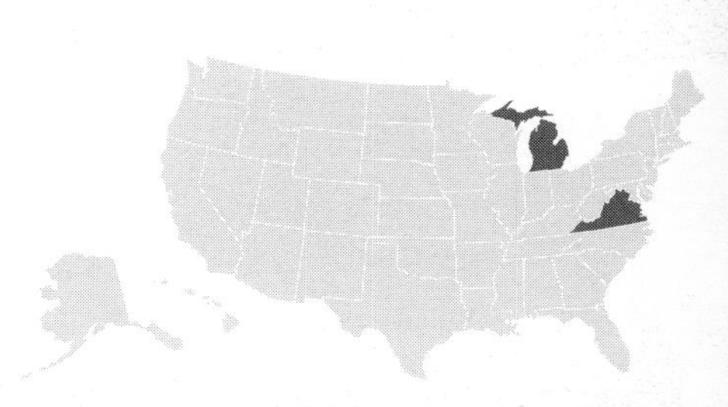

One Semi-Automated Forces (OneSAF)

INVESTMENT COMPONENT

- **Modernization**
- Recapitalization
- Maintenance

MISSION

To provide simulation software that supports constructive and virtual training; mission rehearsal; analysis and research; and embedded solutions for system program managers including the Early Infantry Brigade Combat Team (E-IBCT).

DESCRIPTION

One Semi-Automated Forces (OneSAF) is a next generation, entity-level simulation that supports both computer generated forces and Semi-Automated Forces applications. This enables it to support a wide range of Army brigade-and-below constructive simulations and virtual simulators.

OneSAF is currently being integrated by the Synthetic Environment (SE) Core program as the replacement SAF for virtual trainers such as Aviation Combined Arms Tactical Trainer, Close Combat Tactical Trainer and the Common Gunnery Architecture (CGA). OneSAF will serve as the basis for subsequent modernization activities for simulators across the Army. OneSAF was built to represent the modular and Future Force and provides entities, units, and behaviors across the spectrum of military operations in the contemporary operating environment. OneSAF has been crafted to be uniquely capable of simulating aspects of the contemporary operating environment and its effects on simulated activities and behaviors. OneSAF is unique in its ability to model unit behaviors from fire team to company level for all units—both combat and non-combat operations. Intelligent, doctrinally correct behaviors and improved graphical user interfaces are provided to increase the span of control for workstation operators.

Interoperability support is present for industry standards such as Distributed Interactive Simulation, High Level Architecture, Military Scenario Development Language, Joint Consultation Command and Control Information Exchange Data Model, and Army Battle Command System devices.

OneSAF is a cross-domain simulation suitable for supporting training, analysis, research, experimentation, mission-planning, and rehearsal activities. It provides the latest physics-based modeling and data, enhanced data collection and reporting capabilities.

SYSTEM INTERDEPENDENCIES

OneSAF provides required capabilities for SE Core; OneSAF is a complimentary Tier I program of E-ICBT.

PROGRAM STATUS

- **2QFY08:** OneSAF Version 2.0 released

PROJECTED ACTIVITIES

- **4QFY08:** Release OneSAF (International) Version 2.0
- **1QFY09:** Release OneSAF Version 3.0

ACQUISITION PHASE

Technology Development | Engineering & Manufacturing Development | Production & Deployment | Operations & Support

One Semi-Automated Forces (OneSAF) Objective System

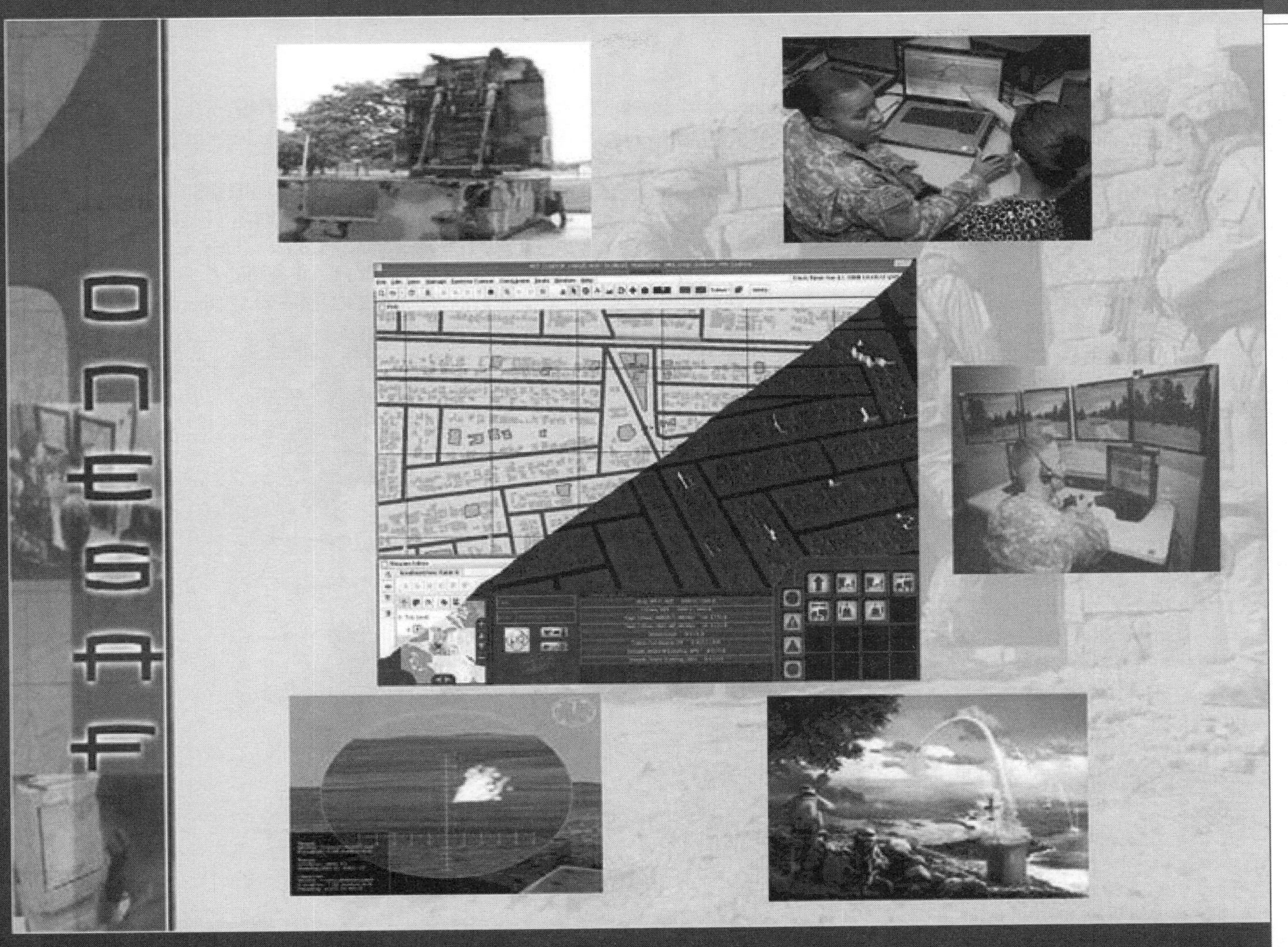

FOREIGN MILITARY SALES
United Kingdom, Canada, Australia, Czech Republic, Bahrain

CONTRACTORS
Science Applications International Corp. (SAIC) (Orlando, FL)
CAE USA (Orlando, FL)
The Aegis Technology Group, Inc. (Orlando, FL)
Northrop Grumman Information Technology (NGIT) (Orlando, FL)
Lockheed Martin (Orlando, FL)

Paladin/Field Artillery Ammunition Supply Vehicle (FAASV)

INVESTMENT COMPONENT

Modernization

Recapitalization

Maintenance

MISSION

To provide the primary artillery support for heavy Brigade Combat Teams and non-divisional heavy fires battalions.

DESCRIPTION

The M109A6 (Paladin) 155mm howitzer is the most technologically advanced self-propelled cannon system in the Army. The field artillery ammunition supply vehicle (FAASV) provides an armored ammunition resupply vehicle in support of the Paladin.

The Paladin Integrated Management (PIM) program supports the fleet management strategy for current Paladins and FAASVs by providing a low-risk and affordable life-cycle solution that addresses obsolescence and ensures long-term sustainment of the fleet through 2050.

PIM uses state-of-the art components to improve the following:

- Survivability: "Shoot and scoot" tactics; improved ballistic and nuclear, biological, and chemical protection on both the Howitzer and FAASV.
- Responsive fires: Capable of firing within 45 seconds from a complete stop with on-board communications, remote travel lock, automated cannon slew capability, and pivot steer technology.
- Accurate fires: On-board position navigator and technical fire control.
- Extended range: 30 kilometers with high-explosive, Rocket-Assisted Projectile (RAP) and Excalibur Projectiles using M203 or M232/M232A1 MACS propellant.
- Increased reliability: Unique chassis built around Bradley Common Powertrain, Track, and Diagnostics.
- Growth capacity (Accept Non-Line of Sight/Future Network capabilities.

Other PIM specifications include the following:

Crew: Paladin, four; FAASV, five
Combat loaded weight: Paladin, 35 tons; FAASV, 28 tons
Paladin on-board ammo: 95 rounds plus 2 Excalibur or Copperhead projectiles
FAASV on-board ammo: 95 rounds
Rates of fire: 4 rounds per minute for first 3 minutes maximum; 1 round per minute sustained
Maximum range: High Explosive Rocket Assisted Projectile (HE/RAP), 22/30 kilometers
Cruising range: Paladin, 186 miles; FAASV, 186 miles
Fire Support Network: Paladin Digital Fire Control System software support Fire Support Network

SYSTEM INTERDEPENDENCIES

AFATDS, Excalibur, PGK, FBCB2

PROGRAM STATUS

- **FY09–1QFY10:** Prototype production

PROJECTED ACTIVITIES

- **1QFY10–3QFY12:** Developmental testing (incudes live fire)
- **1QFY11:** Milestone C
- **2QFY12–3QFY12:** Operational testing

ACQUISITION PHASE

Technology Development | Engineering & Manufacturing Development | Production & Deployment | Operations & Support

Paladin/Field Artillery Ammunition Supply Vehicle (FAASV)

FOREIGN MILITARY SALES
None

CONTRACTORS
BAE Systems (York, PA; Elgin, OK)
Northrop Grumman (Carson, CA)
Anniston Army Depot (Anniston, AL)
Marvin Land Systems (Inglewood, CA)
Kidde Dual Spectrum (Goleta, CA)

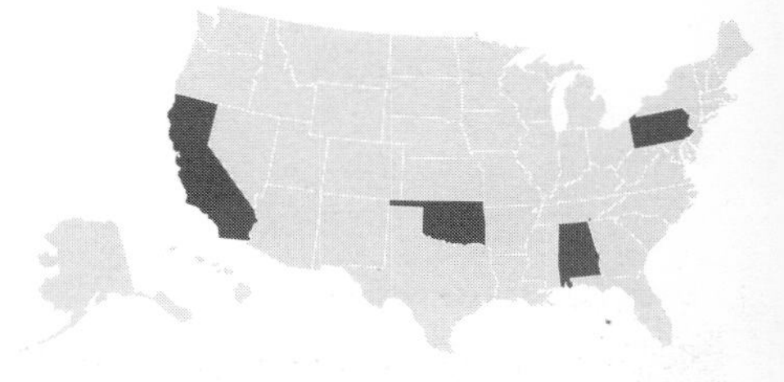

Palletized Load System (PLS) and PLS Extended Service Program (ESP)

INVESTMENT COMPONENT

Modernization
Recapitalization
Maintenance

MISSION

To support combat units by performing cross-country movement of configured loads of ammunition and other classes of supply loaded on flat racks or in containers.

DESCRIPTION

The Palletized Load System (PLS) provides the timely delivery of a high tonnage of ammunition, unit equipment, International Organization for Standardization (ISO) containers/shelters, and all classes of supply to using units and weapon systems as far forward in the maneuver battalion area as the tactical situation allows. The PLS consists of the PLS truck, PLS trailer (PLS–T), and demountable flat racks. The PLS truck is a 10-wheel-drive (10x10) prime mover with an integral onboard load handling system that provides self-loading and unloading capability.

There are two PLS truck variants, the basic PLS truck (M1075) and the PLS truck with material handling crane (M1074). The system also includes the PLS trailer (M1076), Container Handling Unit (CHU) for transporting 20-foot ISO containers, the M3/M3A1 Container Roll-in/Out Platform (CROP), and the M1/M1077A1 flat racks. The PLS has the ability to operate with a degree of mobility commensurate with the supported weapon systems, to facilitate the fighting capabilities of the supported systems and units.

The PLSA1 model is planned for fielding in 2010. It will include: independent front suspension, a new C-15 engine, the Allison 4500 transmission, J-1939 data-bus, and a cab that will be common with the HEMTTA4 and Long Term Armor Strategy (LTAS)-compliant.

The PLS Extended Service Program (ESP) is a recapitalization program that converts high-mileage PLS trucks to 0 miles/0 hours and to the current A0 and future A1 production configurations. The trucks are disassembled and rebuilt with improved technology such as an electronically controlled engine, electronic transmission, air ride seats, four-point seatbelts, bolt-together wheels, increased corrosion protection, enhanced electrical package, and independent front suspension on the A1.

SYSTEM INTERDEPENDENCIES

None

PROGRAM STATUS

- **Current:** To date, fielded approximately 5,500 PLS trucks and 9,081 PLS trailers.

PROJECTED ACTIVITIES

- **2QFY09:** Complete PLSA1 Product verification testing (PVT) at Aberdeen Test Center (ATC), MD
- **2QFY09:** Testing of the Enhanced Container Handling Unit (E–CHU)
- **2–3QFY09:** PLSA1 Logistics demonstration and verification of technical manuals
- **1QFY10:** Type classification/materiel release of Enhanced CHU
- **2QFY10:** Modify existing FHTV3 contract to include production and RECAP of the PLSA1
- **FY10:** PLSA1 TC/MR)
- **FY10:** PLSA1 First unit equipped (FUE)

ACQUISITION PHASE

Technology Development | Engineering & Manufacturing Development | Production & Deployment | Operations & Support

Palletized Load System (PLS) and PLS Extended Service Program (ESP)

	PLS	PLSA1
ENGINE	DDC 8V92 - 500 horse power	CAT C-15 - 600 hp @ 2100 RPM
TRANSMISSION	Allison CLT-755 - 5 Speed	Allison HD 4500 - 6 Speed
TRANSFER CASE	Oshkosh 55,000 - 2 Speed	New Oshkosh - 2 Speed
AXLES - FRONT TANDEM	Rockwell SVI 5MR	Oshkosh / Rockwell
SUSPENSION - FRONT TANDEM	Hendrickson RT-340 - Walking Beam	Oshkosh TAK-4TM Steel Spring
AXLES - REAR TRIDEM	Rockwell SVI 5MR	Rockwell SVI 5MR
SUSPENSION - AXLE #3	Hendrickson-Turner Air Ride	Hendrickson-Turner Air Ride
SUSPENSION - AXLES #4 & #5	Hendrickson RT-400 - Walking Beam	Hendrickson RT-400 - Walking Beam
WHEEL ENDS	Rockwell	Rockwell
CONTROL ARMS	N/A	PLS Block 1 - New
STEERING GEARS - FRONT	492 Master/M110 Slave	M110 Master/M110 Slave
STEERING GEARS - REAR	492	M110
FRAME RAILS	14 inch	14 inch
CAB	PLS	Common Cab
RADIATOR	PLS - Roof Mount	PLS Block 1 - Side Mount
MUFFLER	PLS	PLS Block 1 - New
AIR CLEANER	United Air	United Air
LHS	Multilift MK V	Multilift MK V
CRANE	Grove	Grove
TIRES	Michelin 16.00 R20 XZLT	Michelin 16.00 R20 XZLT
SPARE TIRE	1 - Side Mounted	1 - Roof Mounted
CTI	CMA	CMA
AIR COMPRESSOR	1400 Bendix	922 Bendix
STARTER	Prestolite	Prestolite
ALTERNATOR	12/24V	24 Volt- 260 Amp Niehoff

FOREIGN MILITARY SALES
Turkey, Israel, Jordan

CONTRACTORS
Oshkosh Truck Corp. (Oshkosh, WI)
Detroit Diesel (Emporia, KS; Redford, MI)
Allison Transmissions (Indianapolis, IN)
Michelin (Greenville, SC)
Summa Technologies (Cullman, AL)
GT Machine and Fabrication (Napanee, Ontario, Canada)

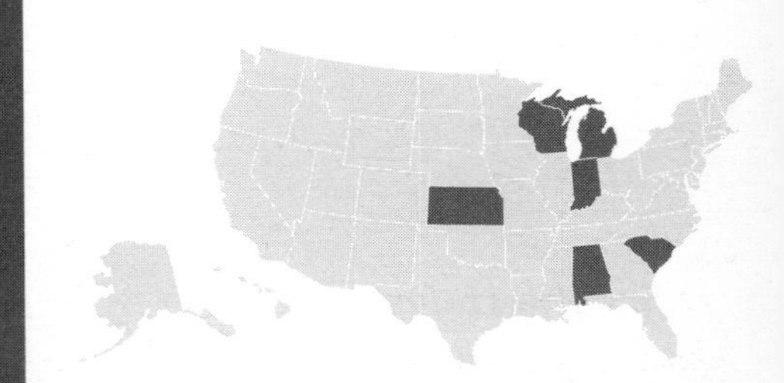

PATRIOT (PAC-3)

INVESTMENT COMPONENT

Modernization

Recapitalization

Maintenance

MISSION

To protect ground forces and critical assets at all echelons from advanced aircraft, cruise missiles, and tactical ballistic missiles.

DESCRIPTION

The PATRIOT Advanced Capability (PAC)-3 program is an air-defense, guided missile system with long-range, medium- to high-altitude, all-weather capabilities designed to counter tactical ballistic missiles (TBMs), cruise missiles, and advanced aircraft. The combat element of the PATRIOT missile system is the fire unit, which consists of a phased array radar set (RS), an engagement control station (ECS), a battery command post, an electric power plant (EPP), an antenna mast group (AMG), a communications relay group (CRG), and launching stations (LS) with missiles.

The RS provides the tactical functions of airspace surveillance, target detection, identification, classification, tracking, missile guidance, and engagement support. The ECS provides command and control. Depending upon configuration, the LS provides the platform for PAC-2 or PAC-3 missiles, sealed in canisters that serve as shipping containers and launch tubes.

The PAC-3 primary mission is to kill maneuvering and non-maneuvering TBMs, and counter advanced cruise missile and aircraft threats. The PAC-3 missile uses hit-to-kill technology for greater lethality against TBMs armed with weapons of mass destruction. The PAC-3 system upgrades have provided improvements that increase performance against evolving threats, meet user requirements, and enhance joint interoperability. PATRIOT's fast-reaction capability, high firepower, ability to track numerous targets simultaneously, and ability to operate in a severe electronic countermeasure environment make it the Army's premier air defense system. The PAC-3 Missile Segment Enhancement (MSE), currently in development, is planned to be used with the PAC-3 system and will be the baseline interceptor for the Medium Extended Air Defense System, which succeeds the PATRIOT system.

SYSTEM INTERDEPENDENCIES

Medium Extended Air Defense System (MEADS), Terminal High Altitude Air Defense (THAAD), Joint Land Attack Cruise Missile Defense Elevated Netted Sensors Systems (JLENS), Integrated Air and Missile Defense (IAMD)

PROGRAM STATUS

- **1QFY09:** Post Deployment Build-6.5 (PDB-6.5) development, test and evaluation

PROJECTED ACTIVITIES

- **3QFY07–1QFY11:** MSE flight testing
- **2QFY10:** Post Deployment Build-6.5 (PDB-6.5) initial operational capability

ACQUISITION PHASE

Technology Development | Engineering & Manufacturing Development | Production & Deployment | Operations & Support

PATRIOT (PAC-3)

FOREIGN MILITARY SALES
Germany, Greece, Israel, Japan, Kuwait, Saudi Arabia, Spain, Taiwan, The Netherlands, Korea, United Arab Emirates

CONTRACTORS
PATRIOT system integrator, ground system modifications, recapitalization program:
Raytheon (Andover, MA; Bedford, MA)
PAC-3 Missile sub-assembly and assembly:
Lockheed Martin (Grand Prairie, TX; Camden, AR; Lufkin, TX)
PAC-3 Missile Seeker sub-contractor:
Boeing (Huntsville, AL)

Precision Guidance Kit

INVESTMENT COMPONENT

- Modernization
- Recapitalization
- Maintenance

MISSION

To improve the accuracy of existing conventional artillery ammunition.

DESCRIPTION

The Precision Guidance Kit (PGK) is an affordable global positioning system (GPS) guidance kit with fuzing functions that is compatible with the existing stockpile of conventional cannon artillery projectiles. The PGK uses an integrated GPS receiver to correct the inherent errors associated with ballistic firing solutions, reducing the number of artillery projectiles required to attack targets. The increase in efficiency and effectiveness offered by PGK provides commanders the operational capability to defeat more targets with the same basic load, while reducing the logistics burden associated with the current mission requirement.

The PGK program is following an incremental program approach. Increment 1, the XM1156 PGK, will be compatible with M107, M549A1, and M795 155mm high explosive (HE) projectiles, and be designed to fire from the M109A6 Paladin and M777A2 Joint Lightweight 155mm Howitzer. Increment 2 will add the M1, M913, M760, and M927 105mm HE projectiles (to be fired from the M119A3 Howitzer), and also improve accuracy. Increment 3 will add 105mm and 155mm cargo projectiles, and implement compatibility with the Early Infantry Brigade Combat Team (E-IBCT) Non Line of Sight–Cannon.

PROGRAM STATUS

- **Current:** Increment 1 program is in engineering and manufacturing development

PROJECTED ACTIVITIES

Increment 1

- **2QFY10:** Initial operations testing and evaluation
- **4QFY10:** Type classification standard and full-materiel release

Increment 2

- **2QFY10:** Milestone B, PGK Increment 2

ACQUISITION PHASE

Technology Development | Engineering & Manufacturing Development | Production & Deployment | Operations & Support

Precision Guidance Kit

FOREIGN MILITARY SALES
None

CONTRACTORS
Increment 1
Prime:
Alliant Techsystems (Plymouth, MN)
Subcontractor:
L-3 Interstate Electronics Corp.
(Los Angeles, CA)

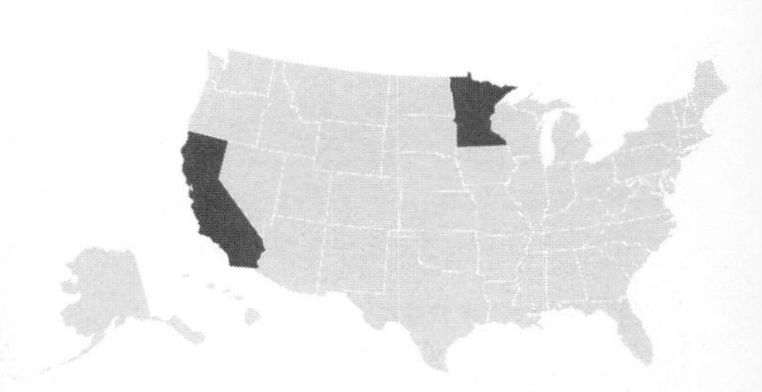

Prophet

INVESTMENT COMPONENT

Modernization

Recapitalization

Maintenance

MISSION

To improve battlespace awareness using electronic support sensors that detect, collect, identify, and locate selected emitters; enhances combat effectiveness using electronic warfare against targeted enemy command and control nodes.

DESCRIPTION

Prophet is a ground-based tactical signals intelligence/electronic warfare sensor that creates a near-real-time electronic picture of the Brigade Combat Team (BCT), Stryker Brigade Combat Team, Armored Cavalry Regiment, and Battlefield Surveillance Brigade battlespace. Prophet provides intelligence support by reporting the location, tracking, and identity of threat emitters. A secondary mission is Electronic Attack (EA) against enemy emitters. The Prophet Spiral I Electronic Support (ES) system is being fielded to Active and Reserve units in support of Operation Enduring Freedom and Operation Iraqi Freedom. The Prophet Spiral 1 ES system provides an increase in capability over the existing Prophet Block I system. Prophet consists of ground collection sensors capable of early entry and airborne insertion. The base dismounted Prophet capability provides force protection information from a man-pack system. Mounted Prophet sensors have an on-the-move collection and reporting capability; they enable Prophet to keep pace with supported units and reposition collection capability easily on the battlefield. Prophet will cross-cue other battlefield sensors and provide additional information that may confirm intelligence from other manned/unmanned battlefield sensors. Prophet EA is packaged in a High Mobility Multipurpose Wheeled Vehicle (HMMWV) trailer, towed behind the ES systems, providing both stationary and on-the-move capabilities. The Prophet Enhanced System will provide an increase in capability over the existing Prophet Spiral 1 ES sensor. Prophet Enhanced production contract was awarded on February 25, 2009. Prophet employs an open systems architecture, modular design, and nonproprietary industry standards, supporting evolutionary growth and expansion via circuit card assemblies and software upgrades. This capability will be used to maintain relevancy on the battlefield and keep pace with technology advancements via a product improvement program to insert planned improvements and new unplanned capabilities into the Spiral 1 ES Sensor and Prophet Enhanced System.

SYSTEM INTERDEPENDENCIES

Trojan lightweight integrated telecommunications equipment, tactical radio communications systems, light tactical vehicles and assured mobility systems, global positioning system, WIN–T for Wideband Beyond-Line-of-Site Communications (WBLOS), Distributed Communications Ground System Army (DCGS–A)

PROGRAM STATUS

- **3QFY08:** Completed Prophet EA limited user test
- **3QFY08–1QFY09:** Continued fielding Prophet ES Spiral 1 systems to Army Transformation BCTs and National Guard
- **3QFY08–1QFY09:** Began defielding Prophet Block I systems
- **4QFY08:** Awarded Engineering Change Proposal to replace VRC-99 with WBLOS satellite communications on Prophet ES Spiral 1

PROJECTED ACTIVITIES

- **2QFY09:** Award Prophet enhanced contract award
- **2QFY09–4QFY09:** Continue Prophet Block I fieldings to National Guard
- **2QFY09–1QFY11:** Continue Prophet ES Spiral 1 fieldings
- **2QFY09–1QFY11:** Continue to defield Prophet Block I systems as Prophet ES Spiral 1 systems are fielded
- **3QFY09:** Prophet ES 1 first unit equipped with WBLOS satellite communications on-the-move
- **1QFY10:** Operational assessment for Prophet ES Spiral 1 with WBLOS SOTM
- **2QFY10:** Prophet Enhanced Quick Reaction Capability first unit equipped
- **2QFY10:** Prophet Enhanced first unit equipped

ACQUISITION PHASE

Technology Development | Engineering & Manufacturing Development | **Production & Deployment** | Operations & Support

Prophet

FOREIGN MILITARY SALES
None

CONTRACTORS
Prophet Enhanced sensor production:
General Dynamics (Scottsdale, AZ)
Prophet ES Spiral 1 sensor and control production:
L-3 Communications (San Diego, CA; Melbourne, FL)

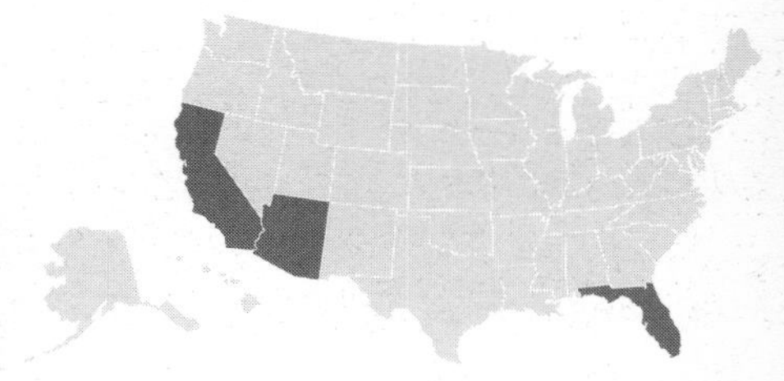

Raven Small Unmanned Aircraft System (SUAS)

INVESTMENT COMPONENT

Modernization

Recapitalization

Maintenance

MISSION

To provide reconnaissance, surveillance, target acquisition and force protection for battalion commanders and below during day/night operations.

DESCRIPTION

The RQ-11B Raven Small Unmanned Aircraft System (SUAS) is a hand-launched, unmanned aircraft system capable of 90 minutes of flight time with an operational range of approximately 10 kilometers. Normal operational altitude is 500 feet or lower. The system, aircraft, and ground control station, are assembled by operators in approximately five minutes. The aircraft, which has a wingspan of 4.5 feet, weighs 4.2 pounds. A small hand controller displays live video and aircraft status. Mission planning is performed on the hand controller or a laptop running flight planning software. Aircraft flight modes include fully manual, altitude holding, fully autonomous navigation, point loiter, and return home. Raven incorporates a secure global positioning system.

The RQ-11B system consists of: three aircraft; two control stations (primary control or remote video monitoring); ten each air vehicle and ground station batteries; two universal battery chargers; two day electro-optical color sensors; three night infrared sensors with laser illuminator; a spare parts kit; and a mission planning laptop.

The Raven is operated by two Soldiers. No specific military occupational specialty is required. Operator training is 10 days in duration.

SYSTEM INTERDEPENDENCIES

None

PROGRAM STATUS

- **System** is currently in the Production and Deployment phase
- **Raven** is currently operational in both Operation Iraqi Freedom and Operation Enduring Freedom

PROJECTED ACTIVITIES

- **FY09–10:** Development, integration, testing, and deployment of Digital Data Link

ACQUISITION PHASE

Technology Development | Engineering & Manufacturing Development | **Production & Deployment** | Operations & Support

Raven Small Unmanned Aircraft System (SUAS)

FOREIGN MILITARY SALES
Denmark

CONTRACTORS
Aerovironment, Inc.
(Simi Valley, CA)
Indigo System Corp. (Goleta, CA)
All American Racers, Inc.
(Santa Ana, CA)
L-3 Communications (San Diego, CA)
Bren-Tronics (Commack, NY)

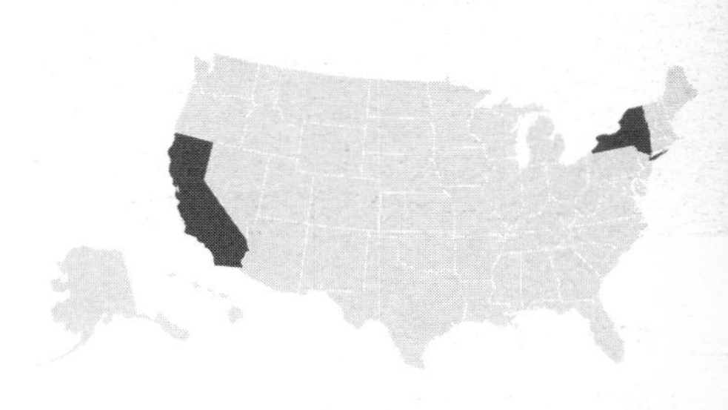

Rough Terrain Container Handler (RTCH)

INVESTMENT COMPONENT

Modernization
Recapitalization
Maintenance

MISSION

To provide a rough-terrain, variable-reach container handler in support of cargo handling operations.

DESCRIPTION

The Rough Terrain Container Handler (RTCH) is a commercial non-developmental item acquired for the cargo handling operation mission requirements worldwide. The vehicle lifts, moves, and stacks both 20-foot and 40-foot long American National Standards Institute/International Organization for Standardization (ANSI/ISO) containers and shelters weighing up to 53,000 pounds. The new Kalmar RTCH, a technological step up from the previous Caterpillar RTCH.

Improvements include the capability to transport by rail, highway, or water in less than 2 1/2 hours and in less than three hours by air (C5 and C17) transport; stack 9-foot, 6-inch containers three high; achieve a forward speed of 23 miles per hour; and an adjustable top handler. The RTCH will operate worldwide, on hard-stand, over-sand terrain, and cross-country during transportation and ordnance ammunition handling operations. Storage, transportation, and deployment will include operations in cold, basic, and hot climates.

The RTCH is not designed to counter or defeat any threat and does not possess lethality capability. During Joint logistics over-the-shore operations (JLOTS) with the absence of a developed port, the top handler variant with the forklift kit installed will be used to position modular causeway sections between the ships and the shore.

SYSTEM INTERDEPENDENCIES

None

PROGRAM STATUS

- **Current:** Ongoing production and fielding

PROJECTED ACTIVITIES

- Continue production and fielding
- **1QFY10:** Follow-on production contract award

ACQUISITION PHASE

Technology Development | Engineering & Manufacturing Development | Production & Deployment | Operations & Support

Rough Terrain Container Handler (RTCH)

FOREIGN MILITARY SALES
None

CONTRACTORS
Kalmar RT Center LLC (Cibolo, TX)

Screening Obscuration Device (SOD)–Visual Restricted (Vr)

INVESTMENT COMPONENT

Modernization

Recapitalization

Maintenance

MISSION

To provide the warfighter with the ability to safely employ short-duration obscuration in the visual, infrared (IR) and millimeter wave (MMW) portions of the electromagnetic spectrum.

DESCRIPTION

The Screening Obscuration Device–Visual Restricted (SOD–Vr), is a member of the Family of Tactical Obscuration Devices Family of Systems. It provides the warfighter commander the capability to rapidly employ small-area, short-duration, screening obscuration effects in the visual through near infrared (IR) spectrum (0.4.–1.2 micron range) during full-spectrum operations. The SOD–Vr is designed for use in restrictive terrain (i.e. urban structures, subterranean locations, and caves). The SOD–Vr degrades proper operation and performance of enemy battlefield weapon systems and enhances friendly capabilities, providing a less hazardous alternative to current non-colored smoke and incendiary hand grenades. The obscurant grenades in the current inventory subject the warfighter to several hazards (i.e. asphyxiation, carcinogen, and fire hazards) if used in restricted terrain. The SOD–Vr is less toxic than current grenades because the fill is non-combustible and non-burning.

SYSTEM INTERDEPENDENCIES

None

PROGRAM STATUS

- **FY09:** Full-rate production decision and fielding

PROJECTED ACTIVITIES

- **FY09:** Continue fielding

ACQUISITION PHASE

Technology Development | Engineering & Manufacturing Development | Production & Deployment | Operations & Support

Screening Obscuration Device (SOD)–Visual Restricted (Vr)

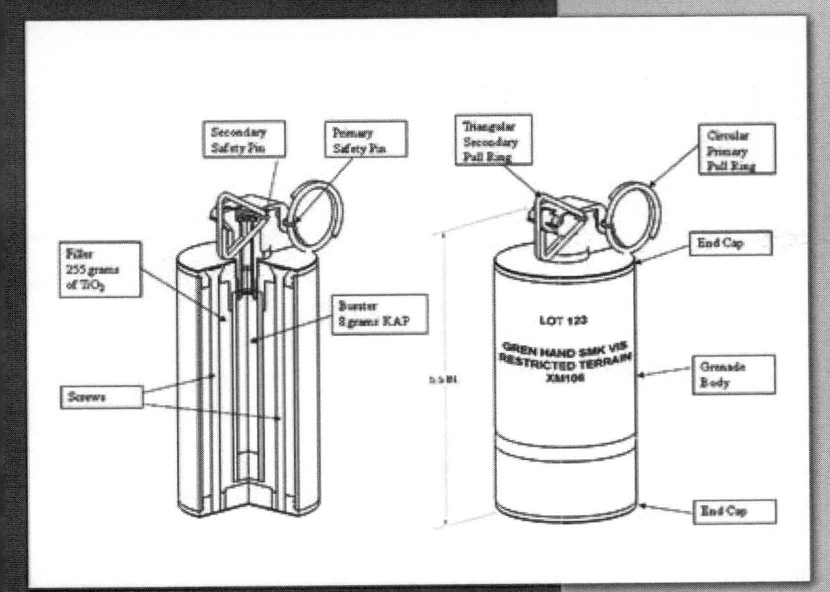

FOREIGN MILITARY SALES
None

CONTRACTORS
Pine Bluff Arsenal (Pine Bluff, AR)

Secure Mobile Anti-Jam Reliable Tactical–Terminal (SMART–T)

INVESTMENT COMPONENT
Modernization
Recapitalization
Maintenance

MISSION

To achieve end-to-end Beyond-Line-of-Sight (BLOS) connectivity that meets joint requirements for command, control, communications, computers, and intelligence protective requirements.

DESCRIPTION

The Secure Mobile Anti-Jam Reliable Tactical–Terminal (SMART–T) is a mobile military satellite communication terminal mounted on a standard High Mobility Multipurpose Wheeled Vehicle (HMMWV). The SMART–T extends the range of current and future tactical communications networks through DoD Milstar communication satellites. The SMART–T's maximum rate for data and voice communications is 1.544 million bits per second (Mbps). It provides the Army with its only protected (anti-jam and low probability of intercept) wideband satellite communication capability. Production is under way to upgrade terminals to communicate with DoD Advanced Extremely High Frequency (AEHF) satellites at a maximum data rate of 8.192 Mbps. The AEHF satellite system will dramatically increase the Army's end-to-end communications throughput capability. The AEHF upgrade to the SMART–T will include up-armoring in compliance with recent changes in Army policy for the deployment of personnel carriers.

SYSTEM INTERDEPENDENCIES

The SMART–T communicates with Milstar military communication satellites, and is being upgraded to communicate with AEHF communication satellites.

PROGRAM STATUS

- **2QFY07–2QFY09:** Fielded 62 SMART–T medium data rate terminals to Army units
- **2QFY07–2QFY09:** Deployed up to 39 SMART–Ts in Southwest Asia; received positive feedback
- **4QFY08:** Awarded up-armor contract change for AEHF upgrade kits production
- **1QFY09:** Initial delivery of the AEHF upgrade kit production for EHF SMART–Ts

PROJECTED ACTIVITIES

- **4QFY09:** Complete fielding of 239 SMART–Ts, including 61 to Army National Guard and Army Reserve Units
- **2QFY10:** Begin installation of AEHF upgrade kits to all EHF SMART–Ts
- **2QFY10:** Award contract to procure 39 AEHF SMART–Ts
- **3QFY10:** Begin fielding AEHF SMART–Ts to operational Army units

ACQUISITION PHASE
Technology Development | Engineering & Manufacturing Development | Production & Deployment | Operations & Support

Secure Mobile Anti-Jam Reliable Tactical–Terminal (SMART–T)

FOREIGN MILITARY SALES
None

CONTRACTORS
AEHF Production:
Raytheon (Largo, FL)
Engineering support:
Lincoln Labs (Lexington, MA)
Hardware:
Teledyne (Lewisburg, TN)
Martin Diesel (Defiance, OH)
Administrative/technical support:
JANUS Research (Eatontown, NJ)
Booz Allen Hamilton (Eatontown, NJ)

Sentinel

INVESTMENT COMPONENT

Modernization

Recapitalization

Maintenance

MISSION

To provide critical air surveillance by automatically detecting, tracking, classifying, identifying, and reporting targets to air defense weapons systems and battlefield commanders.

DESCRIPTION

Sentinel is used with the Army's Forward Area Air Defense Command and Control (FAAD C2) system and provides key target data to Stinger-based weapon systems and battlefield commanders via FAAD C2 or directly, using an Enhanced Position Location Reporting System (EPLRS) or the Single Channel Ground and Airborne Radio System (SINCGARS).

Sentinel consists of the M1097A1 High Mobility Multipurpose Wheeled Vehicle (HMMWV), the antenna transceiver group mounted on a high-mobility trailer, the identification friend-or-foe system (IFF), and the FAAD C2 interface. The sensor is an advanced three-dimensional battlefield X-band air defense phased-array radar with a 40-kilometer range.

Sentinel can operate day and night, in adverse weather conditions, and in battlefield environments of dust, smoke, aerosols, and enemy countermeasures. It provides 360-degree azimuth coverage for acquisition and tracking of targets (cruise missiles, unmanned aerial vehicles, rotary and fixed wing aircraft) moving at supersonic to hovering speeds and at positions from the map of the earth to the maximum engagement altitude of short-range air defense weapons. Sentinel detects targets before they can engage, thus improving air defense weapon reaction time and allowing engagement at optimum ranges. Sentinel's integrated IFF system reduces the potential for engagement of friendly aircraft.

Sentinel modernization efforts include enhanced target range and classification upgrades to engage non-line-of-sight (NLOS) targets; increased detection and acquisition range of targets; enhanced situational awareness; and classification of cruise missiles. The system provides integrated air tracks with classification and recognition of platforms that give an integrated air and cruise missile defense solution for the Air and Missile Defense System of Systems Increment 1 architecture and subsequent increments. Sentinel provides critical air surveillance of the National Capital Region and other areas as part of ongoing homeland defense efforts, and is a component of the counter rocket, artillery, and mortar (C-RAM) batteries in the area of responsibility.

SYSTEM INTERDEPENDENCIES

Forward Area Air Defense (FAAD) Command and Control (C2), Surface Launched Advanced Medium Range Air to Air Missile (SLAMRAAM)

PROGRAM STATUS

- **4QFY09:** First production of 78 Improved Sentinels completed and delivered

PROJECTED ACTIVITIES

- **2QFY09:** Contract award for 14 Improved Sentinels

ACQUISITION PHASE

Technology Development | Engineering & Manufacturing Development | Production & Deployment | Operations & Support

Sentinel

FOREIGN MILITARY SALES
Egypt, Turkish Air Force, Turkish Land Forces, Lithuania

CONTRACTORS
Thales Raytheon Systems (Fullerton, CA; El Paso, TX; Forest, MS; Largo, FL)
CAS, Inc. (Huntsville, AL)

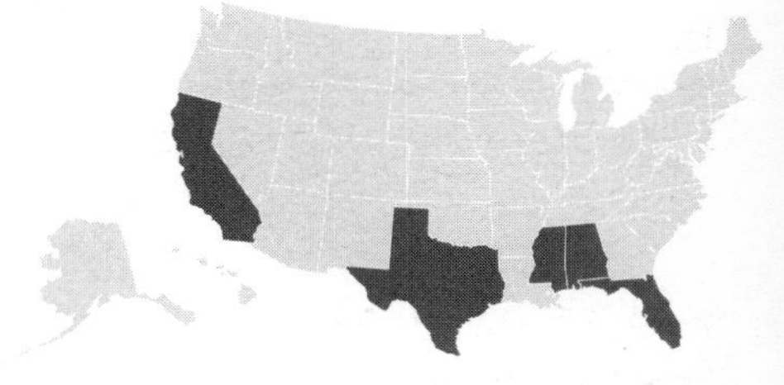

Single Channel Ground and Airborne Radio System (SINCGARS)

INVESTMENT COMPONENT

Modernization

Recapitalization

Maintenance

MISSION

To provide joint commanders with a highly reliable, low-cost, secure, and easily maintained Combat NET Radio (CNR) that has both voice and data handling capability in support of tactical command and control operations.

DESCRIPTION

The Single Channel Ground and Airborne Radio System (SINCGARS) Advanced SINCGARS System Improvement Program (ASIP) radio is the DoD/Army multi-service fielded solution for voice communication for platoon level and above, operating over the 30.000 to 87.975MHz frequency range. This radio provides the capability of establishing two-way communications (including jam-resistance) using the SINCGARS waveform and provides multimode voice and data communications supporting ground, air-to-ground, and ground-to-air line-of-sight communications links. The ASIP radio is the newer version of the SINCGARS radio. It is smaller than the SIP and weighs significantly less, while still maintaining all the functionalities of the SIP for backward compatibility.

Enhancements include the Embedded global positioning system (GPS) Receiver (EGR) and the radio based combat identification/radio based situational awareness (RBCI/RBSA) capability, which provides the warfighter with enhanced situational awareness and identification of friendly forces in targeted areas. RBCI serves as a gap filler for combat identification providing an interrogation/responder capability to satisfy the air-to-ground positive identification of platforms prior to release of weapons to prevent fratricide. RBSA adds a radio beaconing capability for every ASIP-equipped platform to enhance the Blue Force situational awareness picture. The Internet Controller enhancements add improved addressing capabilities in support of tactical internet enhancements being provided by Joint Battle Command–Platform for joint interoperability. Crypto modernization is a programmable communications security (COMSEC) capability for SINCGARS that will allow the radios to continue to provide secure communications to the secret and top secret level of security.

SYSTEM INTERDEPENDENCIES

None

PROGRAM STATUS

- **3QFY08–1QFY09:** Continue to field in accordance with Headquarters Depart of the Army guidance to support the Army Campaign Plan; National Guard, Army Reserve, and Active Army, Operation Enduring Freedom/Operation Iraqi Freedom requirements and urgent Operational Needs Statement
- **3QFY09:** Competitive contract award for procurement of SINCGARS radios to meet approved acquisition objective (AAO) requirement in FY09

PROJECTED ACTIVITIES

- **3QFY09:** Procurement of SINCGARS radios to meet AAO requirement in FY09.
- **2QFY09–4QFY13:** Fielding of SINCGARS to the Global War on Terrorism (GWOT) and Army campaign plan for transformation of the "Total-Army" modular force through FY13

ACQUISITION PHASE

Technology Development | Engineering & Manufacturing Development | Production & Deployment | Operations & Support

Single Channel Ground and Airborne Radio System (SINCGARS)

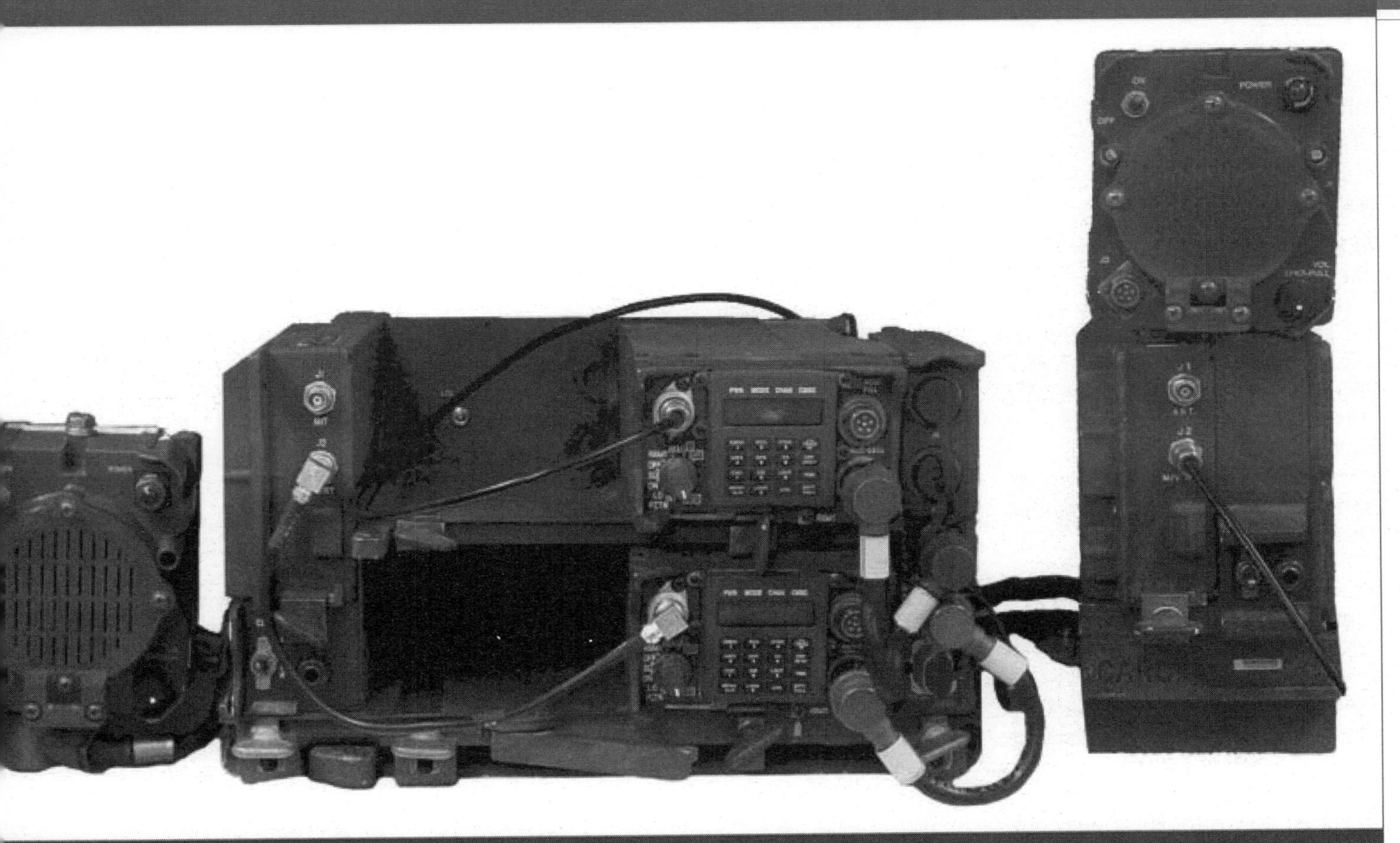

FOREIGN MILITARY SALES

Australia, Bahrain, Croatia, Egypt, Estonia, Finland, Georgia, Greece, Hungary, Ireland, Italy, Korea, Kuwait, Morocco, New Zealand, Portugal, Saudi Arabia, SHAPE Tech Center, Slovakia, Taiwan, Thailand, Ukraine, Uzbekistan, Zimbabwe

CONTRACTORS

Radio design/production:
ITT (Fort Wayne, IN)
Hardware Installation Kits:
UNICOR (Washington, DC)
Engineering Support and Testing:
ITT (Clifton, NJ)
Total Package Fielding:
CACI (Eatontown, NJ)

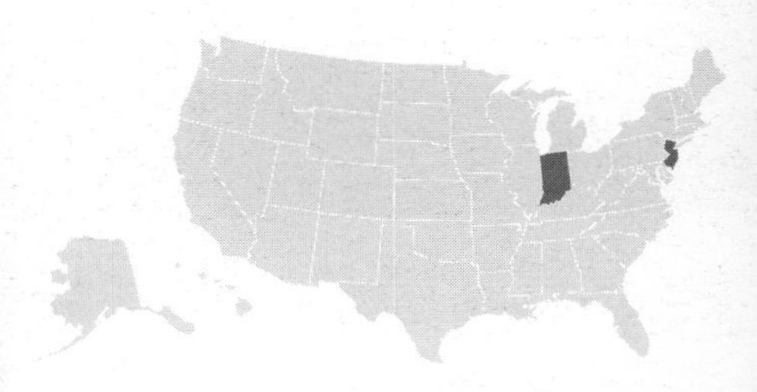

Small Arms–Crew Served Weapons

INVESTMENT COMPONENT

Modernization

Recapitalization

Maintenance

MISSION

To enable warfighters and small units to engage targets with lethal fire to defeat or deter adversaries.

DESCRIPTION

The M240B machine gun is the ground version of the original M240 machine gun. The M240H is used as a defensive armament for the UH-60 Black Hawk and CH-47 Chinook. A lighter weight M240L has been Type Classified for low-rate production, and will replace the M240B in Special Forces/ Ranger, Light Infantry, and Airborne dismounted units.

The M249 Squad Automatic Weapon (SAW) is a lightweight, gas-operated, one-man-portable automatic weapon that delivers substantial, effective fire at ranges out to 1,000 meters. Improved bipods, improved collapsible buttstocks, lightweight ground mounts, and improved combat optics have increased the combat effectiveness of this weapon.

The M2 is a .50 caliber, belt-fed, recoil-operated, air-cooled machine gun. The M2 is capable of single-shot and automatic fire and provides suppressive fire for offensive and defensive purposes against personnel and vehicular targets out to a maximum range of 2,000 yards. It is used primarily for vehicle mounted applications, but can also be tripod mounted for static defensive operations.

The MK19, Mod 3 Grenade Machine Gun is self-powered and air-cooled. It engages point targets up to 1,500 meters and provides suppressive fires up to 2,200 meters. It can be mounted on various tracked and wheeled vehicles, and on the M3 tripod for static defensive operations.

SYSTEM INTERDEPENDENCIES

None

PROGRAM STATUS

- **Current:** M240B/H in production and fielding
- **Current:** M240L Initial production phase
- **Current:** M249 in production and fielding
- **Current:** M2 in production and fielding
- **Current:** MK19 in Production and Fielding

PROJECTED ACTIVITIES

- **FY09:** Continue M240 machine gun production and fielding
- **FY09:** Continue MK19 production and fielding
- **FY09:** Continue M240B Production and fielding; Release Solicitation for a new competitive indefinite delivery, indefinite quantity contract
- **FY09:** Continue M240H production and fielding
- **FY09:** Award M240L low-rate production contract for initial fielding quantities
- **FY09:** Continue M249 production and fielding; release solicitation for a new competitive Indefinite Delivery, Indefinite Quantity contract
- **FY09:** Continue M2 production and fielding; award new competitive indefinite delivery, indefinite quantity contracts
- **FY09:** Continue MK19 production and fielding; conduct first article tests under the recently awarded indefinite delivery, indefinite quantity contracts

ACQUISITION PHASE

Technology Development | Engineering & Manufacturing Development | Production & Deployment | Operations & Support

Small Arms–Crew Served Weapons

FOREIGN MILITARY SALES

M240B Machine Gun: Romania, Afghanistan, Iraq, Panama, Poland, Saudi Arabia, Egypt, Colombia, and Kenya
M249 SAW: Afghanistan, Colombia, Iraq
MK19 Grenade Machine Gun: Canada, Colombia, Djibouti, Hungary, Lebanon, and Poland
M2: Poland

CONTRACTORS

M249 SAW:
Fabrique National Manufacturing, LLC (Columbia, SC)
M240B Machine Gun:
Fabrique National Manufacturing, LLC (Columbia, SC)
MK19 Grenade Machine Gun:
General Dynamics Armament and Technical Products (GDATP) Division (Saco, ME)
Alliant Techsystems, Inc. (Mesa, AZ)

Small Arms–Individual Weapons

INVESTMENT COMPONENT

Modernization

Recapitalization

Maintenance

MISSION

To enable warfighters and small units to engage targets with lethal fire to defeat or deter adversaries.

DESCRIPTION

The M4 Carbine is a compact version of the M16A2 rifle, with a collapsible stock, a flat-top upper receiver, accessory rails, and a detachable rear aperture sight assembly. The M4 achieves more than 85 percent commonality with the M16.

The M320 Grenade Launcher Module attaches to the M4 Carbine and M16A2/M16A4 rifle and fires all existing and improved 40mm low-velocity ammunition. It can also be configured as a standalone weapon.

The M26 Modular Accessory Shotgun System attaches to the M4 carbine rifles and fires all standard lethal, non-lethal, and door-breaching 12-gauge ammunition. It can also be configured as a stand alone weapon.

SYSTEM INTERDEPENDENCIES

None

PROGRAM STATUS

M4 Carbine:

- Army acquisition objective will be bought out in FY10

M320 Grenade Launcher Module:

- Milestone C, operational test, type classified standard, materiel released, full-rate production initiated.

M26 Modular Accessory Shotgun:

- Milestone C, limited user test completed

PROJECTED ACTIVITIES

M4 Carbine: continue production and fielding

M320 Grenade Launcher Module:

- **2QFY09:** First unit equipped

M26 Modular Accessory Shotgun System:

- **3QFY09:** Production qualification test and operational test
- **1QFY10:** Materiel release
- **2QFY10:** First-unit equipped

ACQUISITION PHASE

Technology Development | Engineering & Manufacturing Development | Production & Deployment | Operations & Support

FOREIGN MILITARY SALES

M4 Carbine: Afghanistan, Iraq, Colombia, Fiji, Philippines, Jordan, El Salvador, Panama, Bahrain, Tonga, Honduras, Belize, Suriname, and Kenya

CONTRACTORS

M4 Carbine:
Colt's Manufacturing (Hartford, CT)
M320 Grenade Launcher Module:
Heckler and Koch Defense Inc. (Ashburn, VA)
M26 Modular Accessory Shotgun System:
Vertu Corp. (Manassas, VA)

Small Caliber Ammunition

INVESTMENT COMPONENT
Modernization
Recapitalization
Maintenance

MISSION

To provide America's warfighters with the highest quality, most capable small caliber ammunition for training and combat.

DESCRIPTION

The Small Caliber Ammunition program consists of the following cartridges: 5.56mm, 7.62mm, 9mm, 10-gauge and 12-gauge shotgun, .22 caliber, .30 caliber, and .50 caliber. Small Caliber Ammunition supports the M9 pistol, M16A1/A2/A4 rifle, M4 carbine, M249 squad automatic weapon, M240 machine gun, .50-caliber M2 machine gun, sniper rifles, and a variety of shotguns. The .30 caliber blank cartridge supports veterans service organizations performing veterans' funeral honors.

SYSTEM INTERDEPENDENCIES

Small Caliber Ammunition is dependent on the weapons currently in use.

PROGRAM STATUS

- **Current:** In production

PROJECTED ACTIVITIES

- **FY10:** Produce and deliver 1.4 billion rounds (5.56mm, 7.62mm, and .50 caliber)
- **FY10:** 5.56mm M855 Lead Free Slug (LFS) replacement program initial production
- **FY10:** Lake City Army Ammunition Plant modernization program completes in FY12.

ACQUISITION PHASE
Technology Development | Engineering & Manufacturing Development | Production & Deployment | Operations & Support

6mm

3 M196 M855 M856 M200 M862

M995AP

7.62mm

M62 M80 M82 M118

M993AP

.50 Cal

M33 M17 M8 M20 M1A1 M903 SLAP

.30 Cal Blank

9mm M822

Small Caliber Ammunition

FOREIGN MILITARY SALES

5.56mm, 7.62mm, .50 caliber:
Afghanistan, Columbia, Czech Republic, El Salvador, France, Hungary, India, Iraq, Israel, Japan, Jordan, Kenya, Lebanon, Philippines, Singapore, Thailand, Tunisia, and Yemen

CONTRACTORS

Alliant Techsystems (Independence, MO)
General Dynamics Ordnance and Tactical Systems (St. Petersburg, FL)
Olin Corp. (East Alton, IL)
General Dynamics (Saint Marks, FL)
SNC Technologies (LeGardeur, Canada)

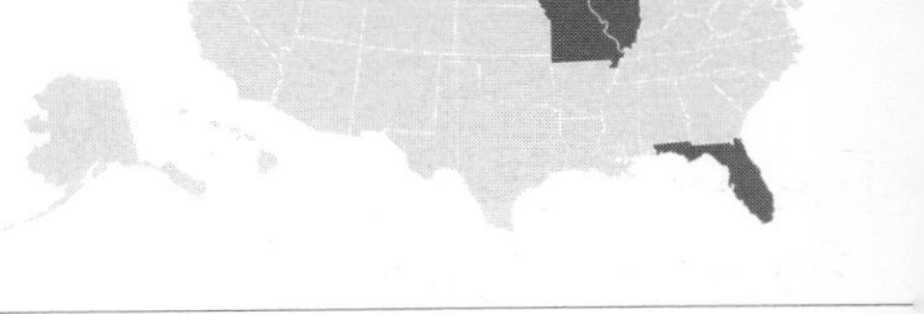

Sniper Systems

INVESTMENT COMPONENT

Modernization

Recapitalization

Maintenance

MISSION

To enable sniper teams to engage targets at extended ranges with lethal force to defeat or deter adversaries.

DESCRIPTION

Sniper systems consist of the following three weapons and their accessories:

- **M107 Semi-Automatic Long Range Sniper Rifle (LRSR)** is a commercial off-the-shelf anti-materiel and counter-sniper semi-automatic .50 caliber rifle. The rifle is a reliable, direct-line-of-sight weapon system, capable of delivering precise rapid fire on targets out to 2,000 meters.
- **M110 Semi-Automatic Sniper System (SASS)** is a commercial off-the-shelf, anti-personnel, 7.62mm semi-automatic sniper rifle that is also effective against light materiel targets. Capable of rapid fire and rapid reload, this suppressed sniper rifle exceeds the rate-of-fire and lethality of the M24 Sniper Weapon System. SASS anti-personnel ranges and accuracy are comparable to the M24. SASS includes an M151 enhanced sniper spotting scope.
- **M24 Sniper Weapon System (SWS)** is a 7.62mm bolt-action six-shot repeating rifle based on the Remington's Model 700. The system consists of the rifle, day optic site, iron sights, bipod, deployment kit, cleaning kit, soft rifle carrying case, optic case, system case, and operators manual. Components include the M144 spotting scope, laser filter, flash blast suppressor, and anti-reflection device. The primary round of ammunition is the M118 long-range cartridge. The maximum effective range is 800 meters.

SYSTEM INTERDEPENDENCIES

None

PROGRAM STATUS

- **M107:** Production completed; XM107 to M107 maintenance work order nearly complete
- **M110:** In production and fielding
- **M24:** Sustainment

PROJECTED ACTIVITIES

- **M107:** Procurements complete: upgrades and fielding near completion
- **M110:** Production and fielding ongoing
- **M24 SWS:** Sustain and turn-in upon M110 transition

ACQUISITION PHASE

Technology Development | Engineering & Manufacturing Development | Production & Deployment | Operations & Support

Sniper Systems

FOREIGN MILITARY SALES
M107: Thailand, Colombia, Kenya, Austria, Poland

CONTRACTORS
M107:
Barrett Firearms Manufacturing (Murfreesboro, TN)
M110:
Knight's Armaments Co. (Titusville, FL)
M24:
Remington (Ilion, NY)

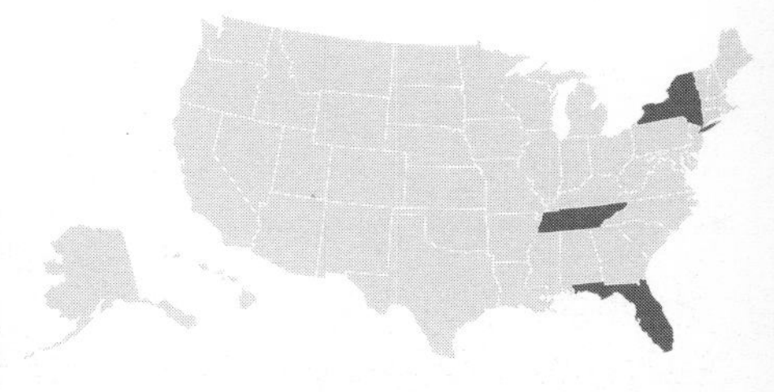

Spider

INVESTMENT COMPONENT

Modernization

Recapitalization

Maintenance

MISSION

To provide the commander with a new capability to shape the battlefield, protect the force, and respond to changing battlefield environments in a graduated manner while minimizing risk to friendly troops and non-combatants.

DESCRIPTION

Spider is a hand-emplaced, remotely-controlled, Man-In-The-Loop (MITL), anti-personnel munition system. Spider provides munition field effectiveness, but does so without residual life-threatening risks after hostilities end or when warring factions depart. The fielding of this system, with its sensors, communications, and munitions, changes the way Soldiers operate in an otherwise unpredictable battlefield. Each munition is controlled by a remotely stationed Soldier who monitors its sensors, allowing for more precise (non-lethal to lethal) responses—a significant advancement and advantage. The system's design allows for safe and rapid deployment, reinforcement, and recovery as well as safe passage of friendly forces. Spider eliminates the possibility of an unintended detonation through early warning and selective engagement of enemy forces. Spider is designed for storage, transport, rough handling, and use in worldwide military environments.

SYSTEM INTERDEPENDENCIES

Interface with Tactical Internet through Force XXI Battle Command Brigade-and-Below (FBCB2) and obstacle positioning through GPS.

PROGRAM STATUS

- **2QFY09:** Fielding of urgent materiel release (UMR) Hardware to Operation Enduring Freedom (OEF)
- **4QFY09:** Award contract modification to procure up to an additional 125 systems

PROJECTED ACTIVITIES

- **1QFY10:** Materiel release/type classification standard
- **1QFY10:** Full-rate production decision
- **3QFY10:** Full-rate production contract award

ACQUISITION PHASE

Technology Development | Engineering & Manufacturing Development | Production & Deployment | Operations & Support

Spider

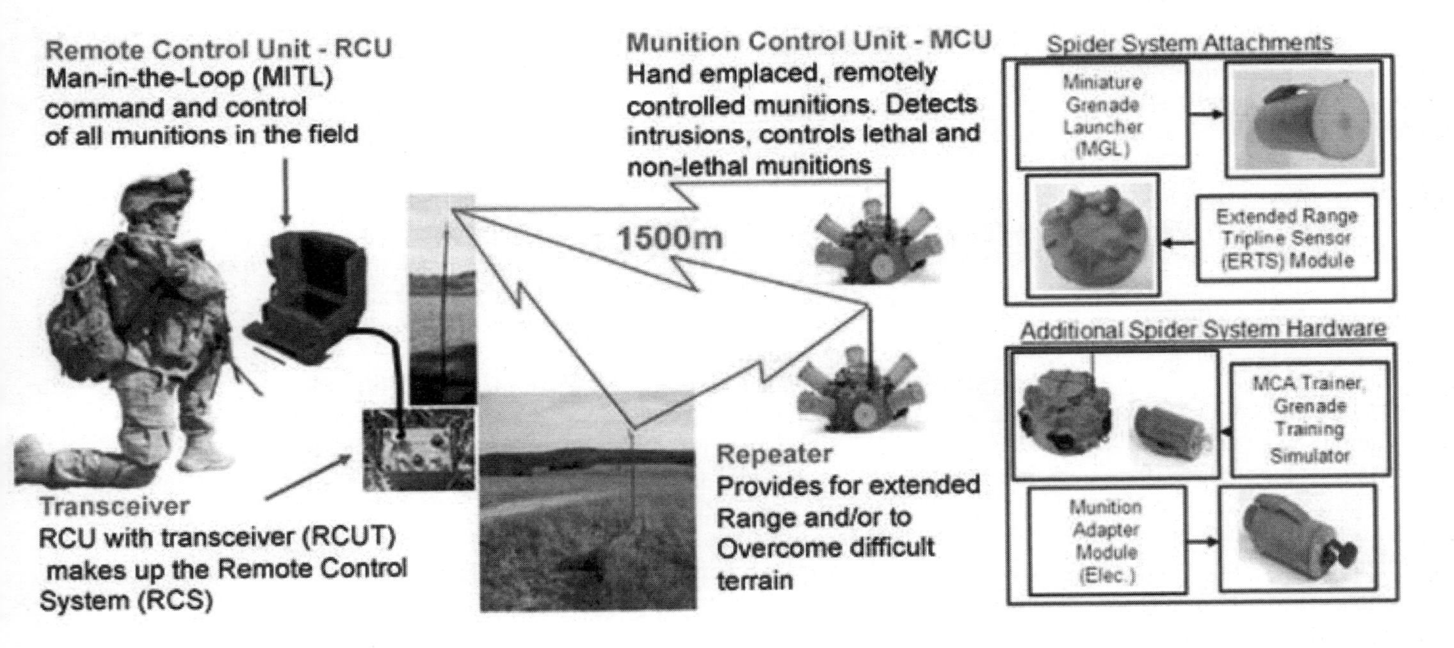

FOREIGN MILITARY SALES
None

CONTRACTORS
Prime:
Textron Defense Systems
(Wilmington, MA)
Alliant Techsystems (Plymouth, MN)
Subcontractors:
Alliant Techsystems (Rocket Center, WV)
BAE/Holston (Kingsport, TN)
American Ordnance (Milan, TN)

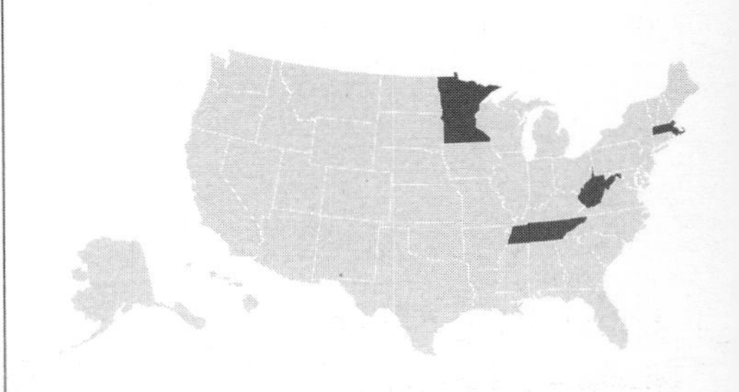

Stryker

INVESTMENT COMPONENT

Modernization

Recapitalization

Maintenance

MISSION

To enable the Army to respond immediately to urgent operational requirements anywhere in the world using readily deployable, combat-ready support vehicles capable of rapid movement.

DESCRIPTION

Stryker is a family of eight-wheeled armored vehicles that combine high battlefield mobility, firepower, survivability, and versatility with reduced logistics requirements. It includes two types of vehicles: the Infantry Carrier Vehicle (ICV) and the Mobile Gun System (MGS). The ICV, a troop transport vehicle, can carry nine infantry Soldiers, their equipment, and a crew of two: driver and vehicle commander. The MGS, designed to support infantry, has a 105mm turreted gun and autoloader system to breach bunkers and concrete walls.

Eight other configurations based on the ICV support combat capabilities: Reconnaissance Vehicle (RV), Mortar Carrier (MC), Commander's Vehicle (CV), Fire Support Vehicle (FSV) Engineer Squad Vehicle (ESV), Medical Evacuation Vehicle (MEV), Anti-Tank Guided Missile (ATGM) vehicle, Nuclear, Biological, and Chemical Reconnaissance Vehicle (NBCRV)

The ICV (excluding the MEV, ATGM, FSV, and RV) is armed with a remote weapons station supporting an M2 .50 caliber machine gun or MK19 automatic grenade launcher, the M6 grenade launcher, and a thermal weapons sight. Stryker supports a communications suite integrating the Single Channel Ground and Airborne Radio System (SINCGARS); Enhanced Position Location Reporting System (EPLRS); Force XXI Battle Command Brigade-and-Below (FBCB2); GPS; and high-frequency and near-term digital radio systems. In urban terrain, Stryker gives 360-degree protection against 14.5mm armor piercing threats. It is deployable by C-130 aircraft and combat-capable on arrival.

The Stryker program leverages non-developmental items with common subsystems and components to quickly field these systems. Stryker integrates government furnished materiel subsystems as necessary. Stryker stresses performance and commonality to reduce the logistics footprint and minimize costs. Since October 2003, Strykers in Iraq have logged over 24 million miles and maintained operational readiness above 90 percent. SBCTs require 332 Stryker vehicles. The current program requires more than 3,616 Strykers to field seven SBCTs and meet additional requirements. Funding has increased Stryker ready-to-fight fleet requirements to 150 Strykers, and Repair Cycle Floats to 322 Strykers.

SYSTEM INTERDEPENDENCIES

None

PROGRAM STATUS

- **4QFY08:** Acquisition Decision Memorandum approving additional limited-rate production of 62 MGS Strykers
- **1QFY09:** Completed fielding to all seven SBCTs with exception of NBCRV and MGS
- **1QFY09:** Configuration Steering Board addressing scope of requirements for MGS
- **4QFY09:** Configuration Steering Board to address Stryker modernization

PROJECTED ACTIVITIES

- **2QFY09–4QFY10:** NBCRV reliability growth testing
- **4QFY10:** NBCRV operational test and evaluation II
- **2QFY11:** NBCRV Milestone III for full-rate production

ACQUISITION PHASE

Technology Development

Engineering & Manufacturing Development

Production & Deployment

Operations & Support

Stryker

FOREIGN MILITARY SALES
None

CONTRACTORS
General Dynamics (Anniston, AL;
Sterling Heights, MI; Lima, OH)
Interstate Electronics (Anaheim, CA)
Composix (Newark, OH)
Mittal (Coatesville, PA)

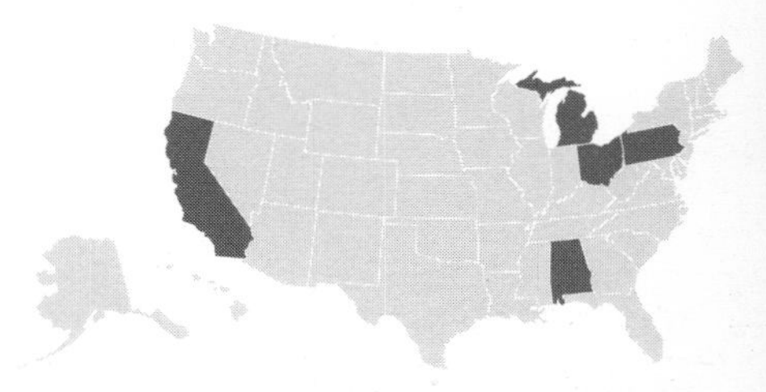

Surface Launched Advanced Medium Range Air-to-Air Missile (SLAMRAAM)

INVESTMENT COMPONENT

Modernization
Recapitalization
Maintenance

MISSION

To defeat aerial threats (Cruise Missiles, Unmanned Aerial Vehicles [UAVs], fixed and rotary wing aircraft) to current and future maneuver forces and critical assets; the missile will support a variety of missions at the tactical, operational, and strategic levels of warfare.

DESCRIPTION

The Surface Launched Advanced Medium Range Air-to-Air Missile (SLAMRAAM) will defend designated critical assets and maneuver forces against aerial threats. It is a key component of the Integrated Air and Missile Defense (IAMD) Composite Battalion and will replace the Avenger in the Army's Air and Missile Defense forces. SLAMRAAM is a lightweight, day-or-night, and adverse weather, non-line-of-sight system for countering cruise missiles and unmanned air vehicle threats with engagement capabilities in excess of 18 kilometers. The system is comprised of an Integrated Fire Control Station (IFCS) for command and control, integrated sensors, and missile launcher platforms. While SLAMRAAM uses Sentinel as its organic radar to provide surveillance and fire control data, the system will also receive data from other joint and Army external sensors when available. The SLAMRAAM launcher is a mobile platform with common joint launch rails, launcher electronics, on-board communication components, and four to six AIM-120-C Advanced Medium Range Air-to-Air Missiles (AMRAAMs).

SYSTEM INTERDEPENDENCIES

SLAMRAAM will provide engagement operations interoperability by participating on the Joint Data Network (JDN) (Link 16) Internal and External communication. The IFCS will manage engagement operations, and the Advanced Sentinel radar will be the initial sensor. In addition, SLAMRAAM will be supported by the Joint Land Attack Cruise Missile Defense Elevated Netted Sensor (JLENS) System.

PROGRAM STATUS

- **2QFY09:** Long lead decision
- **2QFY09:** Begin developmental testing/limited user testing

PROJECTED ACTIVITIES

- **1QFY10:** Completion of developmental testing/limited user testing
- **1QFY10:** Milestone C decision for low-rate initial production
- **1QFY10:** Low-rate initial production
- **1QFY11:** Begin initial operational test and evaluation (IOTE)
- **2QFY11** Complete IOTE

ACQUISITION PHASE

Technology Development | **Engineering & Manufacturing Development** | Production & Deployment | Operations & Support

Surface Launched Advanced Medium Range Air-to-Air Missile (SLAMRAAM)

FOREIGN MILITARY SALES
None

CONTRACTORS
Raytheon (Fullerton, CA; Andover, MA; Tewksbury, MA)
CAS, Inc. (Huntsville, AL)
Boeing (Huntsville, AL)

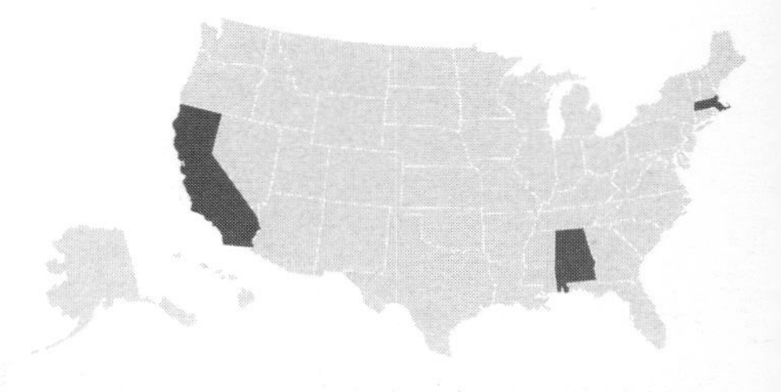

Shadow Tactical Unmanned Aerial Vehicle (TUAV)

INVESTMENT COMPONENT

Modernization

Recapitalization

Maintenance

MISSION

To provide the tactical maneuver commander near real-time reconnaissance, surveillance, target acquisition, and force protection during day/night and adverse weather conditions.

DESCRIPTION

The RQ-7B Shadow Tactical Unmanned Aerial Vehicle (TUAV) has a wingspan of 14 feet and a payload capacity of approximately 60 pounds; gross takeoff weight exceeds 380 pounds and endurance is more than six hours on-station at a distance of 50 kilometers. The system is compatible with the All Source Analysis System, Advanced Field Artillery Tactical Data System, Joint Surveillance Target Attack Radar System Common Ground Station, Joint Technical Architecture-Army, and the Defense Information Infrastructure Common Operating Environment. The One System Ground Control Station (OSGCS) is also the only joint-certified GCS in the DoD. The RQ-7B Shadow can be transported by six Air Force C-130 aircraft. Shadow is currently operational in both the Army and Marine Corps.

The RQ-7B Shadow configuration, fielded in platoon sets, consists of:

- Four air vehicles with electro-optical/infrared imaging payloads including infrared illuminators
- Two ground control station (GCS) shelters mounted on High Mobility Multipurpose Wheeled Vehicles (HMMWV) and their associated ground data terminals; one portable GCS and one portable ground data terminal
- Two air vehicle transport HMMWV, one of which tows a trailer-mounted hydraulic launcher
- Two HMMWV with trailers for operations/maintenance personnel and equipment transport
- One HMMWV with Maintenance Section Multifunctional (MSM) shelter and trailer
- One HMMWV with Mobile Maintenance Facility (MMF) shelter
- Two automatic Take-off and Landing Systems (TALS)
- Four One System Remote Video Terminals (OSRVT) and antennas

The Shadow is manned by a platoon of 22 soldiers and, typically, two contractors. The Soldier platoon consists of a platoon leader, platoon sergeant, UAV warrant officer, 12 Air Vehicle Operators (AVOs)/Mission Payload Operators (MPO), four electronic warfare repair personnel and three engine mechanics supporting launch and recovery. The MSM is manned by Soldiers who also transport spares and provide maintenance support. The MMF is manned by contractor personnel located with the Shadow platoon to provide logistics support to include "off system support" and "maintenance by repair."

The Shadow also has an early entry configuration of 15 Soldiers, one GCS, the air vehicle transport HMMWV, and the launcher trailer, which can be transported in three C-130s. All components can be slung under a CH-47 or CH-53 helicopter for transport.

SYSTEM INTERDEPENDENCIES

HMMWV, SINCGARS, MEP803A 10kW

PROGRAM STATUS

- **Current:** System is in the Production and Deployment phase
- **Current:** Since achieving Initial Operating Capability, the Shadow has flown more than 386,000 hours in support of combat operations in Operation Iraqi Freedom and Operation Enduring Freedom

PROJECTED ACTIVITIES

- **FY08–09:** Continue fielding Shadow platoons in support of Army Modularity, Integrate Laser Designation systems; develop and field numerous reliability improvements; develop and field a larger wing

ACQUISITION PHASE

Technology Development | Engineering & Manufacturing Development | Production & Deployment | Operations & Support

Shadow Tactical Unmanned Aerial Vehicle (TUAV)

FOREIGN MILITARY SALES
None

CONTRACTORS
Air Vehicle/Ground Data Terminal:
AAI Corp. (Hunt Valley, MD)
GCS, Portable GCS:
CMI (Huntsville, AL)
Auto-land system:
Sierra Nevada Corp. (Sparks, NV)
Ground Data Terminal Pedestal:
Tecom (Chatsworth, CA)
MMF/MSM shelter:
General Dynamics (Marion, VA)
Avionics:
Rockwell Collins (Warrenton, VA)
Ground Station Software:
CDL Systems (Calgary, Canada)
Aircraft Engine:
UAV Engines Limited (Shenstone, UK)

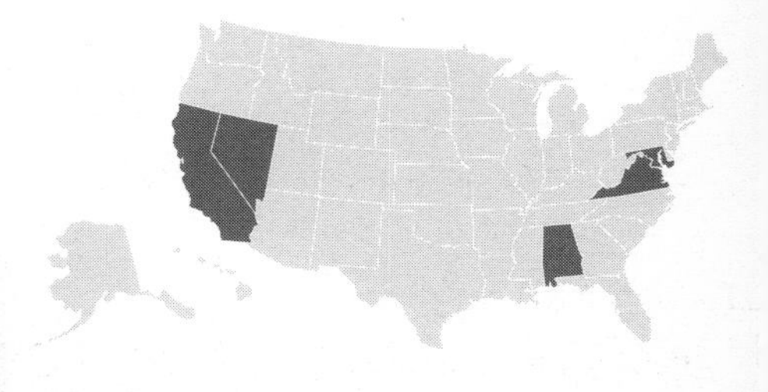

Tactical Electric Power (TEP)

INVESTMENT COMPONENT

Modernization

Recapitalization

Maintenance

MISSION

To provide modernized tactical electric power sources for all military services.

DESCRIPTION

The Tactical Electric Power (TEP) program consists of Small [2–3 kilowatt (kW)], Medium (5–60kW), and Large (100–840kW) electrical power generating systems, trailer mounted power units and power plants, and electrical distribution equipment that provide standardized power management solutions to all DoD agencies and numerous allied nations. The Project Manager-Mobile Electric Power systems:

- Maximize fuel efficiency
- Increase reliability (500–600 hours average time between failure), maintainability, and transportability through standardization
- Minimize weight and size while meeting all user requirements
- Operate at rated loads in all military environments
- Reduce infrared signature and noise (less than 70 A-weighted decibels at 7 meters)
- Are survivable in chemical, biological, and nuclear environments
- Meet power generation and conditioning standards in accordance with military standards (Mil-STD 1332)
- Provide quality electric power for command posts; command, control, communications, computers, intelligence, surveillance, and reconnaissance (C4ISR) systems; weapon systems; and other battlefield support equipment

SYSTEM INTERDEPENDENCIES

None

PROGRAM STATUS

- **FY08–09:** Production and fielding ongoing for 2 kilowatt Military Tactical Generator (MTG), 3kW, 5kW, 10kW, 15kW, 30kW, 60kW, 100kW and 200kW Tactical Quiet Generator (TQG)
- **FY08–09:** Continue assembly and fielding of power units and power plants (trailer-mounted generator sets)
- **FY08–09:** Advanced Medium Mobile Power Sources (AMMPS), the next generation of medium TEP sources, continues in Phase 2, System Demonstration
- **FY08–09:** Final Deployable Power Generation and Distribution System (DPGDS) Army "B" models delivered to the 249th Engineer Battalion (Prime Power)
- **FY08–09:** Production and fielding ongoing for Power Distribution Illumination System Electrical (PDISE)
- **FY08–09:** Fielded the 15kW "B" model TQG

PROJECTED ACTIVITIES

- **FY09–10:** Continue production and fielding of PDISE.
- **FY09–10:** Continue production and fielding of MTG/TQG generator sets, power units, and power plants
- **2QFY09:** First delivery of AMMPS pre-production test models
- **3QFY09:** AMMPS Developmental and Operational Testing
- **3QFY10:** AMMPS Milestone C and full-rate production contract
- **3QFY09:** AMMPS Milestone C and full-rate production contract

ACQUISITION PHASE

Technology Development | Engineering & Manufacturing Development | Production & Deployment | Operations & Support

Tactical Electric Power (TEP)

2kW Military Tactical Generator

3kW Tactical Quiet Generator

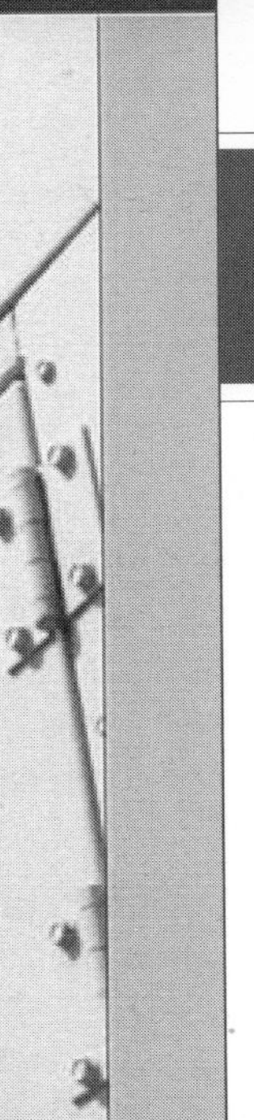

5kW Tactical Quiet Generator

10kW Tactical Quiet Generators

15kW Tactical Quiet Generator

30kW Tactical Quiet Generator

60kW Tactical Quiet Generator

100kW Tactical Quiet Generator

840kW Deployable Power Generation and Distribution System

FOREIGN MILITARY SALES
Tactical quiet generators have been purchased by Egypt, Israel, Korea, Kuwait, Saudi Arabia, Turkey, United Arab Emirates, and 11 other countries.

CONTRACTORS
3kW, 5kW, 10kW, 15kW, 100kW and 200kW TQG:
DRS Fermont (Bridgeport, CT)
30kW, 60kW TQG:
L-3 Westwood (Tulsa, OK)
2 kilowatt MTG:
Dewey Electronics (Oakland, NJ)
DPGDS:
DRS Technical Solutions (Herndon, VA)
PDISE:
Fidelity Technologies Corp. (Reading, PA)
Trailers for power units and power plants:
Schutt Industries (Clintonville, WI)
Advanced Medium Mobile Power Sources 5–60kW:
Cummins Power Generation (Minneapolis, MN)

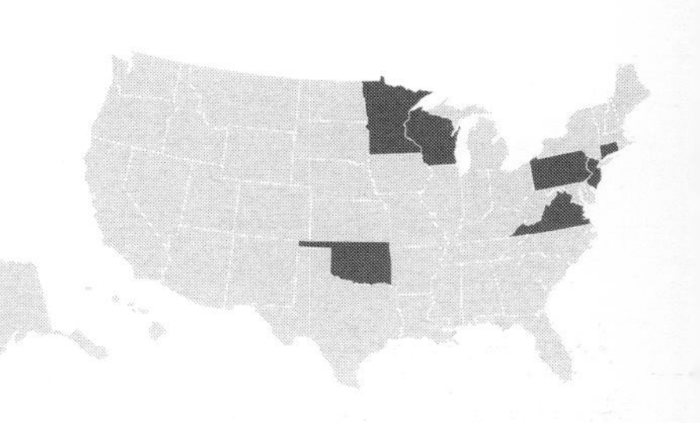

Tank Ammunition

INVESTMENT COMPONENT

- **Modernization**
- Recapitalization
- Maintenance

MISSION

To provide overwhelming lethality overmatch in direct-fire tank ammunition for use in current ground combat weapons platforms.

DESCRIPTION

The current 120mm family of tactical tank ammunition consists of fourth generation kinetic energy, multipurpose, and canister ammunition. Kinetic Energy ammunition lethality is optimized by firing a maximum-weight sub-caliber projectile at the greatest velocity possible. The M829A3 is the only kinetic energy cartridge currently in production for U.S. use, while the older M829A1 and M829A2 remain in inventory. For foreign sales, the Kinetic Energy–Tungsten (KE–W) munition is in production. Multipurpose ammunition uses a high-explosive warhead to provide blast, armor penetration, and fragmentation effects. There are three high-explosive cartridges in the current inventory: M830A1, M830, and M908 Obstacle Reduction. The M830A1 completed production in FY09. The shotgun shell-like M1028 canister cartridge provides the Abrams tank with effective, rapid, lethal fire against massed assaulting infantry and is also used in training. The 120mm family has two dedicated training cartridges in production: M865 and M1002. The M831A1, a dedicated 120mm training cartridge, is also in inventory.

To support the Stryker force, the 105mm Mobile Gun System uses the new M393A3 high-explosive and M1040 canister cartridges. The M393A3 cartridge destroys hardened enemy bunkers and creates openings through which infantry can pass. The M1040 canister cartridge provides rapid, lethal fire against massed assaulting infantry at close range and is also used in training. Also in production is the M467A1 training cartridge, which is ballistically matched to the M393A3 cartridge. Tactical rounds in inventory include the M900 kinetic energy (KE) and the M456A2 High Explosive Anti-Tank (HEAT) cartridges. Training rounds in inventory include the M724 and M490A1 cartridges.

SYSTEM INTERDEPENDENCIES

The Abrams Main Battle Tank can fire 120mm ammunition; The Stryker Mobile Gun System can fire 105mm ammunition.

PROGRAM STATUS

- **FY08:** M829A3, M830, M830A1, M1002 and M908, M1028, M1040, M393A3, M467A1 are all fielded

PROJECTED ACTIVITIES

- **FY10:** The M829A3, M830A1, M467A1, and M1028 will no longer be in production

ACQUISITION PHASE

Technology Development | Engineering & Manufacturing Development | **Production & Deployment** | Operations & Support

Tank Ammunition

FOREIGN MILITARY SALES

M831A1 and M865: Iraq

CONTRACTORS

M830A1, M1002, M1028, M865, and M467A1:
Alliant Techsystems (Plymouth, MN)

M1002, M865, and KEW:
General Dynamics Ordnance and Tactical Systems (St. Petersburg, FL)

M393A3 and M1040:
L-3 Communications (Lancaster, PA)

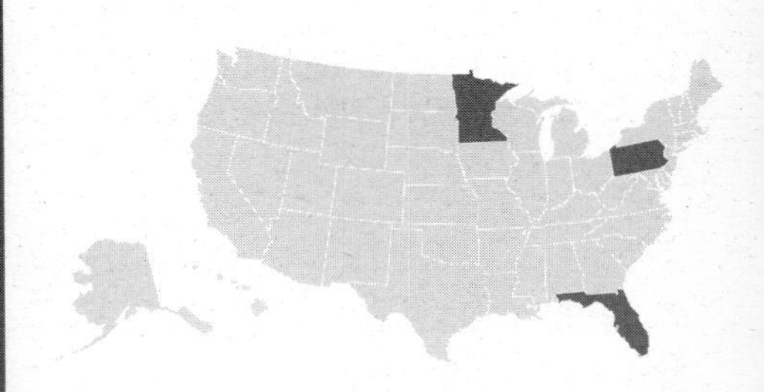

Test Equipment Modernization (TEMOD)

INVESTMENT COMPONENT
- **Modernization**
- Recapitalization
- Maintenance

MISSION

To improve readiness of Army weapon systems and minimize general purpose electronic test equipment proliferation and obsolescence, thereby reducing operations and support costs.

DESCRIPTION

The Test Equipment Modernization (TEMOD) program procures general purpose electronic test equipment that is essential to the continuing support of weapon systems required by Current/Future Forces. Acquisitions are commercial items that have significant impact on readiness, power projection, safety, and training operations of the United States Army, Army Reserve, and National Guard. The TEMOD program has procured 38 products replacing over 334 models. Projects are prioritized as a result of an annual working group composed of the materiel developer, the combat developer, and the user community.

Radar Test Set Identification Friend or Foe Mode 5 Upgrade Kit and Radar Test Set with Mode 5 (TS-4530A/UPM) This option is used to perform pre-flight checks of aviation and missile transponders and interrogators to alleviate potential fratricide concerns. It is also required to ensure Army aircraft are in compliance with European and Federal Aviation Administration mandates. Mode 5 encryption is required as a result of the National Security Administration's decertification of Mode 4 encryption.

Signal Generators, low frequency (SG-1364/U) and high frequency (SG-1366/U) These are signal sources to test electronic receivers and transmitters of all types throughout the Army and provide standards to compare signals. They generate a known signal into radios to test receiver sensitivity and ensure that battlefield commanders can communicate in adverse conditions. The signal generators will be integrated into aviation facilities, systems particular to ground support missiles, and special weapons facilities.

Telecommunication System Test Set (TS-4544/U) This test set analyzes signal quality between communication systems to ensure accurate data exchange. It measures and displays various bit-data information as related to digital transmission.

SYSTEM INTERDEPENDENCIES

None

PROGRAM STATUS

Low Frequency Signal Generator
- **3QFY08:** Contract awarded
- **1QFY09:** Low-rate initial production (LRIP)

Radar Test Set with Friend or Foe Mode 5
- **3QFY08:** Issued request for proposals
- **2QFY09:** Contract awarded

PROJECTED ACTIVITIES

Low Frequency Signal Generator
- **2QFY09:** Product verification testing (PVT)
- **1QFY10:** Full-rate production (FRP)

Radar Test Set with Friend or Foe Mode 5
- **2QFY10:** LRIP
- **2QFY10:** PVT
- **4QFY10:** FRP

High Frequency Signal Generator
- **2QFY09:** Issue Letter Request for Bid Samples
- **2QFY10:** Contract Award
- **3QFY10:** LRIP and PVT
- **4QFY11:** FRP

Telecommunication System Test Set
- **4QFY09:** Issue Letter Request for Bid Samples
- **3QFY10:** Contract award
- **1QFY11:** LRIP

ACQUISITION PHASE

Technology Development	Engineering & Manufacturing Development	**Production & Deployment**	Operations & Support

Test Equipment Modernization (TEMOD)

TS-4530 Operation

TS-4530
Radar Test Set

FOREIGN MILITARY SALES

Radar Test Set with Friend or Foe Mode 5:
United Kingdom, Greece, Singapore, Kuwait, Hungary, Azerbaijan, Portugal, Netherlands, Norway, Saudi Arabia

CONTRACTORS

Low Frequency Signal Generator:
Rohde & Schwarz (Columbia, MD)
High Frequency Signal Generator:
To be determined
Telecommunication System Test Set:
To be determined
Radar Test Set with Friend or Foe Mode 5:
To be determined

SG-1364U Signal Generator Operation

SG-1364U
Signal Generator

Thermal Weapon Sight

INVESTMENT COMPONENT

- Modernization
- Recapitalization
- Maintenance

MISSION

To enable combat forces to acquire and engage targets with small arms during day, night, obscurant, no-light, and adverse weather conditions.

DESCRIPTION

The AN/PAS-13 Thermal Weapon Sight (TWS) Generation II family enables individual and crew-served gunners to see deep into the battlefield, to increase situational awareness and target acquisition range, and to penetrate obscurants, day or night. TWS II systems use forward-looking infrared (FLIR) technology and provide a standard video output for training, image transfer or remote viewing. TWS II systems are silent, lightweight, compact, durable, battery-powered thermal sights powered by commercial Lithium AA batteries. TWS II systems offer a minimum 20 percent longer range at roughly two-thirds the weight and with 50 percent power savings over the legacy TWS systems.

The TWS family comprises three variants:

- AN/PAS-13(V)1 Light Weapon Thermal Sight (LWTS) for the M16 and M4 series rifles and carbines as well as the M136 Light Anti-Armor Weapon.
- AN/PAS-13(V)2 Medium Weapon Thermal Sight (MWTS) for the M249 and M240B series medium machine guns.
- AN/PAS-13(V)3 Heavy Weapon Thermal Sight (HWTS) for the squad leaders weapon M16 and M4 series rifles and carbine, M24 and M107 sniper rifles, M2 HB machine gun and MK19 grenade machine gun.

SYSTEM INTERDEPENDENCIES

M2, M4, M16, M249. M240B, MK19, M24, M107

PROGRAM STATUS

- **Current:** TWS II in production and being fielded

PROJECTED ACTIVITIES

- **Continue:** Production and fielding in accordance with Headquarters, Department of the Army (HQDA) G8 priorities

ACQUISITION PHASE

Technology Development | Engineering & Manufacturing Development | Production & Deployment | Operations & Support

Thermal Weapon Sight

FOREIGN MILITARY SALES
None

CONTRACTORS
BAE Systems (Lexington, MA)
DRS Technologies (Melbourne, FL; Dallas, TX)
Raytheon (McKinney, TX)

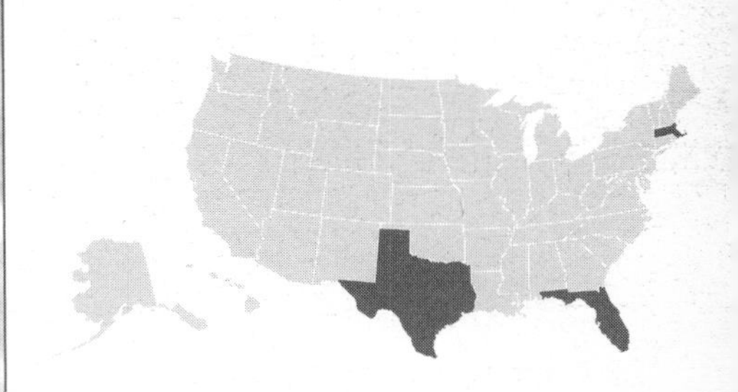

Transportation Coordinators' Automated Information for Movement System II (TC-AIMS II)

INVESTMENT COMPONENT

- Modernization
- Recapitalization
- Maintenance

MISSION

To facilitate movement management and control of personnel, equipment, and supplies from a home station to a theater of operations and back; to provide in-theater support for onward movement, sustainment planning requirements, and source in transit visibility data.

DESCRIPTION

The Transportation Coordinators' Automated Information for Movement System II (TC-AIMS II) is a service migration system. Characteristics include: source feeder system to Joint Force Requirements Generation II, Joint Planning and Execution System, Global Transportation Network, and Services' command and control systems; common user interface to facilitate multi-service user training and operations; commercial off-the-shelf hardware/software architecture; net-centric implementation with breakaway client-server and/or stand alone/workgroup configurations; incremental, block upgrade developmental strategy.

SYSTEM INTERDEPENDENCIES

None

PROGRAM STATUS

- **2QFY07–1QFY09:** Continued Block 2 fielding
- **2QFY07–1QFY09:** Completed development of Block 3, which provides combatant commanders a reception, staging, onward movement, and integration capability, directly supporting in-theater transportation movement activities
- **2QFY07–1QFY09:** Successful test of Block 3, favorable milestone decision to field Block 3
- **2QFY07–1QFY09:** Began fielding of Block 3

PROJECTED ACTIVITIES

- **2QFY09–2QFY11:** Complete fielding Block 2 and Block 3

ACQUISITION PHASE

Technology Development | Engineering & Manufacturing Development | Production & Deployment | Operations & Support

Transportation Coordinators' Automated Information for Movement System II (TC-AIMS II)

FOREIGN MILITARY SALES
None

CONTRACTORS
Systems integration:
Engineering Research and Development Command (Vicksburg, MS)
Future Research Corp. (Huntsville, AL)
Apptricity Corp. (Dallas, TX)
Program support:
L-3 Communications (Titan Group) (Newington, VA)

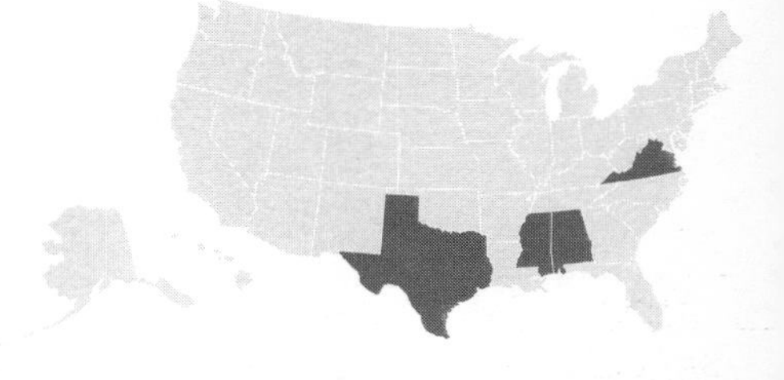

Tube-Launched, Optically-Tracked, Wire-Guided (TOW) Missiles

INVESTMENT COMPONENT

Modernization

Recapitalization

Maintenance

MISSION

To provide long-range, heavy anti-tank and precision assault fire capabilities to Army and Marine forces.

DESCRIPTION

The Close Combat Missile System–Heavy (CCMS–H) TOW (Tube-Launched, Optically-Tracked, Wire-Guided) is a heavy anti-tank/precision assault weapon system, consisting of a launcher and a missile. The missile is six inches in diameter (encased, 8.6 inches), and 49 inches long. The gunner defines the aim point by maintaining the sight cross hairs on the target. The launcher automatically steers the missile along the line-of-sight toward the aim point via a pair of control wires or a one-way radio frequency (RF) link, which links the launcher and the missile.

TOW missiles are employed on the High Mobility Multipurpose Wheeled Vehicle (HMMWV)-mounted Improved Target Acquisition System (ITAS), HMMWV-mounted M220A4 launcher (TOW 2), Stryker Anti-Tank Guided Missile (ATGM) Vehicles, and Bradley Fighting Vehicles (A2/A2ODS/A2OIF/A3) within the Infantry, Stryker, and Heavy Brigade Combat Teams respectively. TOW missiles are also employed on the Marine HMMWV-mounted ITAS, HMMWV-mounted M220A4 launcher (TOW 2), LAV–ATGM Vehicle, and AH1W Cobra attack helicopter. TOW is also employed by allied nations on a variety of ground and airborne platforms.

The TOW 2B Aero is the most modern and capable missile in the TOW family, with an extended maximum range to 4,500 meters. The TOW 2B Aero has an advanced counter active protection system capability and defeats all current and projected threat armor systems. The TOW 2B Aero flies over the target (offset above the gunner's aim point) and uses a laser profilometer and magnetic sensor to detect and fire two downward-directed, explosively-formed penetrator warheads into the target. The TOW 2B Aero's missile weight is 49.8 pounds (encased, 65 pounds).

The TOW Bunker Buster is optimized for performance against urban structures, earthen bunkers, field fortifications, and light-skinned armor threats. The missile impact is at the aim point. It has a 6.25 pound, 6-inch diameter high-explosive, bulk-charge warhead, and its missile weighs 45.2 pounds. The TOW BB has an impact sensor (crush switch) located in the main-charge ogive and a pyrotechnic detonation delay to enhance warhead effectiveness. The PBXN-109 explosive is housed in a thick casing for maximum performance. The TOW BB can produce a 21- to 24-inch diameter hole in an 8-inch thick, double-reinforced concrete wall at a range of 65 to 3,750 meters.

SYSTEM INTERDEPENDENCIES

M1121/1167 HMMWV, Stryker ATGM, ITAS

PROGRAM STATUS

TOW 2B Aero and Bunker Buster (BB)

- **2QFY10:** TOW 2B Aero RF and TOW BB production award
- **2QFY10–2QFY12:** TOW 2B Aero RF and TOW BB production deliveries
- **3QFY05:** Awarded contract option for TOW Bunker Buster production for Army and U.S. Marine Corps
- **4QFY06:** Awarded FY06–09 multi-year contract for TOW Missile production

PROJECTED ACTIVITIES

- **Continue:** Production of TOW 2B Aero and Bunker Buster
- **2QFY11:** TOW 2B Aero RF and TOW BB production award

ACQUISITION PHASE

Technology Development | Engineering & Manufacturing Development | Production & Deployment | Operations & Support

Tube-Launched, Optically-Tracked, Wire-Guided (TOW) Missiles

FOREIGN MILITARY SALES

The TOW weapon system has been sold to more than 43 allied nations over the life of the system

CONTRACTORS

TOW 2B Aero and TOW BB

Prime:
Raytheon Missile Systems (Tucson, AZ)

Control Actuator, Shutter Actuator:
Moog (Salt Lake City, UT)

Warheads:
Aerojet General (Socorro, NM)

Gyroscope:
BAE Systems (Cheshire, CT)

Sensor (TOW 2B only):
Thales (Basingstoke, UK)

Launch Motor:
ATK (Radford, VA)

Flight Motor:
ATK (Rocket Center, WV)

Machined/Fabricated Parts:
Klune (Spanish Fork, UT)

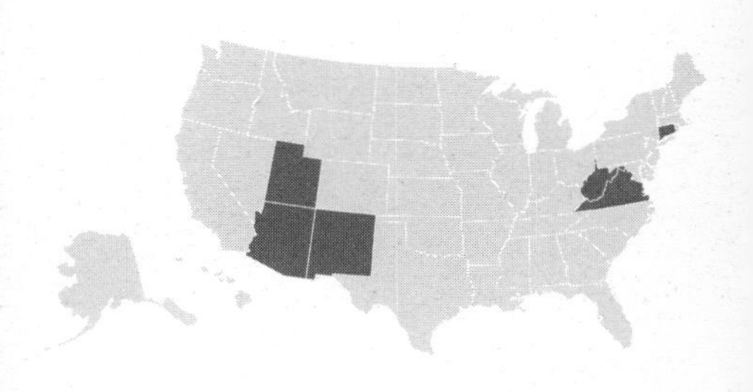

Unit Water Pod System (Camel)

INVESTMENT COMPONENT
Modernization
Recapitalization
Maintenance

MISSION

To provide the Army with the capability to receive, store, and dispense potable water to units at all echelons throughout the battlefield.

DESCRIPTION

The Unit Water Pod System (Camel) replaces the M107, M149, and M1112 series water trailers. It consists of an 800–900 gallon capacity baffled water tank with integrated freeze protection and all hoses and fittings necessary to dispense water by gravity flow. The acquisition strategy consists of two increments: Increment 1 is the basic system with freeze protection. Increment 2 will provide modular component(s) to give the Camel water chilling, pumping, circulation, and on-board power generation as add-on capabilities. The Camel sits on a M1095 Trailer that allows for better transportability on and off the road by utilizing the Family of Medium Tactical Vehicle Truck. It holds a minimum of 800 gallons of water and provides one day of supply of potable water for drinking and other purposes. If the unit has another source of drinking water, such as bottled water, then the Camel can provide two days of supply (DOS) of potable water for other purposes. It is operational from -25 to +120 degrees Fahrenheit. The system also contains six filling positions for filling canteens and five gallon water cans.

SYSTEM INTERDEPENDENCIES

- M1095 Medium Tactical Vehicle Trailer
- Family of Medium Tactical Vehicle Truck

PROGRAM STATUS

- **Current:** Combined Arms Support Command (CASCOM), the Sustainment Center of Excellence, is staffing the Capability Production Document (CPD).

PROJECTED ACTIVITIES

- **FY09:** CPD approval
- **FY10:** Award Camel contract
- **FY11:** Complete production verification testing (PVT)

ACQUISITION PHASE
Technology Development
Engineering & Manufacturing Development
Production & Deployment
Operations & Support

Unit Water Pod System (Camel)

FOREIGN MILITARY SALES
None

CONTRACTORS
To be determined

Warfighter Information Network–Tactical (WIN–T) Increment 1

INVESTMENT COMPONENT

Modernization

Recapitalization

Maintenance

MISSION

To provide "networking at-the-halt" capability down to battalion level using high-speed, high-capacity voice, data, and video communications in the area of operations that employ internet standards.

DESCRIPTION

Warfighter Information Network–Tactical (WIN–T) Increment 1 represents a generational leap forward in allowing widely dispersed, highly maneuverable units to communicate. Increment 1 is a converged tactical communications network providing voice, data, and video capability to connect the battalion-level warfighter, allowing greater flexibility of troop movement. It is divided into two sub increments defined as Increment 1a "extended networking at-the-halt" and Increment 1b "enhanced networking-at-the-halt." Increment 1 is a rapidly deployable, early entry system housed in a Lightweight Multipurpose Shelter (LMS) and mounted on an Expanded Capacity High Mobility Multipurpose Wheeled Vehicle (HMMWV) for roll-on/roll-off mobility.

Increment 1a upgrades the former Joint Network Node satellite capability to access the K_a-band defense Wideband Global Satellite, reducing the reliance on commercial K_u-band satellites.

WIN–T Increment 1b introduces the Net Centric Waveform, a dynamic wave form that optimizes bandwidth and satellite utilization. It also introduces a colorless core security architecture, which meets Global Information Grid Information Assurance security compliance requirements

WIN–T Increment 1 is a Joint compatible communications package that allows the warfighter to use advanced networking capabilities, retain interoperability with Current Force systems and keep in step with future increments of WIN–T.

SYSTEM INTERDEPENDENCIES

None

PROGRAM STATUS

- **1QFY09:** Increment 1a initial operational test
- **2QFY09:** Increment 1b limited user test

PROJECTED ACTIVITIES

- **4QFY11:** Increment 1b operational test

ACQUISITION PHASE

Technology Development | Engineering & Manufacturing Development | Production & Deployment | Operations & Support

Joint Network Node (JNN)

Battalion Command Post Node

Tactical Hub

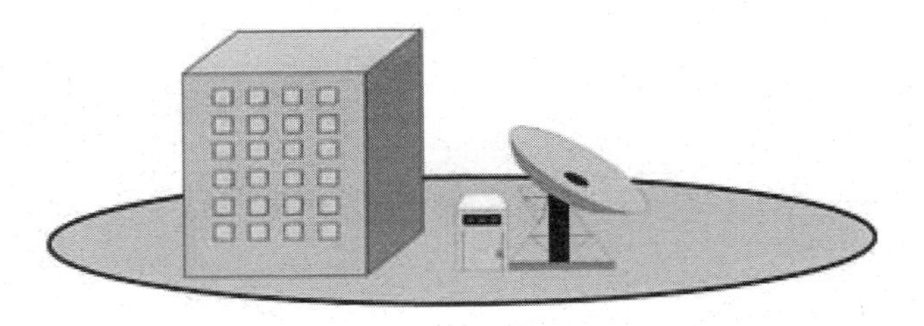

Fixed Regional Hub Node (FRHN)

Satellite Terminal Trailer (STT)

Warfighter Information Network–Tactical (WIN–T) Increment 1

FOREIGN MILITARY SALES
None

CONTRACTORS
General Dynamics C4 Systems (Taunton, MA)
General Dynamics SATCOM Tech (Duluth, GA)
Data Path, Inc. (Duluth, GA)

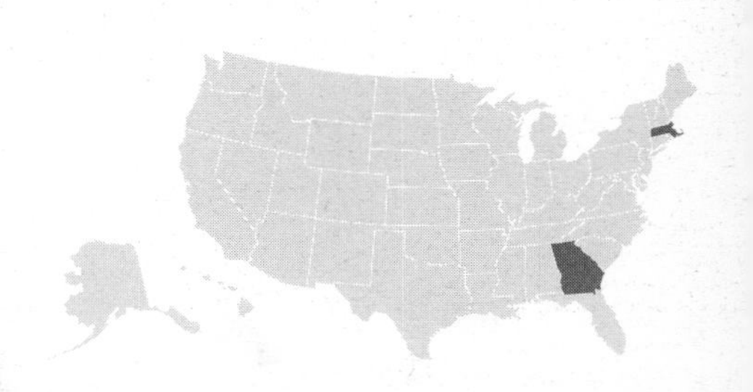

Warfighter Information Network–Tactical (WIN–T) Increment 2

INVESTMENT COMPONENT

Modernization

Recapitalization

Maintenance

MISSION

To provide "initial networking on-the-move" as a converged tactical communications and transport layer network leveraging proven commercial and government technology, enabling joint land forces to engage enemy forces deeper and more effectively, while incurring fewer losses.

DESCRIPTION

WIN–T Increment 2 accelerates delivery of a self-forming, self-healing mobile communication/transport layer network via commercial off-the-shelf and government off-the-shelf technologies. Increment 2 leverages an early release of the objective Highband Networking Waveform running on the Highband Networking Radio to provide high throughput line-of-sight communications and leverages an early release of the objective Net Centric Waveform on a MPM-1000 modem for on-the-move satellite communications enabling greater situational awareness and command and control. Multiple configuration items tailor capability from division down to company. It provides an accelerated delivery of network operations capability that allows management, prioritization, and protection of information while reducing organizational and operational support.

Increment 2 network operations includes automated planning, on-the-move node planning, automated link planning for currently fielded systems, initial automated spectrum management, initial quality of service planning and monitoring, and over-the-air network management and configuration of WIN-T radios. Additionally, Increment 2 network operations automates the initial Internet Protocol planning and routing configurations.

SYSTEM INTERDEPENDENCIES

None

PROGRAM STATUS

- **1QFY09:** Developmental test
- **2QFY09:** Limited user test

PROJECTED ACTIVITIES

- **1QFY10:** Milestone C, entering low rate initial production
- **3QFY10:** Production qualification test (contractor)
- **2QFY11:** Logistics demonstration
- **2QFY11:** Product qualification test (government)
- **2QFY11:** Cold Region Test
- **4QFY11:** Initial operational test
- **1QFY12:** Full rate production decision review
- **2QFY12:** First unit equipped
- **4QFY12:** Initial operational capability

ACQUISITION PHASE

Technology Development | **Engineering & Manufacturing Development** | Production & Deployment | Operations & Support

WIN-T Increment 2 - Initial Networking On The Move

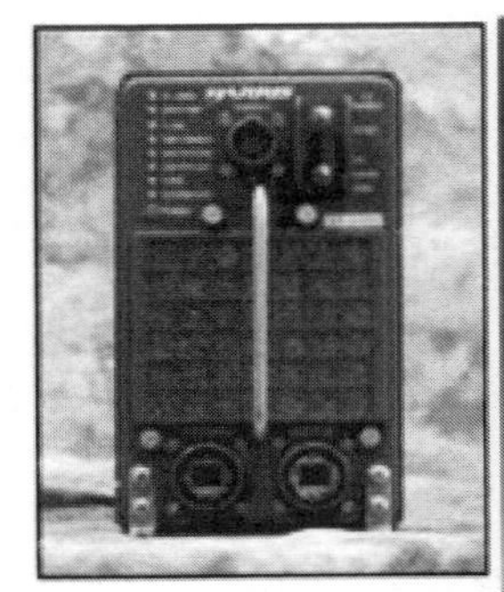

Highband Networking Line of Sight Radio

Tactical Communications Node

On the move Satellite Communications Antenna (Satellite Radio Frequency Unit Antenna)

Highband Networking Line of Sight Antenna (Higband Radio Frequency Unit Antenna)

Initial Network Operations Capability

MPM 1000 Modem for satellite Communications

FOREIGN MILITARY SALES
None

CONTRACTORS
General Dynamics C4 Systems (Taunton, MA)
Lockheed Martin Mission Systems (Gaithersburg, MD)
Harris Corp. (Palm Bay, FL)
BAE Systems (Wayne, NJ)
L-3 Communications (San Diego, CA)

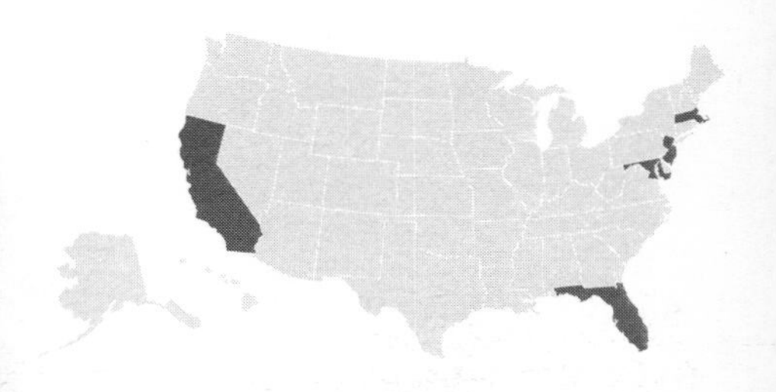

Warfighter Information Network–Tactical (WIN–T) Increment 3

INVESTMENT COMPONENT

Modernization

Recapitalization

Maintenance

MISSION

To provide "full networking on-the-move" to combat/maneuver Army and Future Combat Systems as a mobile, multi-tiered, tactical communications/transport layer network, enabling joint land forces to engage enemy forces deeper and more effectively, incurring fewer losses.

DESCRIPTION

Warfighter Information Network–Tactical (WIN–T) Increment 3 enables the full-objective mobile, tactical network distribution of command, control, communications, computers, intelligence, surveillance, and reconnaissance information via voice, data, and real-time video. Building on previous increments, Increment 3 provides more robust connectivity and greater network access via military specification radios, higher bandwidth satellite communications (SATCOM) and line of sight (LOS) waveforms, an air tier (LOS airborne relay), and integrated network operations. It manages, prioritizes, and protects information through network operations (network management, quality of service and information assurance) while reducing organizational and operational support. It ensures communications interoperability with Joint, Allied, Coalition, Current Force, and commercial voice and data networks. Using communications payloads mounted on Unmanned Aerial Systems, Increment 3 introduces an air tier to increase network reliability and robustness with automatic routing between LOS and SATCOM. This extends connectivity and provides increased warfighter mobility, providing constant mobile communications.

SYSTEM INTERDEPENDENCIES

None

PROGRAM STATUS

- **1QFY09:** Engineering field test to demonstrate technology maturity

PROJECTED ACTIVITIES

- **1QFY12:** Critical design review

ACQUISITION PHASE

Technology Development | Engineering & Manufacturing Development | Production & Deployment | Operations & Support

Warfighter Information Network–Tactical (WIN–T) Increment 3

WIN-T Increment 3 - Full Networking On The Move

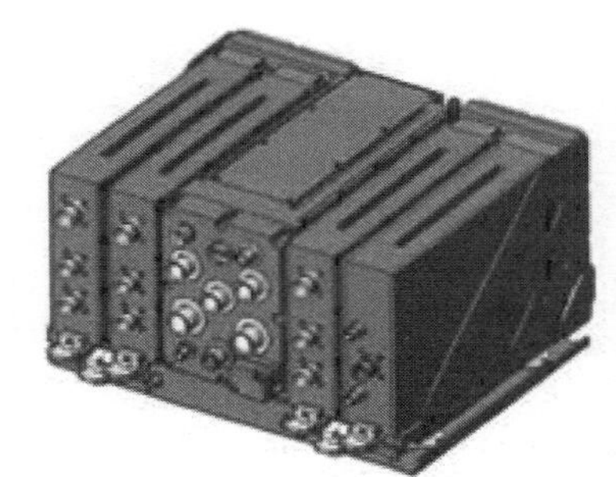

4 Channel Air Cooled Radio

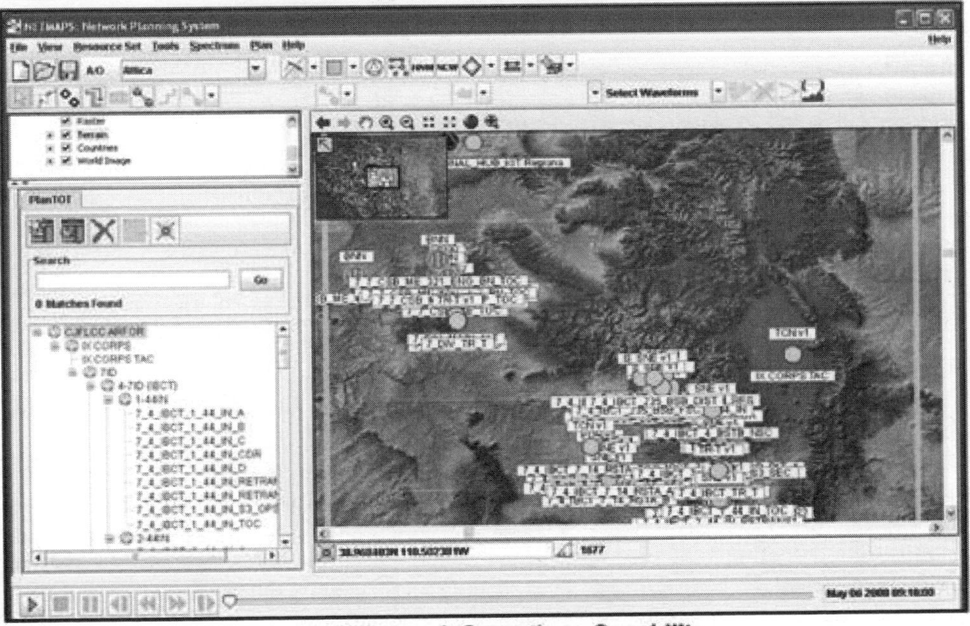

Full Network Operations Capability

2 Channel Air Cooled Radio

Dual Band LOS Antenna

Aerial Tier

FOREIGN MILITARY SALES

None

CONTRACTORS

General Dynamics C4 Systems (Taunton, MA)
Lockheed Martin Mission Systems (Gaithersburg, MD)
Harris Corp. (Palm Bay, FL)
BAE Systems (Wayne, NJ)
L-3 Communications (San Diego, CA)

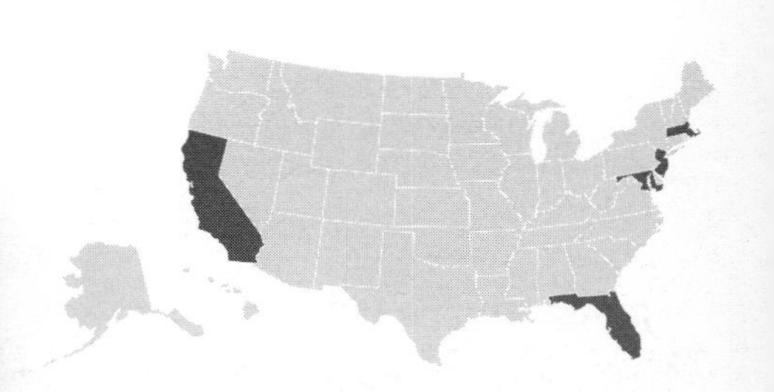

Weapons of Mass Destruction Elimination

INVESTMENT COMPONENT

Modernization

Recapitalization

Maintenance

MISSION

To enable Weapons of Mass Destruction–Civil Support Teams (WMD–CSTs) to perform on-site analysis of unknown samples in support of first responders with a mobile laboratory. The system also provides voice and data communications to enhance assessment of and response to WMD events.

DESCRIPTION

Analytical Laboratory System (ALS) Increment 1 is a mobile analytical laboratory that provides the CST capabilities for detecting and identifying chemical, biological, or radiological contamination. ALS Increment 1 is a system enhancement program (SEP) to replace the current Mobile ALS and interim Dismounted Analytical Platform. It provides advanced technologies with enhanced sensitivity and selectivity in the detection and identification of biological and chemical warfare agents and toxic industrial chemicals and materials.

The Unified Command Suite (UCS) vehicle is a self-contained, stand-alone, C-130 air mobile communications platform that provides both voice and data communications capabilities to CST commanders. The UCS consists of a combination of commercial and existing government off-the-shelf communications equipment (both secure and non-secure data) to provide the full range of communications necessary to support the CST mission. It is the primary means of reach-back communications for the ALS for the CSTs and acts as a command and control hub to deliver a common operational picture for planning and fulfilling an incident response. It provides:

- Digital voice and data over satellite network
- Secure Internet Protocol Router Network (SIPRNET) and Non-Secure (NIPRNET)
- Radio remote and intercom with cross-banding
- Over-the-horizon communication interoperable interface with state emergency management and other military units

The Common Analytical Laboratory System (CALS) provides a common CBRNE analytical capability across multiple domain spaces. Developed in both a mobile platform (light) as well as a semi-fixed site platform (heavy), the CALS is a modular design that provides the necessary array of analytical, diagnostic, and investigative capabilities tailored for a specified mission or contingency operation.

SYSTEM INTERDEPENDENCIES

UCS

PROGRAM STATUS

ALS-1:

- **4QFY09:** Full operational capability

20th SUPCOM Light Lab:

- **4QFY09:** Full operational capability

PROJECTED ACTIVITIES

CALS:

- **1QFY10:** Materiel development decision

20th SUPCOM Heavy Lab:

- **4QFY10:** Full operational capability

ACQUISITION PHASE

Technology Development | Engineering & Manufacturing Development | Production & Deployment | Operations & Support

ALS
ANALYTICAL LABORATORY SYSTEM

UCS
UNIFIED COMMAND SUITE

Heavy Lab (20th SUPCOM)

Light Lab (20th SUPCOM)

FOREIGN MILITARY SALES
None

CONTRACTORS
ALS:
Wolf Coach, Inc., an L-3 Communications Company (Auburn, MA)
UCS Vehicle:
Wolf Coach, Inc., an L-3 Communications Company (Auburn, MA)
UCS Communications system integrator:
Naval Air Warfare Center Aircraft Division (Patuxent River, MD)
20th SUPCOM Heavy Lab:
ECBC (Edgewood, MD)

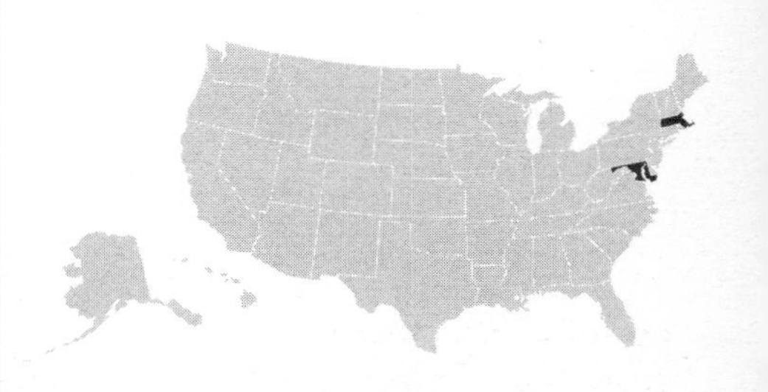

Science & Technology

The Army Science and Technology (S&T) strategy (Figure 1) supports the Army's goals to restore balance between current and future demands by providing new technologies to enhance and modernize systems in the Current Force and to enable new capabilities in the Future Force. This strategy is enabled through a portfolio with three types of investments, each providing different results in distinct timeframes.

Figure 1: Strategy—Develop and mature technology to enable the Future Force while seeking to enhance the Current Force.

The three types of S&T investment are: far-term, funding basic research for discovery and understanding of phenomena; mid-term, funding applied research laboratory concept demonstrations; and near-term, funding advanced technology development demonstrations in relevant environments outside the laboratory (Figure 2). The technology demonstrations prove technology concepts and their military utility to inform the combat developments process and provide the acquisition community with evidence of technologies' readiness to satisfy system requirements. This portfolio supports the overseas contingency operations in three ways: 1) Soldiers benefit today from technologies that emerged from our past investments; 2) we exploit transition opportunities by accelerating mature technologies derived from ongoing S&T efforts; and 3) we leverage the expertise of our scientists and engineers to develop solutions to unforeseen problems encountered during current operations such as the armor applied to Mine-Resistant Ambush Protected (MRAP) combat vehicles for enhanced protection from rocket propelled grenades (RPGs). The entire S&T program is adaptable and responsive as evidenced in its support of the Army Modernization Strategy.

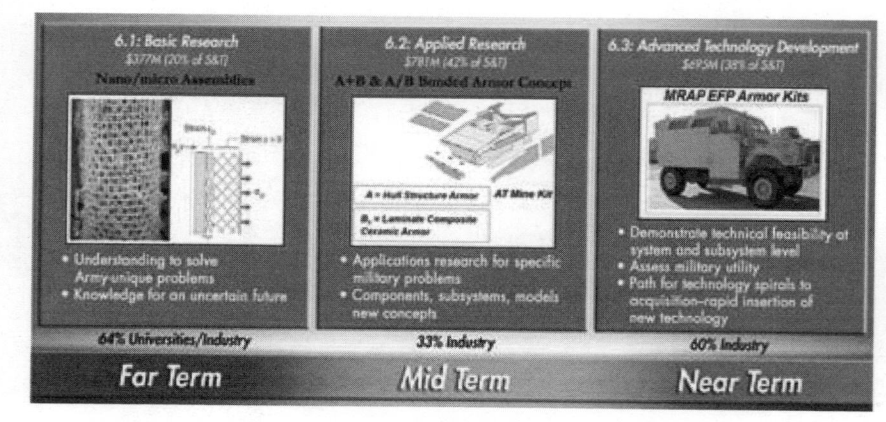

Figure 2: The S&T portfolio consists of three types of investments.

S&T INVESTMENT—FUTURE FORCE TECHNOLOGY AREAS

The diverse S&T portfolio is characterized in terms of Future Force Technology Areas. The investments in these areas are shown on the next page in a color depiction (Figure 3) that approximates their proportionate dollar value in FY2010 by Technology Area. Training and Doctrine Command/Army Capabilities Integration Center (TRADOC/ARCIC) represents the warfighter in the S&T process, and articulates the warfighter's needs to the S&T community through the development, staffing, and coordination of the TRADOC "Warfighter Outcomes" to guide the S&T investment. The Deputy Assistant Secretary of the Army for Research and Technology consolidated the 37 Tier 1 TRADOC "Warfighter Outcomes" into 10 Comprehensive Warfighter Outcomes.

Within these Technology Areas, the highest priority efforts are designated by Headquarters Department of the Army (HQDA) as Army Technology Objectives (ATOs). We do not designate ATOs within the basic research area since these investments fund sciences (discovery and understanding) not technology. The ATOs are co-sponsored by the S&T developer and the warfighter's representative, TRADOC. The ATOs are focused efforts that develop specific S&T products within the cost, schedule, and performance metrics assigned when they are approved. The goal is to mature technology within ATOs to transition to program managers for system development and demonstration and, subsequently, to acquisition.

This S&T section of the *U.S. Army Weapon Systems 2010* handbook is organized by Future Force Technology Area. Selected ATOs are described within most of the Technology Areas. The complete portfolio of 97 ATOs is described in the 2009 Army Science and Technology Master Plan (distribution limited to government and current government contractors).

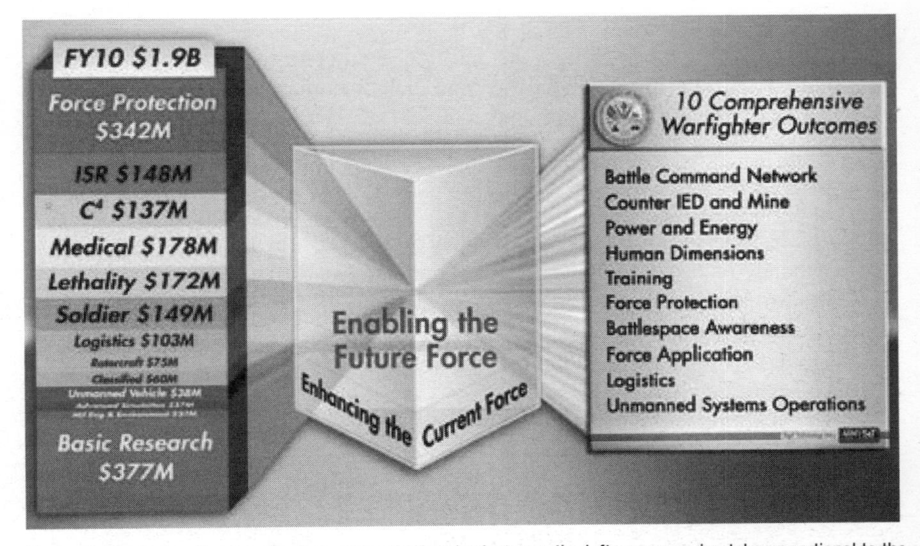

Figure 3: The Future Force technology area color bands shown on the left are approximately proportional to the financial investment within the Army's requested FY10 S&T budget. The specific technologies funded in these investment areas are aligned to the 10 Comprehensive Warfighter Outcomes. The 10 CWOs are the consolidation of TRADOC's Tier I Warfighter Outcomes, which include their "Big 5" Integrated Warfighter Outcomes. The Warfighter Outcomes articulate the warfighter capability needs.

Short descriptions of Future Force Technology Areas:

- **Force Protection** technologies enable Soldiers and platforms to avoid detection, acquisition, hit, penetration, and kill. These technologies include advanced armor, countermine, and counter improvised explosive devices (IEDs) detection and neutralization, and counter rocket, artillery, and mortars (CRAM) aircraft survivability and active protection systems.
- **Intelligence, Surveillance, and Reconnaissance (ISR)** technologies enable persistent and integrated situational awareness and understanding to provide actionable intelligence that is specific to the needs of the Soldier across the range of military operations.
- **Command, Control, Communications, and Computers (C4)** technologies provide capabilities for superior decision making, including intelligent network decision agents and antennas to link Soldiers and leaders into a seamless battlefield network.

- **Lethality** technologies enhance the ability of Soldiers and platforms to provide overmatch against threat capabilities and include nonlethal technologies enabling tailorable lethality options.
- **Medical** technologies protect and treat Soldiers to sustain combat strength, prevent or treat infectious diseases, reduce casualties, improve clinical care and rehabilitative medicine, and save lives. It includes technologies to enhance Soldier performance in extremely demanding environments imposed by battlefield physical and psychological demands as well as extremes in topography and climate.
- **Unmanned Systems** technologies enhance the effectiveness of unmanned air and ground systems through improved perception, cooperative behaviors, and increased autonomy.
- **Soldier Systems** technologies provide materiel solutions that protect, network, sustain, and equip Soldiers, and non-materiel solutions that enhance human performance. Together these solutions enable Soldiers to adapt and dominate against any threat.
- **Logistics** technologies enhance strategic response and reduce logistics demand. Focus is on technologies that increase efficiency of systems or subsystems or sustainment processes that enable production of consumables closer to the point of use, that conserve or reduce demand for consumables (such as fuel and water), and that enhance the nation's assurance of sufficient energy for Army missions.
- **Military Engineering and Environment** technologies enhance deployability and sustainability. These technologies also enable sustainment of training and testing range activities.
- **Advanced Simulation** technologies provide increasingly realistic training and mission rehearsal environments to support battlefield operations, system acquisition, and requirements development.
- **Rotorcraft** technologies enhance the performance and effectiveness of current and future rotorcraft while seeking to reduce operational and sustainment costs.
- **Basic Research** investments seek to develop new understanding to enable revolutionary advances or paradigm shifts in future operational capabilities.

FORCE PROTECTION

Kinetic Energy Active Protection System

The Kinetic Energy Active Protection System ATO provides the additional capability to defeat tank-fired kinetic energy rounds to the chemical energy system that currently defines the Brigade Combat Team (BCT) Modernization Point-of-Departure Active Protection System. This program develops warhead and interceptor chassis designs and conducts robust component testing. These components support the hit-avoidance suite designed to enhance the protection of BCT against tank-fired threats.

Figure 4: Kinetic Energy Active Protection System.

Tactical Wheeled Vehicle Survivability

Tactical Wheeled Vehicle Survivability ATO identifies, analyzes, develops, demonstrates, and transitions an integrated suite of advanced survivability technologies for the protection of crew and passengers in current and future tactical wheeled vehicle (TWV) fleets. For TWV platforms, both traditional and nontraditional armor approaches do not independently defeat objective threats within the system's weight, power, and cost constraints. Integrated survivability technology suites will be determined through trade-off analyses to balance payload, performance, and protection at a reasonable cost. Technologies included are high-performance, lighter weight ballistic materials; active protection systems; electronic warfare; and signature management for both Army and

Marine Corps TWVs. The armor technologies developed in this program are designed as a B-kit solution and follow the interface requirements of the current Long-Term Armor Strategy (LTAS) A–B kit configuration. Similarly, the non-armor survivability technologies are designed to easily interface with the LTAS A-kit configuration in terms of size, weight, power, and cooling considerations. When integrated, these technologies (both armor and non-armor) offer an upgradable, modular approach to protection, thus supporting the warfighter's need for mission-adaptable survivability concepts for TWVs.

Threat and Minefield Detection Payload for Shadow Tactical Unmanned Aerial Vehicle

This ATO matures and demonstrates a tactical unmanned aerial vehicle (TUAV) payload incorporating multi/hyper-spectral imaging sensors, adaptive spectral detection, and change detection algorithms. The TUAV payload will demonstrate real-time detection of roadside threats, threat deployment activity, and minefields at realistic mission altitudes. It also provides an advanced reconnaissance, surveillance, and target acquisition capability for detection of difficult targets, including home-made explosives.

Detection for In-Road Threats

This ATO matures and demonstrates an advanced mine and threat detection capability to address a broader spectrum of in-road threats—including those deeply buried—at higher rates of advance for modular engineer platforms and the Early Infantry Brigade Combat Team (E-IBCT). In order to meet current and Future Force needs, this effort matures and then integrates ground penetrating radar and metal detection technologies onto vehicles to detect the evolving underbelly threat on primary and secondary roads. The technologies demonstrated include an optimized metal detector, signal processing, a downward-looking ground penetrating radar, and algorithms optimized for both shallow and deep targets.

Extended Area Protection & Survivability (EAPS) Integrated Demo

Enhanced Area Air Defense System (EAADS) is the Army's objective maneuver Air and Missile Defense system. EAADS will be a deployable maneuver capability that leverages the best combination of directed energy and/or kinetic energy technology against the aerial threat. The most technologically challenging element of the EAADS mission is the protection against rocket, artillery, and mortar (RAM) attack. The Counter RAM (C-RAM) multi-pillar system of systems is a successful, quickly fielded, initial capability against the near- and medium-term RAM threat. The C-RAM intercept pillar does not, however, meet objective EAADS criteria, including effectiveness at required range, multiple simultaneous engagements, 360-degree coverage, and ability to control collateral damage. This ATO further matures missile and bullet technologies, and integrates these technologies for hardware demonstrations to bridge the gap between the initial C-RAM capability and the objective EAADS.

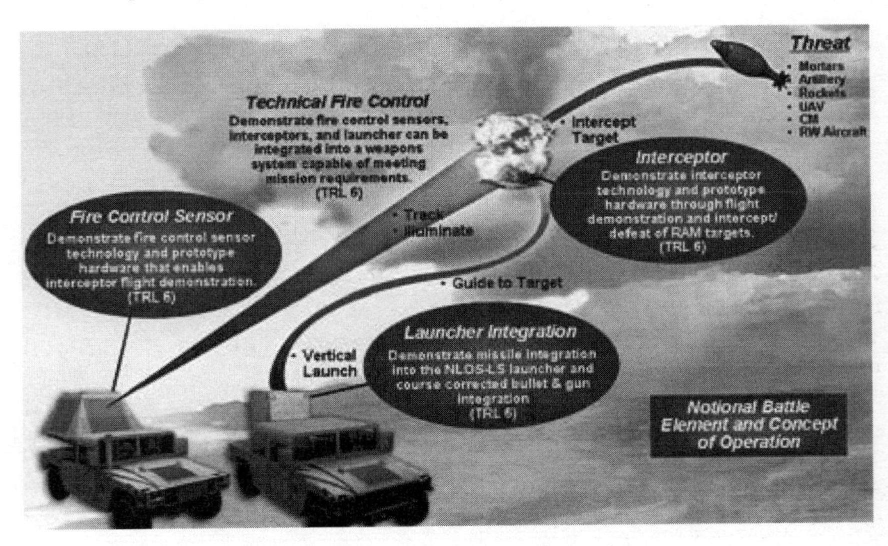

Figure 5: Extended Area Protection & Survivability (EAPS) Integrated Demo

INTELLIGENCE, SURVEILLANCE, RECONNAISSANCE

All-Terrain Radar for Tactical Exploitation of Moving Target Indicator and Imaging Surveillance (ARTEMIS)

This ATO matures and demonstrates an airborne, all-weather, all-terrain ground moving target indication (GMTI), tracking, and cueing system for a Class IV unmanned aerial system (UAS). Unlike most tactical radars, this system will be able to track both mounted and dismounted threats moving in open terrain or using cover for concealment. Additionally, ARTEMIS incorporates synthetic aperture radar (SAR) capability that is able to image vehicle-sized threats in foliated and open terrains, as well as smaller threats that are shallowly buried or in the open. The effort builds a smaller multifunction foliage penetration radar system that satisfies Class IV UAS requirements. The efforts under demonstration are: persistent SAR and GMTI surveillance against mounted and dismounted threats; reduced susceptibility to camouflage concealment and deception measures; and detection of surface/sub-surface roadside threats.

Figure 6: All-Terrain Radar for Tactical Exploitation of Moving Target Indicator and Imaging Surveillance

Battlespace Terrain Reasoning Awareness—Battle Command

This ATO provides integrated battle command capabilities to create and utilize actionable information from terrain, atmospheric, and weather effects on systems, platforms, and Soldiers. This will enable agile, integrated ground and air operations in all operational environments. In FY10, an initial spiral of urban-based technologies from the Network-Enabled Command and Control ATO program will be incorporated. The resulting capability will provide net-centric, n-tier, terrain reasoning services and embedded battle command applications.

Figure 7: Battlespace Terrain Reasoning Awareness—Battle Command

This effort is working with key transformational battle command programs and TRADOC schools to (1) conduct controlled demonstrations to gain insight into effectively integrating actionable terrain, atmospheric, and weather information into battle command system-of-systems (SoS), staffs, processes, and functions; (2) improve, extend, and mature terrain- and weather-based information products and embedded applications within battle command SoS; (3) transition capabilities to the Distributed Common Ground System–Army (DCGS–A), BCT, and commercial joint mapping toolkit; and (4) support the development of a geo-battle management language that extends the current model to include representation of actionable terrain, weather, and atmospheric information.

Target Location Designation System

This ATO demonstrates an improved, man-portable, target acquisition and laser designation system with reduced size, weight, and power. The effort enables real-time target identification and acquisition, laser designation, and precision target location of distant targets in a very lightweight, low-power, cost-effective, and high-performance package. This effort will produce: (1) an improved mid-wave infrared focal plane array; (2) a common designator module using end-pumped, mono-block laser technology; and (3) precision target location with improved global positioning, gyroscope, and magnetometer. The results of this effort will demonstrate to the warfighter improvements in target acquisition, precision target location, and laser designation capabilities to thus increase combat effectiveness and lethality. The increased target acquisition range will provide a greater standoff range and increase Soldier survivability; the reduced weight will achieve greater Soldier mobility.

Figure 8: Flexible Display Technology for Soldiers and Vehicles

Flexible Display Technology for Soldiers and Vehicles

This ATO will develop flexible display technologies for affordable, lightweight, rugged, low-power, and reduced-volume displays in conjunction with the development of human factors parameters for systems utilizing flexible displays. Flexible displays have reduced weight and are inherently rugged with ultra-low power electro-optic technologies as compared to traditional liquid-crystal, glass-based displays. The development of displays on flexible substrates will enable novel applications that cannot be achieved by glass-based technologies (e.g., wearable and conformal for Soldier applications, conformal for vehicle and cockpit applications, and compact display that can be rolled out for multiuser applications). This ATO program is coordinated with human factors studies to optimize design trade-offs, and will produce flexible, 4-inch diagonal displays (greater than 320 x 240 resolution), as well as technology for color emissive and reflective displays. Benefits to the warfighter include a 60 percent weight reduction of display components compared to glass displays, and a 30 to 90 percent power reduction compared to liquid crystal displays.

Multi-Spectral Threat Warning

Ultra-violet (UV) sensors utilized in aircraft threat warning systems are limited in their ability to accurately distinguish Man-Portable Air Defense System threats from false alarm sources. This ATO investigates and quantifiably measures the benefits to aircraft protection of integrating currently fielded, UV-based Missile Warning System with infrared (IR) and acoustic sensors. Specifically, Multi-Spectral Threat Warning seeks to enhance the current system's probability of detection and reduce its false alarm rate through correlating IR signature data with the UV data. Additionally, providing acoustic spectra to the current UV-based system's Hostile Fire Indication algorithms increases the probability of detection for non-tracer rounds.

Figure 9: Multi-Spectral Threat Warning

COMMAND, CONTROL, COMMUNICATIONS, AND COMPUTERS (C4)

Network-Enabled Command and Control

The Network-Enabled Command and Control (NEC2) ATO develops, integrates, and transitions technologies, products, and software services that provide network-centric command and control capabilities to the current and Future Force. Transition of these products and services are focused on current, transitional, and future battle command systems throughout all environments and phases of operations. NEC2 will develop advanced software and algorithms that tailor and manage the flow of battle command information and command and control services across current and Future Force systems. This will enable the commander and his staff to effectively use vast amounts of information horizontally and vertically throughout the theater of operations for decision and information superiority. Technology efforts under NEC2 focus on applications in complex and urban terrain; battle command planning, execution, and replanning products for unmanned systems and sensors; and decision making tools that account for political, religious, and cultural factors, and expand the commander's reach to other government and nongovernment experts. An Unmanned Systems Capstone Experiment will: (1) evaluate unmanned software services for air and ground systems performance across tactical application scenarios; and (2) collect and process communications characterization data and deliver refined unmanned software services to the BCT modernization program.

Tactical Mobile Networks

This ATO develops, matures, and demonstrates communications and networking technologies that optimize throughput, bandwidth usage, size, energy, and network prediction of tactical voice and data networks. Tactical Mobile Networks address emerging Future Force requirements through (1) proactive diverse link selection (PAD–LS) algorithms to optimize use of available communications links within multilink nodes (vehicles, TOCs, etc.); (2) multiband, multimode tactical voice and data network communications services for dismounted Soldiers and manned and unmanned systems (sensors, munitions, etc.) through the development of a Joint Tactical Radio System (JTRS) Software Communications

Architecture (SCA) v2.2 Soldier Radio Waveform (SRW); and (3) software tools to dynamically predict and visualize on-the-move communications network performance.

The Tactical Mobile Networks ATO conducts modeling and simulation to verify the functional and performance characteristics of PAD–LS algorithms during development and develops implementations, and conducts demonstrations of the link selection algorithms under controlled environment. The effort matures the network management tools, incorporating increasing number of networking waveforms, entities, processing speeds, network topologies, and network visualization (network statistics and user priorities).

Collaborative Battlespace Reasoning and Awareness

The Collaborative Battlespace Reasoning and Awareness (COBRA) ATO develops and demonstrates multiplatform, cross community applications and software services that support the integration and synchronization of intelligence and operations functions through the design, development and implementation of information interoperability, and through collaborative management and decision support technologies. This ATO also develops and demonstrates systems that will improve mission execution success by providing software to more tightly couple operations and intelligence and to better facilitate collaboration. Research and development will be focused on mapping intelligence and geospatial information requirements to military tasks. This effort will make possible faster and higher quality decision cycles and increased battle command unification through collaboration and real-time sharing, exploitation, and analysis to support the operational mission, tasks, and desired effects.

RF Adaptive Technologies Integrated with Communications and Location (RADICAL)

This ATO develops and demonstrates Radio Frequency (RF) dynamic spectrum technologies for tactical communications and improved position determination in Global Positioning System (GPS)-degraded environments (Figure 10). ATO efforts include a software module that enables spectrum policy management for dynamic spectrum access-enabled radios, architecture development to integrate and enhance disruption tolerant networking (DTN) in the tactical environment, and a software module that improves position determination based on net-assisted GPS and RF ranging technologies. RADICAL will leverage the Defense Advanced Research Projects Agency (DARPA) Wireless Network After Next (WNAN) program to provide consistent dynamic spectrum policy management using software implementation, ensure reliable message delivery in a disruptive communications environment by enhancing and extending the DTN technology into tactical networks, mitigate multipath through RF ranging, and improve GPS performance through net-assisted GPS technologies.

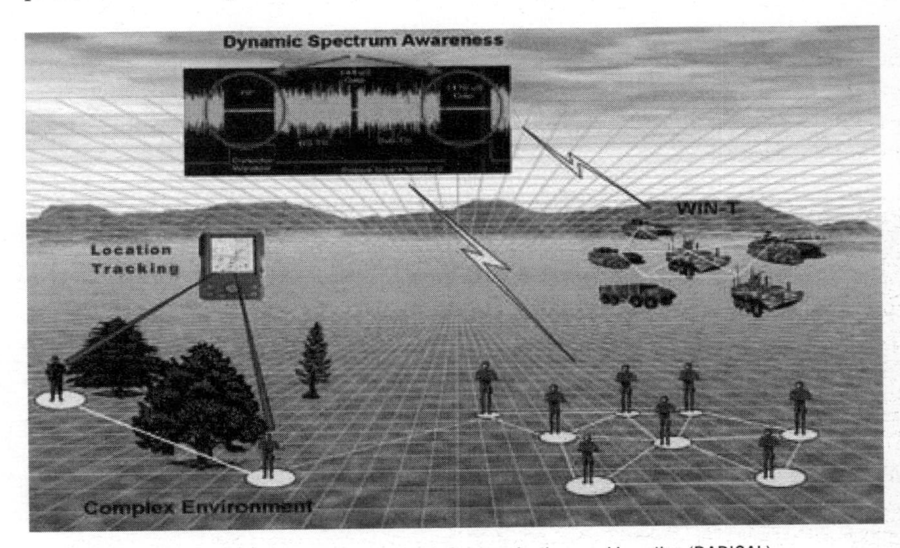

Figure 10: RF Adaptive Technologies Integrated with Communications and Location (RADICAL)

LETHALITY

Non Line of Sight–Launch System Technology

The Non Line of Sight–Launch System (NLOS–LS) Technology ATO is developing and maturing improved components and subsystem technologies for the NLOS–LS missile system, a core program of the BCT modernization effort.

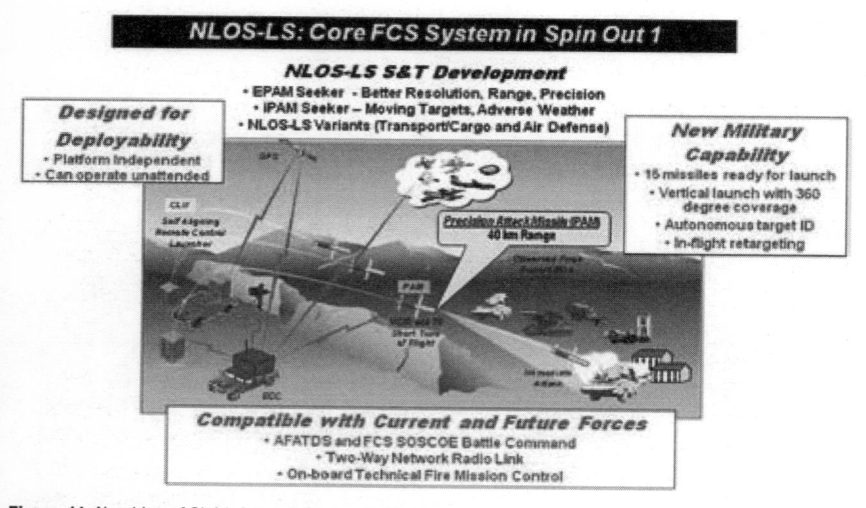

Figure 11: Non Line of Sight–Launch System Technology

The ATO supports the NLOS–LS development by transitioning affordable, mature components that enhance the threshold performance through a subsystem maturation effort; continuing critical component development efforts for future performance enhancements and integrated missile variants providing mission flexibility launched from the NLOS–LS. This effort has developed and successfully transitioned enhanced seeker technology for the Precision Attack Missile (PAM) baseline seeker that provides better resolution and acquisition range at a reduced average unit production cost to the NLOS–LS program manager and prime contractor. Other technology development efforts have been focused on maturation through the development, fabrication, and testing of critical subsystems including semi-active laser (SAL)/laser radar (LADAR) seeker; controllable rocket motor propulsion; high efficiency turbine engine technology; multi-purpose warhead technologies; an improved multi-mode seeker (iPAM); and miniature electronics. Modeling and simulation efforts have included the linkage of physics-based engineering models, hardware and software-in-the-loop (HWIL/SWIL) designs, constructive analysis, and virtual prototype development and exercise.

Advanced Lasers and Unmanned Aerial System Payloads

This ATO develops, integrates, and demonstrates a 7-pound advanced sensor payload with laser rangefinding and laser designating capabilities to address the reconnaissance, surveillance, and target acquisition mission requirements for the BCT Class I unmanned aerial system (UAS). New multifunction lightweight lasers, optical receiver components, and electronics will be developed suitable for UAS and other Soldier applications. The new laser components will be integrated with a compact, small-pixel, uncooled infrared imaging sensor into a two-axis pointing platform (gimbal) to enable an airborne organic laser designation capability for the lower echelon warfighter. The advanced lasers and UAS sensor payload will enable Soldiers to quickly see and characterize potential targets as well as nontarget objects that are in the open or in complex and urban terrain, and support beyond-line-of-sight situational awareness, targeting, and engagement with precision weapons. A parallel ManTech effort seeks to develop an optimized manufacturing process for a universal, monoblock laser designator module component that can be integrated into a wide variety of laser applications.

Applied Smaller, Lighter, Cheaper Munitions Components

Affordably reducing space, weight, and power at the component level remains essential to increasing precision munition lethality for full spectrum operations, particularly military operations on urban terrain (MOUT) (Figure 12). This ATO focuses on developing increasingly smaller, lighter, cheaper components and subsystems that will enhance current system capabilities against asymmetric threats and mature technologies for next-generation small precision munitions. Primary investment areas include: nano/advanced composite structures and new fabrication techniques to save weight while maintaining or enhancing structural and thermal properties; miniaturized electronics to reduce size and weight, and support increased processing demands for capability enhancements like image stabilization; sensor/image processing for MOUT environments, including people tracking; and warhead safe and arm integration for precision lethality against expanded target sets in urban terrain. Major warfighter payoffs will be enhanced precision lethality and cost savings through common components.

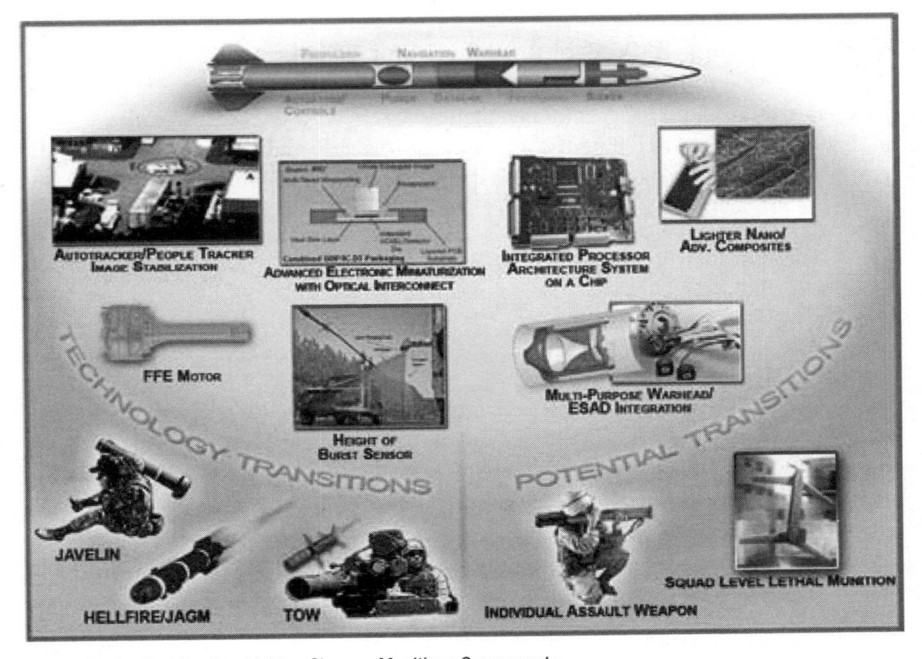

Figure 12: Applied Smaller, Lighter, Cheaper Munitions Components

Scalable Technology for Adaptive Response

The Scalable Technology for Adaptive Response (STAR) ATO is maturing and demonstrating new energetic materials, fuzes, and warhead technologies that can provide selectable and scalable effects against platforms and personnel. The STAR ATO will demonstrate technologies for selectable lethal effects in large-, medium-, and small-diameter munitions and missiles; and development of controlled lethal effects, multipurpose energetics and formulations, reactive materials, and advanced fuzing and power technologies. The STAR ATO will demonstrate 250mm (Guided Multiple Launch Rocket System), 155mm (Excalibur), and 30mm (M789/Mk238); enable improved weapon effectiveness and lethality; and reduce collateral damage and logistics.

MEDICAL

Psychological Resetting after Combat Deployment: Advanced Battlemind

This ATO develops and validates an advanced unit-training program to reduce combat-related psychological problems, including symptoms related to mild traumatic brain injury (mTBI) and post-traumatic stress disorder (PTSD) during the post-deployment resetting phase. The goal is to facilitate recovery from psychological injuries related to combat, build individual and unit resilience in preparation for subsequent deployments, reduce the incidence of debilitating symptomatic problems, and reduce risk-taking behaviors that have the greatest impact on a Soldier's mental health, well-being, relationships, and job performance. An in-depth six session Battlemind Training Package will be developed that integrates state-of-the-art cognitive-behavioral approaches to traumatic stress, while maintaining the focus on Soldier strengths, unit cohesion, leadership skills, and individual cognitive skill building. The package will also incorporate cognitive education strategies shown to be effective in reducing symptoms from mTBI, which often overlaps with PTSD.

Damage Control Resuscitation

This ATO pursues the best combination and optimal use of alternatives to whole blood (plasma, red blood cells, blood clotting agents, etc.) to prevent bleeding and maintain oxygen delivery and nutrients to tissue (Figure 13). These products will likely enhance survival of casualties after severe blood loss, which is the leading cause of death to injured warfighters. Recent data from the battlefield suggests that blood clotting disorders and immune system activation, which damages normal cellular metabolic processes, commonly occur in severely injured patients. Therefore, a priority is to maintain blood clotting capability and oxygen and nutrient delivery to tissues by using the best resuscitation products that can be administered at far forward locations.

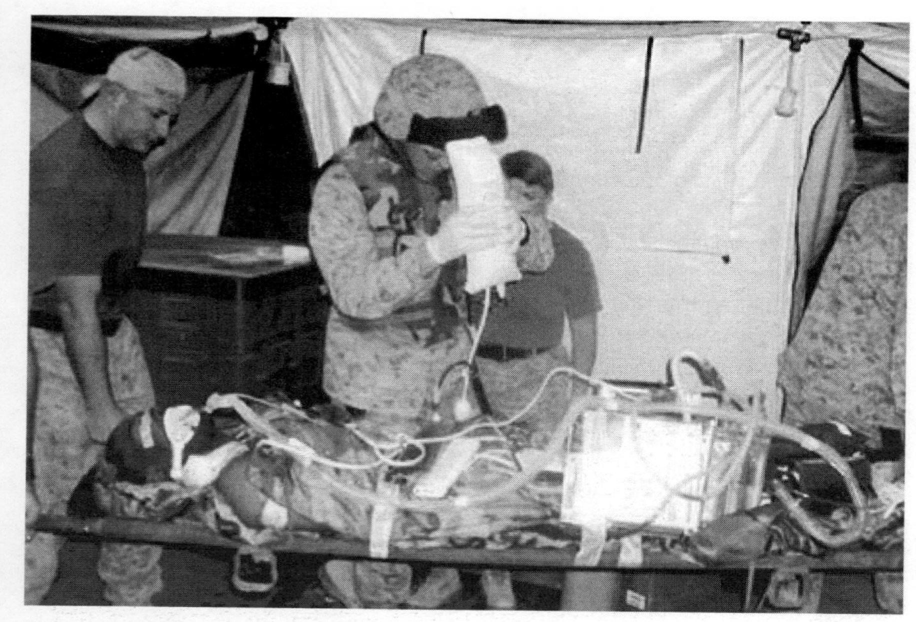

Figure 13: Damage Control Resuscitation

Drug for the Treatment of Traumatic Brain Injury (TBI)

This ATO is testing a candidate drug to treat TBI to determine its safety and effectiveness in 200 human subjects that have suffered TBI. It is estimated that 15 to 25 percent of all injuries in recent conflicts are to the head. TBI survivors often have physical and cognitive impairment, memory loss, and mood and personality disorders. There are currently no drugs to treat or reduce brain related injuries.

Prophylactic Drugs to Prevent Drug Resistant Malaria

This ATO develops candidate antimalarial prophylactic drugs and test these candidates in animals Successful completion of this ATO will allow clinical testing of candidates and potentially may lead to a safe and efficacious replacement antimalarial prophylactic drug. The goals in developing these candidate prophylactic drugs are to replace current drugs that are becoming less effective due to the development of drug resistance in the malaria parasite, to reduce or eliminate unwanted side effects that impact Soldier's use of the drugs, and to allow for a less critical and more convenient dosing schedule for deployed Soldiers. This drug would also increase operational readiness by maintaining a healthy force, as well as reduce the logistical and combat health support burden associated with treatment in theatre or after evacuation.

UNMANNED SYSTEMS

Robotic Vehicle Technologies Control Architecture for BCT Modernization

The Robotic Vehicle Technologies Control Architecture (RVCA) for BCT modernization develops an unmanned ground vehicle (UGV) end-to-end control architecture to reduce future integration risk and demonstrate the viability of autonomous UGV operations in a relevant environment using representative system-of-systems hardware and software components. This program will enhance UGV program viability and reduce program risks through the maturation, integration, and test representative hardware and software onto a surrogate UGV platform. The Crusher vehicle developed by DARPA under its UGCV PerceptOR Integration program will serve as the initial integration platform and be used for test and experimentation. Phase I of the effort integrates the autonomous navigation system onto the Crusher platform to provide autonomous maneuver capabilities, and will also mature and integrate the mission execution, computer operating environment, and vehicle management system hardware and software necessary for unmanned vehicle control. A series of engineering evaluations and a Soldier operational exercise to measure system performance and effectiveness from both the technical and operational contexts will be conducted. Phase II of the effort will see the migration/integration of hardware and software from the Crusher vehicle to a new test platform developed under the Tank-Automotive Research, Development and Engineering Center (TARDEC) Autonomous Platform Demonstrator program, and the task will finalize hardware and software implementations. It will conduct another series of engineering evaluations and conduct a final Soldier operational exercise in military significant environment.

Safe Operations of Unmanned Systems for Reconnaissance in Complex Environments

Safe Operations of Unmanned systems for Reconnaissance in Complex Environments (SOURCE) ATO develops, integrates, and demonstrates robust robotic technologies required for Future Modular Force unmanned systems (figure 14). The ATO will advance the state of the art in perception and control technologies to permit unmanned systems (UMS) to autonomously conduct missions in populated, dynamic urban environments while adapting to changing conditions; develop initial tactical/mission behavior technologies to enable a group of heterogeneous UMS to maneuver in collaboration with mounted and dismounted forces; optimize soldier operation of UMS; and provide improved situational awareness for enhanced survivability. Modeling and simulation will be used to develop, test, and evaluate the unmanned systems technologies (e.g. tactical behaviors and perception algorithms). Test bed platforms will be integrated with the software and associated hardware developed under this program, as well as appropriate mission modules, to support warfighter experiments in a militarily significant environment in conjunction with TRADOC.

Figure 14: Safe Operation of Unmanned Systems for Reconnaissance in Complex Environments

SOLDIER SYSTEMS

Soldier Planning Interfaces & Networked Electronics

This ATO develops a government-owned, Soldier-borne electronic equipment architecture that incorporates a National Security Agency-approved wireless personal area network subsystem (Figure 15). Soldier Planning Interfaces & Networked Electronics (SPINE) will reduce the soldier-borne footprint and system weight by 30 percent through the loss of wires and connectors. The wireless network will be powered by a conformal battery currently under development which will increase power by 50percent for 24 hour period. Additionally, it will utilize emerging software services to enable Soldier connectivity and data exchange to current and future tactical radio networks and battle command systems. Throughout this effort, capability demonstrations will be conducted at the C4ISR On The Move (OTM) test bed at Ft. Dix, NJ.

Figure 15: Soldier Planning Interfaces & Networked Electronics

Soldier Blast and Ballistic Protective System Assessment and Analysis Tools

This ATO provides analysis tools and test protocols to aid development and assessment of ballistic and blast protective systems. It will provide a benchmark of current capability, and develop system and component test protocols and devices with an initial focus on primary blast lung and facial/ocular injury. Models such as the Integrated Casualty Estimation Methodology (ICEM) model will be enhanced and exploited to begin characterizing blast effects to mounted and dismounted Soldiers for an improved Soldier armor design prototype. The payoff will be technology for improved Soldier armor and blast protection systems.

Enhanced Performance Personnel Armor Technology

Existing personnel armor systems cover less than 50 percent of the Soldier's body. This ATO will consider materials technology and tools to provide armor protection to the head, face, and extremities and will consider the penalties associated with that protection. ATO products will include new materials concepts for expanded Soldier body armor protection against blast and ballistic threats; improved materials models for predicting blast and ballistic performance; and full scale, high-fidelity modeling and diagnostic tools to guide technology development. The technologies and tools will transition to advanced technology development efforts in FY10 or earlier as options mature to TRL 5.

High-Definition Cognition (HD-COG) In Operational Environments

This ATO researches real-time understanding of brain function in operational environments to allow matching of Soldier capabilities and advanced technologies. For example, vehicle crewstations could cue Soldiers based on how their brains process what they see, hear, and feel. Such neuro-ergonomic designs can exploit how the brain functions, providing tremendous Soldier performance improvements. This program will develop technologies to assess Soldier neuro-cognitive processes in operational environments, as well as techniques to use them for neuro-ergonomic design. Technology development will focus on solutions to cognition, visual scanning, and platform control for mounted and dismounted operations. Approximately three experiments will be performed each year to look at ATO-developed technologies in a motion-based simulation environment.

LOGISTICS

Power for the Dismounted Soldier

This ATO matures and demonstrates technologies to provide small, lightweight, low-cost power sources. It demonstrates batteries what are half the size and twice the energy of C4ISR primary batteries (e.g., SINCGARS ASIP); conformal rechargeable Soldier system batteries; a soldier-mission-extending hybrid fuel cell; and a JP8-powered Soldier-portable power source for tactical battery recharging. Resulting efforts include: reduction in weight (~50 percent) for Soldier power; extended mission times in Soldier and sensor applications; reduction in resupply quantity, weight and costs; and increased Soldier mobility, sustainability, survivability and deployability by providing higher energy sources and recharging capability.

Wheeled Vehicle Power and Mobility

Wheeled Vehicle Power and Mobility ATO addresses the mobility and power requirements for the Army's current and future wheeled vehicles. With fleet modernization, wheeled vehicles require enhanced power and suspension capabilities to power more electronic components, transport payloads, support armor upgrades, and increase fuel efficiency. The ATO will demonstrate commercial engines adapted to military requirements that provide better fuel economy and lower heat rejection; compact, reliable, safe, and lightweight hybrid electric technology; incorporation of SiC power electronics; and a TRL 6 demonstration of an advanced magneto-rheological suspension system. The ATO provides wheeled vehicle platforms with power generation and control to include hybrid electric drive systems as well as an advanced suspension system for improved vehicle ride stability. The ATO provides the warfighter with enhanced vehicle mobility and safety to accomplish future missions.

High Performance Lightweight Track

This ATO will provide two high-performance lightweight track system options for 30–40 ton class vehicles: a Segmented Band Track and Lightweight Metallic Track for platform weights of 30–40 tons. Future combat vehicles need lightweight track with acceptable maintainability, durability, and survivability. The current lightweight track ATO developed a 16.5" wide segmented band track for a 25-ton vehicle. Requirements growth for BCT MGV has caused critical demand for a higher capacity, more survivable lightweight track. Lightweight track systems are challenged by increased vehicle weights and performance requirements and require innovative materials and design improvements to meet high strength, durability, and survivability targets. The program will improve/optimize lightweight segmented track technology through utilization of "Best in Class" high-performance elastomers and designs to enhance durability and survivability. This ATO seeks to develop and refine Lightweight Metallic Track through optimized and innovative designs and materials that deliver performance, maintainability, and survivability at 30–40 tons.

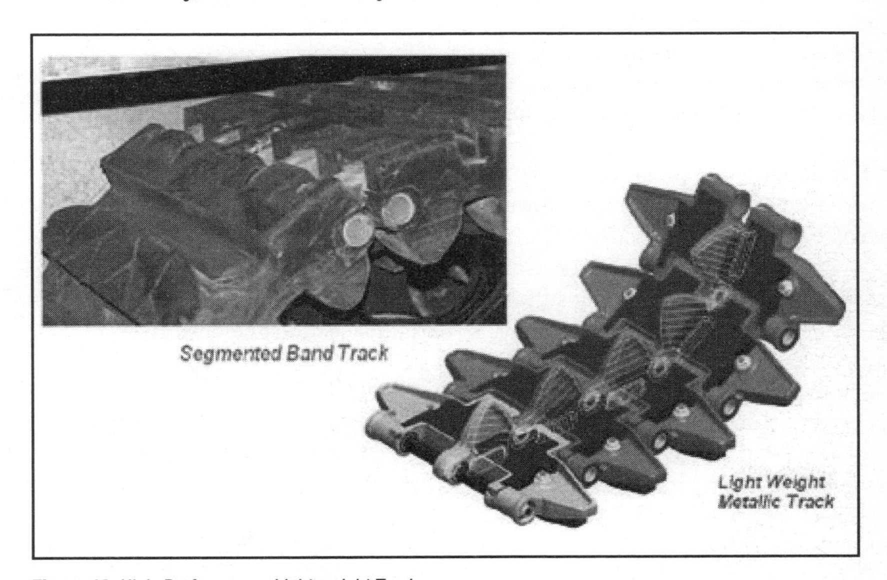

Figure 16: High-Performance, Light weight Track

Prognostics and Diagnostics for Operational Readiness and Condition-Based Maintenance

Near-term and future systems readiness and maintainability rely on the ability to detect health status and performance, and the environmental conditions that limit component lifetime. Improvements to current detection and prediction capabilities would reduce component losses and the logistics train, resulting in improved mission completion. This ATO develops prognostic software and application-specific sensors for remote health detection and prediction of vehicle and weapons systems component performance. Part of this capability is comprised by diagnostic sensors that enable health assessment. Prediction of remaining lifetime also requires holistic interpretation of the data, and is a function of both the component and data quality. To achieve both, it is imperative that commanders and logisticians be able to access the data expeditiously with a minimum of effort. This effort will develop a core "tag" with embedded sensors and processing that can be wirelessly interrogated. The system component's sensor history data will be analyzed by both on-board and post-processed prognostics algorithms developed in this ATO, in order to assess immediate readiness and remaining time to maintenance or lifetime. Resultant data will yield actionable information for both commander and logistician leading to increased readiness, enhanced awareness of materiel condition, increased confidence of mission completion, and smaller logistics footprint through condition-based maintenance.

JP-8 Reformation for Alternate Power Sources

The JP-8 Reformer for Alternate Power Sources ATO provides the research and development required to convert JP-8 fuel into a hydrogen-rich alternate fuel for downstream power generation. This program will develop a JP-8 reformer brassboard capable of removing sulfur and other aromatic contaminants that are detrimental to fuel cell operation. Careful selection and design of desulphurization, reformer, thermal, water, and sensor technologies are paramount. The design, tests, and operation of the JP-8 reformer brassboard will be highly dependent on the fuel cell system design. The brassboard will be designed to provide from 12–120 liters/minute of a high-grade, low-sulfur (<1.0 parts per million) hydrogen-rich fuel for continuous operation. This reformed fuel will power a commercially available fuel cell platform. This critical front-end reforming step will be an integral technology development enabling 10 kilowatt of available power for silent watch and other power requirements in the theater. The success of this reformer program is designed to complement parallel developments in suitable fuel cell architectures under development within the Department of Defense, Department of Energy, and commercial fuel cell developers.

ADVANCED SIMULATION

Research for Scalable Embedded Training and Mission Rehearsal

Embedded training (ET), a key performance parameter for Future Force vehicles and Soldiers, is also required by Abrams, Bradley, and Stryker vehicles, but has been slow to evolve. The Scalable Embedded Training and Mission Rehearsal ATO will support a common implementation strategy and address known technology shortfalls in ET across current and Future Force systems. This ATO will accelerate ET and mission rehearsal implementation; develop tactical engagement simulation sensors for dismounted Soldier training, size, power, and accuracy requirements; and provide ET risk mitigation for GSS, Heavy Brigade Combat Team, and Stryker Brigade Combat Team. The ATO will be completed in FY09 with field demonstrations of mission rehearsal and live, virtual, and constructive ET using Current Force combat vehicles and dismounted Soldiers as the experimental force.

Figure 17: Soldier Planning Interfaces & Networked Electronics

Simulated Severe Trauma for Medical Simulation

The Severe Trauma Simulation ATO researches technologies that prepare Soldiers physically and psychologically for the severe injuries encountered on the battlefield. Many Soldiers are not prepared for the shock of treating severe trauma and are less effective in the use of their medical skills. Experience shows that training traditional techniques for treating injuries can be improved. Realistic simulated trauma will allow Soldiers to master their skills and equipment before entering the battlefield. This effort will investigate battlefield injuries and evaluate them against current training; research effects of severe trauma on patients and caregivers; and design a methodology to support combat medic training with realistic battlefield injuries, including compartmentalized trauma, physiology, transfer of care, and time milestones of care. The ATO will provide prototype simulations with advances in materials (realistic skin, flesh, blood, bone, fluids, and organs), sensor technologies, and simulated fluid loss. The prototypes will function as standalone training systems, as well as external accessories for patient simulators and actors. In FY09, the ATO will evaluate the developed components in the current program of instruction to assess training effectiveness at military training venues.

BASIC RESEARCH

Basic research investments are a critical hedge in acquiring new knowledge in areas that hold great promise in advancing new and technically challenging Army capabilities and concepts to enable revolutionary advances and paradigm-shifting future operational capabilities. Areas of emerging interest and focus in basic research are: Neuroscience, Autonomous Systems, Quantum Information Science, Immersive Technology, Biotechnology, Nanotechnology, and Network Science. Investment in basic research within the Army provides insurance against an uncertain future and guards against technological surprise. And if we are successful, these investments will make it possible to conduct ever more complex military operations, with greater speed and precision, to devastate any adversary on any battlefield. The following is a brief summary of the areas of investment, the synergy among them, and some of the capabilities they may provide.

1 Neuroscience—Understanding how the human brain works

Fundamental to the conduct of military operations is superior Soldier performance. Understanding how the human brain works, i.e. determining the brain's "software," is key to developing these capabilities. When embedded into a wide range of military platforms, this "software" will provide superior training methods and human system interfaces that will be tuned to an individual's characteristics, thereby resulting in superior Soldier performance. Research in this area will also dramatically advance our ability to prevent and treat those suffering from various types of battlefield brain injury.

2 Autonomous Systems—Extending the operational effectiveness of Soldiers through robotic systems

A major military objective is to totally frustrate and defeat our adversaries across a wide spectrum of conflicts while dramatically increasing the survivability of our Soldiers by keeping them out of harm's way. Autonomous systems of extraordinary capability can fulfill this objective; however, they must be completely safe and secure while operating in highly complex operational environments. Achieving such levels of capability will require significant investments in highly sophisticated sense, response and processing systems approaching that of biological systems; major advances in artificial intelligence; the development of intelligent agents approaching human- performance levels; and advances in machine learning, swarming, and actuation and control.

3 Quantum Information Science—Overcoming the limitations of Moore's Law
Increasing demands for information to support rapid and effective decision-making on the battlefield require advanced sensor systems to collect relevant data, as well as the means for processing it into actionable forms. Major advancements in processing power are required to cope with the demand to process ever larger amounts of data. Investments in this area will exploit the massive parallelism of the quantum world to create computers that will dwarf the capabilities of the most powerful computers today, making them look like pocket calculators. The development of such computational systems will enable the embedding of high-performance computing in all military platforms including the Soldier's uniform.

4 Immersive Technology—The path to virtual reality training
The evolving threat environment continues to put increasing demand on the diversity and effectiveness of Soldier skills. To meet this demand, superior training tools and methods are needed. Virtual worlds can provide this capability; however, we are currently at primitive stages in their realization. With advances in computational processing and steady progress in understanding the brain's "software" comes the possibility of creating highly realistic virtual training environments inhabited by humanlike avatars. Such environments will provide a paradigm shift in the way we provide training, while achieving low-cost, safe, low-environmental impact, highly variable simulation environments for the future training of our soldiers.

5 Biotechnology—Leveraging four billion years of evolution
The increasing importance and demands for wide-area persistent surveillance create significant challenges for sensor systems, real-time processing of vast amounts of data, the real-time interpretation of information for decision-making and challenging power and energy requirements to support such demanding systems. Through four billion years of evolution, biological systems have engineered solutions to some of these challenges. We seek to leverage research in these areas for improving the performance of our Soldiers. Major investments in this area through reverse engineering will lead to totally new sensing systems, new ways for the rapid processing of data into information, the development of novel sense and response systems, and biologically inspired power and energy solutions for our Soldiers.

6 Network Science—Managing complex military operations with greater speed and precision
Networks tie together the following: highly distributed sensor systems for reconnaissance and surveillance, information for decision-making, Soldiers, and the execution of fast distributed precision fires. Better functioning networks are essential to advancing our ability to conduct complex military operations with greater speed and precision. However, our state of knowledge of these networks is relatively primitive and, as such, significantly impairs our ability to fully realize the potential that networks can provide on current and future battlefields. A new multidisciplinary approach is being implemented that combines communications, information and the social/human component of networks, and that changes the way we address the challenges associated with optimizing the use of networks. This new research effort will allow us to predict and optimize network performance through the creation of totally new design tools before we build them.

S&T ROLE IN FORMAL ACQUISITION MILESTONES

The Army S&T community role in acquisition involves not only technology development and transition, but also formal participation in milestone decisions for acquisition programs of record. As the component S&T executive, the Deputy Assistant Secretary of the Army (DASA) for Research and Technology (R&T) is responsible for conducting a technology readiness assessment (TRA) at milestone B and C decision points for major defense acquisition programs (MDAPs). This assessment has become even more important with recent statutory requirements for the Milestone Decision Authority (MDA) to certify to Congress that the technologies of an MDAP have been demonstrated in a relevant environment prior to making a milestone B decision. The TRA serves as the gauge of this readiness for the MDA's certification at both Army and Office of the Secretary of Defense levels. The TRA process is a collaborative effort carried out among the program office, the S&T community, and (for acquisition category (ACAT) 1D programs) the Office of the Undersecretary of Defense USD Acquisition Technology & Logistics (AT&L).

SUMMARY

The technological sophistication required for 21st century operations constantly increases with the broadening nature of threats and the greater availability of technology to our adversaries. Much of the Army's ongoing research is characterized as high-risk, high-payoff—the type that the private sector is not likely to sustain over the long haul because there is no linkage to acquisition programs at the outset of research. This high-risk research is essential if we are to achieve the technological breakthroughs for dramatic performance improvements in the Army's systems. One such breakthrough in guidance and control technology led to the Excalibur precision artillery munition that has virtually eliminated collateral damage to noncombatants. Today's Current Force has significant technology-enabled advantages as a result of the Army's past investments in S&T, particularly in night vision, precision munitions, and individual Soldier protection. Scientists and engineers continue to expand the limits of our understanding to provide technology to our Soldiers in the systems they use to achieve transformational capabilities required for decisive victories.

Appendices

Army Combat Organizations

Glossary of Terms

Systems by Contractors

Contractors by State

Points of Contact

Army Combat Organizations

Army organizations are inherently built around people and the tasks they must perform. Major combat organizations are composed of smaller forces, as shown here.

Squad
- Leader is a sergeant
- Smallest unit in Army organization
- Size varies depending on type: Infantry (9 Soldiers), Armor (4 Soldiers), Engineer (10 Soldiers)
- Three or four squads make up a platoon

Platoon
- Leader is a lieutenant
- Size varies: Infantry (40 Soldiers), Armor (4 tanks, 16 Soldiers)
- Three or four platoons make up a company

Company
- Leader is a captain
- Usually up to 220 Soldiers
- Artillery unit of this size is called a battery
- Armored Cavalry or Air Cavalry unit is called a troop
- Basic tactical element of the maneuver battalion or cavalry squadron
- Normally five companies make up a battalion

Battalion
- Leader is a lieutenant colonel
- Tactically and administratively self-sufficient
- Armored Cavalry and Air Cavalry equivalents are called squadrons
- Two or more combat battalions make up a brigade

Brigade
- Leader is a colonel
- May be employed on independent or semi-independent operations
- Combat, combat support, or service support elements may be attached to perform specific missions
- Normally three combat brigades are in a division

Division
- Leader is a major general
- Fully structured division has own brigade-size artillery, aviation, engineer, combat support, and service elements
- Two or more divisions make up a corps commanded by a lieutenant general

Glossary of Terms

Acquisition Categories (ACAT)

ACAT I programs are Milestone Decision Authority Programs (MDAPs [see also Major Defense Acquisition Program]) or programs designated ACAT I by the Milestone Decision Authority (MDA [see also Milestone Decision Authority]).

Dollar value: estimated by the Under Secretary of Defense (Acquisition and Technlogy) (USD [A&T]) to require an eventual total expenditure for research, development, test and evaluation (RDT&E) of more than $365 million in fiscal year (FY) 2000 constant dollars or, for procurement, of more than $2.190 billion in FY 2000 constant dollars. ACAT I programs have two sub-categories:

1. **ACAT ID**, for which the MDA is USD (A&T). The "D" refers to the Defense Acquisition Board (DAB), which advises the USD (A&T) at major decision points.
2. **ACAT IC**, for which the MDA is the DoD Component Head or, if delegated, the DoD Component Acquisition Executive (CAE). The "C" refers to Component. The USD (A&T) designates programs as ACAT ID or ACAT IC.

ACAT IA programs are MAISs (see also Major Automated Information System (MAIS) Acquisition Program), or programs designated by the Assistant Secretary of Defense for Command, Control, Communications, and Intelligence (ASD [C3I]) to be ACAT IA.

Estimated to exceed: $32 million in FY 2000 constant dollars for all expenditures, for all increments, regardless of the appropriation or fund source, directly related to the AIS definition, design, development, and deployment, and incurred in any single fiscal year; or $126 million in FY 2000 constant dollars for all expenditures, for all increments, regardless of the appropriation or fund source, directly related to the AIS definition, design, development, and deployment, and incurred from the beginning of the Materiel Solution Analysis Phase through deployment at all sites; or $378 million in FY 2000 constant dollars for all expenditures, for all increments, regardless of the appropriation or fund source, directly related to the AIS definition, design, development, deployment, operations and maintenance, and incurred from the beginning of the Materiel Solution Analysis Phase through sustainment for the estimated useful life of the system.

ACAT IA programs have two sub-categories:

1. **ACAT IAM**, for which the MDA is the Chief Information Officer (CIO) of the DoD, the ASD (C3I). The "M" refers to Major Automated Information System Review Council (MAISRC). (Change 4, 5000.2-R)
2. **ACAT IAC**, for which the DoD CIO has delegated milestone decision authority to the CAE or Component CIO. The "C" refers to Component.

ACAT II programs are defined as those acquisition programs that do not meet the criteria for an ACAT I program, but do meet the criteria for a major system, or are programs designated ACAT II by the MDA.

ACAT III programs are defined as those acquisition programs that do not meet the criteria for an ACAT I, an ACAT IA, or an ACAT II. The MDA is designated by the CAE and shall be at the lowest appropriate level. This category includes less-than-major AISs.

Acquisition Phase

All the tasks and activities needed to bring a program to the next major milestone occur during an acquisition phase. Phases provide a logical means of progressively translating broadly stated mission needs into well-defined system-specific requirements and ultimately into operationally effective, suitable, and survivable systems. The acquisition phases for the systems described in this handbook are defined below:

Technology Development Phase

The purpose of this phase is to reduce technology risk, determine and mature the appropriate set of technologies to be integrated into a full system, and to demonstrate critical technology elements on prototypes. Technology Development is a continuous technology discovery and development process reflecting close collaboration between the Science and Technology (S&T) community, the user, and the system developer. It is an iterative process designed to assess the viability of technologies while simultaneously refining user requirements. Entrance into this phase depends on the completion of the Analysis of Alternatives (AOA), a proposed materiel solution, and full funding for planned Technology Development Phase activity.

Engineering and Manufacturing Development (EMD) Phase
(Statutes applicable to the Systems Development and Demonstration phase shall be applicable to the EMD phase.)

The purpose of the EMD phase is to develop a system or an increment of capability; complete full system integration (technology risk reduction occurs during Technology Development); develop an affordable and executable manufacturing process; ensure operational supportability with particular attention to minimizing the logistics footprint; implement human systems integration (HSI); design for producibility; ensure affordability; protect critical program information by implementing appropriate techniques such as anti-tamper; and demonstrate system integration, interoperability, safety, and utility. The Capability Development Document, Acquisition Strategy, Systems Engineering Plan, and Test and Evaluation Master Plan (TEMP) shall guide this effort. Entrance into this phase depends on technology maturity (including software), approved requirements, and full funding. Unless some other factor is overriding in its impact, the maturity of the technology shall determine the path to be followed.

Production and Deployment Phase
The purpose of the Production and Deployment phase is to achieve an operational capability that satisfies mission needs. Operational test and evaluation shall determine the effectiveness and suitability of the system. The MDA shall make the decision to commit the DoD to production at Milestone C and shall document the decision in an Acquisition Decision Memorandum. Milestone C authorizes entry into low rate initial production (for MDAPs and major systems), into production or procurement (for non-major systems that do not require LRIP) or into limited deployment in support of operational testing for MAIS programs or software-intensive systems with no production components. The tables in Enclosure 4 identify the statutory and regulatory requirements that shall be met at Milestone C. Entrance into this phase depends on the following criteria: acceptable performance in developmental test and evaluation and operational assessment (OSD OT&E oversight programs); mature software capability; no significant manufacturing risks; manufacturing processes under control (if Milestone C is full-rate production); an approved Initial Capabilities Document (ICD) (if Milestone C is program initiation); an approved Capability Production Document (CPD); a refined integrated architecture; acceptable interoperability; acceptable operational supportability; and demonstration that the system is affordable throughout the life cycle, fully funded, and properly phased for rapid acquisition. The CPD reflects the operational requirements, informed by EMD results, and details the performance expected of the production system. If Milestone C approves LRIP, a subsequent review and decision shall authorize full-rate production.

Operations and Support Phase
The purpose of the Operations and Support phase is to execute a support program that meets materiel readiness and operational support performance requirements, and sustains the system in the most cost-effective manner over its total life cycle. Planning for this phase shall begin prior to program initiation and shall be documented in the Life-Cycle Sustainment Plan (LLSP). Operations and Support has two major efforts: life-cycle sustainment and disposal. Entrance into the Operations and Support Phase depends on meeting the following criteria: an approved CPD; an approved LCSP; and a successful Full-Rate Production (FRP) Decision.

Acquisition Program

A directed, funded effort designed to provide a new, improved or continuing weapons system or AIS capability in response to a validated operational need. Acquisition programs are divided into different categories that are established to facilitate decentralized decision-making, and execution and compliance with statutory requirements.

Advanced Concept Technology Demonstrations (ACTDs)

ACTDs are a means of demonstrating the use of emerging or mature technology to address critical military needs. ACTDs themselves are not acquisition programs, although they are designed to provide a residual, usable capability upon completion. If the user determines that additional units are needed beyond the residual capability and that these units can be funded, the additional buys shall constitute an acquisition program with an acquisition category generally commensurate with the dollar value and risk of the additional buy.

Automated Information System (AIS)
A combination of computer hardware and software, data, or telecommunications, that performs functions such as collecting, processing, transmitting, and displaying information. Excluded are computer resources, both hardware and software, that are physically part of, dedicated to, or essential in real time to the mission performance of weapon systems.

Commercial and Non-Developmental Items
Market research and analysis shall be conducted to determine the availability and suitability of existing commercial and non-developmental items prior to the commencement of a development effort, during the development effort, and prior to the preparation of any product description. For ACAT I and IA programs, while few commercial items meet requirements at a system level, numerous commercial components, processes, and practices have application to DoD systems.

Demilitarization and Disposal
At the end of its useful life, a system must be demilitarized and disposed of. During demilitarization and disposal, the program manager shall ensure materiel determined to require demilitarization is controlled and shall ensure disposal is carried out in a way that minimizes DoD's liability due to environmental, safety, security, and health issues.

Developmental Test and Evaluation (DT&E)
DT&E shall identify potential operational and technological capabilities and limitations of the alternative concepts and design options being pursued; support the identification and description of design technical risks; and provide data and analysis in support of the decision to certify the system ready for operational test and evaluation.

Joint Program Management
Any acquisition system, subsystem, component or technology program that involves a strategy that includes funding by more than one DoD component during any phase of a system's life cycle shall be defined as a joint program. Joint programs shall be consolidated and collocated at the location of the lead component's program office, to the maximum extent practicable.

Live Fire Test and Evaluation (LFT&E)
LFT&E must be conducted on a covered system, major munition program, missile program, or product improvement to a covered system, major munition program, or missile program before it can proceed beyond low-rate initial production. A covered system is any vehicle, weapon platform, or conventional weapon system that includes features designed to provide some degree of protection to users in combat and that is an ACAT I or II program. Depending upon its intended use, a commercial or non-developmental item may be a covered system, or a part of a covered system. (Change 4, 5000.2-R) Systems requiring LFT&E may not proceed beyond low-rate initial production until realistic survivability or lethality testing is completed and the report required by statute is submitted to the prescribed congressional committees.

Low Rate Initial Production (LRIP)
The objective of this activity is to produce the minimum quantity necessary to provide production-configured or representative articles for operational tests; establish an initial production base for the system; and permit an orderly increase in the production rate for the system, sufficient to lead to full-rate production upon successful completion of operational testing.

Major Automated Information System (MAIS) Acquisition Program
An AIS acquisition program that is (1) designated by ASD (C3I) as a MAIS, or (2) estimated to require program costs in any single year in excess of $32 million in FY 2000 constant dollars, total program costs in excess of $126 million in FY 2000 constant dollars, or total life-cycle costs in excess of $378 million in FY 2000 constant dollars. MAISs do not include highly sensitive classified programs.

Major Defense Acquisition Program (MDAP)
An acquisition program that is not a highly sensitive classified program (as determined by the Secretary of Defense) and that is: (1) designated by the USD (A&T) as an MDAP, or (2) estimated by the USD (A&T) to require an eventual total expenditure for research, development, test and evaluation of more than $365 million in FY 2000 constant dollars or, for procurement, of more than $2.190 billion in FY 2000 constant dollars.

Major Milestone
A major milestone is the decision point that separates the phases of an acquisition program. MDAP milestones include, for example, the decisions to authorize entry into the engineering and manufacturing development phase or full rate production. MAIS milestones may include, for example, the decision to begin program definition and risk reduction.

Major Systems
Dollar value: estimated by the DoD Component Head to require an eventual total expenditure for RDT&E of more than $140 million in FY 2000 constant dollars, or for procurement of more than $660 million in FY 2000 constant dollars.

Materiel Solution Analysis Phase
The purpose of this phase is to assess potential materiel solutions and to satisfy the phase-specific entrance criteria for the next program milestone designated by the MDA. Entrance into this phase depends upon an approved ICD resulting from the analysis of current mission performance and an analysis of potential concepts across the DoD components, international systems from allies, and cooperative opportunities.

Milestone Decision Authority (MDA)
The individual designated in accordance with criteria established by the USD (A&T), or by the ASD (C3I) for AIS acquisition programs, to approve entry of an acquisition program into the next phase.

Modifications
Any modification that is of sufficient cost and complexity that it could itself qualify as an ACAT I or ACAT IA program shall be considered for management purposes as a separate acquisition effort. Modifications that do not cross the ACAT I or IA threshold shall be considered part of the program being modified, unless the program is no longer in production. In that case, the modification shall be considered a separate acquisition effort. (Added from 5000.2-R)

Operational Support
The objectives of this activity are the execution of a support program that meets the threshold values of all support performance requirements and sustainment of them in the most life-cycle cost-effective manner. A follow-on operational testing program that assesses performance and quality, compatibility, and interoperability, and identifies deficiencies shall be conducted, as appropriate. This activity shall also include the execution of operational support plans, to include the transition from contractor to organic support, if appropriate. (Added from 5000.2-R)

Operational Test and Evaluation (OT&E)
OT&E shall be structured to determine the operational effectiveness and suitability of a system under realistic conditions (e.g., combat) and to determine if the operational performance requirements have been satisfied. The following procedures are mandatory: threat or threat representative forces, targets, and threat countermeasures, validated in coordination with Defense Intelligence Agency (DIA), shall be used; typical users shall operate and maintain the system or item under conditions simulating combat stress and peacetime conditions; the independent operational test activities shall use production or production representative articles for the dedicated phase of OT&E that supports the full-rate production decision, or for ACAT IA or other acquisition programs, the deployment decision; and the use of modeling and simulation shall be considered during test planning. There are more mandatory procedures (9 total) in 5000.2-R.

For additional information on acquisition terms, or terms not defined, please refer to AR 70-1, Army Acquisition Policy, available on the Internet at http://www.army.mil/usapa/epubs/pdf/r70_1.pdf; or DA PAM 70-3, Army Acquisition Procedures, available on the Internet at http://www.dtic.mil/whs/directives/corres/pdf/500002p.pdf.

Systems by Contractors

AAI Corp.
Extended Range Multipurpose (ERMP) Sky Warrior Unmanned Aircraft System (UAS)
Shadow Tactical Unmanned Aerial Vehicle (TUAV)

AAR Mobility Systems
Family of Medium Tactical Vehicles (FMTV)
Mobile Maintenance Equipment Systems (MMES)

Accenture
General Fund Enterprise Business Systems (GFEBS)
Global Command and Control System–Army (GCCS–A)

Action Manufacturing
2.75" Family of Rockets

ADSI
High Mobility Engineer Excavator (HMEE)

Aerial Machine and Tool, Inc.
Air Warrior (AW)

Aerojet
Guided Multiple Launch Rocket System (GMLRS)

Aerojet General
Tube-Launched, Optically-Tracked, Wire-Guided (TOW) Missiles

Aerovironment Inc.
Raven Small Unmanned Aircraft System (SUAS)

Agilent Technologies, Inc.
Calibration Sets Equipment (CALSETS)

Airborne Systems North America
Joint Precision Airdrop System (JPADS)

Airflyte Electronics Co.
Armored Knight

Agilent Technologies Inc.
Calibration Sets Equipment (CALSETS)

Alenia Aeronautica
Joint Cargo Aircraft

All American Racers Inc.
Raven Small Unmanned Aircraft System (SUAS)

Alliant Techsystems Inc.
2.75" Family of Rockets
Artillery Ammunition
Excalibur (XM982)
HELLFIRE Family of Missiles
Medium Caliber Ammunition
Precision Guidance Kit
Small Arms–Crew Served Weapons
Small Caliber Ammunition
Spider
Tank Ammunition

Allison
Family of Medium Tactical Vehicles (FMTV)

Allison Transmissions
Heavy Expanded Mobility Tactical Truck (HEMTT)/HEMTT Extended Service Program (ESP)
Palletized Load System (PLS) and PLS Extended Service Program (ESP)

AM General (AMG)
High Mobility Multipurpose Wheeled Vehicle (HMMWV)
Improved Ribbon Bridge (IRB)

American Eurocopter
Light Utility Helicopter (LUH)

American Ordnance
Artillery Ammunition
Spider

American Science & Engineering, Inc.
Non-Intrusive Inspection Systems (NII)

AMT
Mortar Systems

AMTEC Corp.
Medium Caliber Ammunition

Anniston Army Depot (ANAD)
Abrams Upgrade
Paladin/Field Artillery Ammunition Supply Vehicle (FAASV)

ANP Technologies
Joint Chemical Biological Radiological Agent Water Monitor (JCBRAWM)

Apptricity Corp.
Transportation Coordinators' Automated Information for Movement System II (TC-AIMS II)

ArgonST Radix
Guardrail Common Sensor (GR/CS)

Armacel Armor
Interceptor Body Armor

Armtec Defense
Artillery Ammunition

ATK
Tube-Launched, Optically-Tracked, Wire-Guided (TOW) Missiles

Atlantic Inertial Units
Excalibur (XM982)

Austal USA
Joint High Speed Vessel (JHSV)

Avon Protection Systems
Joint Service General Purpose Mask (JSGPM)

BAE Systems
Air Warrior (AW)
Airborne Reconnaissance Low (ARL)
Armored Security Vehicle (ASV)
Bradley Upgrade
Family of Medium Tactical Vehicles (FMTV)
Heavy Loader

High Mobility Artillery Rocket System (HIMARS)
High Mobility Multipurpose Wheeled Vehicle (HMMWV)
Interceptor Body Armor
Joint Tactical Ground Stations (JTAGS)
Joint Tactical Radio System Airborne, Maritime/Fixed Station (JTRS AMF)
Joint Tactical Radio System Ground Mobile Radios (JTRS GMR)
Joint Tactical Radio System Handheld, Manpack, and Small Form Fit (JTRS HMS)
Lightweight 155mm Howitzer (LW155)
Mine Protection Vehicle Family (MPVF)
Multiple Launch Rocket System (MLRS) M270A1
Paladin/Field Artillery Ammunition Supply Vehicle (FAASV)
Thermal Weapon Sight
Tube-Launched, Optically-Tracked, Wire-Guided (TOW) Missiles
Warfighter Information Network–Tactical (WIN–T) Increment 2
Warfighter Information Network–Tactical (WIN–T) Increment 3

BAE Systems Bofors Defense (teamed with Raytheon)
Excalibur (XM982)

BAE Systems Land & Armaments
Joint Light Tactical Vehicle (JLTV)

BAE Systems Land & Armaments, Ground Systems Division
Mine Resistant Ambush Protected Vehicles (MRAP)

BAE/Holston
Spider

BAE-TVS
Mine Resistant Ambush Protected Vehicles (MRAP)

Barrett Firearms Manufacturing
Sniper Systems

Bell Helicopter
Kiowa Warrior

Berg Companies, Inc.
Force Provider (FP)

Binary Group
General Fund Enterprise Business Systems (GFEBS)

Boeing
Early Infantry Brigade Combat Team (E-IBCT) Capabilities
Chinook/CH-47 Improved Cargo Helicopter (ICH)
Joint Air-to-Ground Missile (JAGM)
Joint Tactical Radio System Ground Mobile Radios (JTRS GMR)
Joint Tactical Radio System Network Enterprise Domain (JTRS NED)
Longbow Apache
PATRIOT (PAC-3)
Surface Launched Advanced Medium Range Air-to-Air Missile (SLAMRAAM)

Booz Allen Hamilton
Distributed Common Ground System (DCGS–Army)
AcqBusiness

Secure Mobile Anti-Jam Reliable Tactical–Terminal (SMART–T)

Bracco Diagnostics, Inc.
Joint Service Personnel/Skin Decontamination System (JSPDS)

Bren-Tronics
Raven Small Unmanned Aircraft System (SUAS)

Bruhn New-Tech
Joint Warning and Reporting Network (JWARN)

CACI
Aerial Common Sensor (ACS)
Biometric Family of Capabilities for Full Spectrum Operations (BFCFSO)
Airborne Reconnaissance Low (ARL)
Army Key Management System (AKMS)
Biometric Enterprise Core Capability (BECC)
Combat Service Support Communications (CSS Comms)
Single Channel Ground and Airborne Radio System (SINCGARS)

CACI Technologies
Nuclear Biological Chemical Reconnaissance Vehicle (NBCRV)–Stryker

CAE
One Semi-Automated Forces (OneSAF) Objective System

CAO USA
Light Utility Helicopter (LUH)

Carleton Technologies, Inc.
Air Warrior (AW)

CAS, Inc.
Joint Land Attack Cruise Missile Defense Elevated Netted Sensor System (JLENS)
Sentinel
Surface Launched Advanced Medium Range Air-to-Air Missile (SLAMRAAM)

Casteel Manufacturing
Line Haul Tractor

Caterpillar
Family of Medium Tactical Vehicles (FMTV)
Heavy Expanded Mobility Tactical Truck (HEMTT)/HEMTT Extended Service Program (ESP)

Caterpillar Defense and Federal Products (OEM)
Heavy Loader

CDL Systems
Shadow Tactical Unmanned Aerial Vehicle (TUAV)

CDW-G
Medical Communications for Combat Casualty Care (MC4)

CECOM Software Engineering Center
Maneuver Control System (MCS)

Ceradyne, Inc.
Interceptor Body Armor

Charleston Marine Containers
Force Provider (FP)

Chenega
Armored Security Vehicle (ASV)

Cisco
Common Hardware Systems (CHS)

CMI
Shadow Tactical Unmanned Aerial Vehicle (TUAV)

Colt's Manufacturing
Small Arms–Individual Weapons

Composix
Stryker

Computer Sciences Corp. (CSC)
AcqBusiness
Advanced Field Artillery Tactical Data System (AFATDS)
Global Combat Support System–Army (GCSS–Army)
Installation Protection Program (IPP) Family of Systems
Medical Simulation Training Center (MSTC)

COMTECH
Force XXI Battle Command Brigade-and-Below (FBCB2)

COMTECH Mobile Datacom
Movement Tracking System (MTS)

Critical Solutions International, Inc.
Mine Protection Vehicle Family (MPVF)

CSS
Army Key Management System (AKMS)

Cubic Defense Systems
Instrumentable–Multiple Integrated Laser Engagement System (I–MILES)

Cummins Power Generation
Tactical Electric Power (TEP)

Cummins Mid-South LLC
Armored Security Vehicle (ASV)

Daimler Truck, North America/ Freightliner
Line Haul Tractor

Data Link Solutions
Multifunctional Information Distribution System (MIDS)–Joint Tactical Radio System (JTRS)

Data Path Inc.
Warfighter Information Network–Tactical (WIN–T) Increment 1

Defiance
High Mobility Multipurpose Wheeled Vehicle (HMMWV)

DELL
Common Hardware Systems (CHS)

Deloitte LLP
AcqBusiness

Detroit Diesel
Line Haul Tractor
Palletized Load System (PLS) and PLS Extended Service Program (ESP)

Dewey Electronics
Tactical Electric Power (TEP)

DRS Fermont
Tactical Electric Power (TEP)

DRS Mobile Environmental Systems
Close Combat Tactical Trainer (CCTT)

DRS Optronics Inc.
Kiowa Warrior

DRS Sustainment Systems, Inc. (DRS-SSI)
Armored Knight
Modular Fuel System (MFS)

DRS Tactical Systems
Armored Knight

DRS Technical Solutions
Tactical Electric Power (TEP)

DRS Technologies
Bradley Upgrade
Chemical Biological Protective Shelter (CBPS)
Common Hardware Systems (CHS)
Force XXI Battle Command Brigade-and-Below (FBCB2)
Integrated Family of Test Equipment (IFTE)
Joint Service Transportable Decontamination System (JSTDS)–Small Scale (SS)
Thermal Weapon Sight

DSE (Balimoy) Corp.
Medium Caliber Ammunition

Ducommun AeroStructures
Longbow Apache

DynCorp
Fixed Wing

Dynetics, Inc.
Calibration Sets Equipment (CALSETS)

Dynetics Millennium Davidson (DMD)
Integrated Air and Missile Defense (IAMD)

DynPort Vaccine
Chemical Biological Medical Systems–Prophylaxis

EADS North America
Light Utility Helicopter (LUH)

ECBC
Weapons of Mass Destruction Elimination

ECS
Common Hardware Systems (CHS)

E.D. Etnyre and Co.
Modular Fuel System (MFS)

EG&G
Force Protection Systems

Elbit Systems
Common Hardware Systems (CHS)

Elbit Systems of America
Bradley Upgrade
Kiowa Warrior
Mortar Systems

Engineering Professional Services
Advanced Field Artillery Tactical Data System (AFATDS)

Engineering Research and Development Command
Transportation Coordinators' Automated Information for Movement System II (TC-AIMS II)

Engineering Solutions & Products Inc.
Force XXI Battle Command Brigade-and-Below (FBCB2)
Global Command and Control System–Army (GCCS–A)

FBM Babcock Marine
Improved Ribbon Bridge (IRB)

Fabrique National Manufacturing, LLC
Small Arms–Crew Served Weapons

Fairfield
Distributed Learning System (DLS)

FASCAN International
Countermine

Fidelity Technologies Corp.
Tactical Electric Power (TEP)

FLIR Systems, Inc.
Lightweight Laser Designator Range Finder

Fluke Corp.
Calibration Sets Equipment (CALSETS)

Force Protection Industries, Inc.
Mine Protection Vehicle Family (MPVF)
Mine Resistant Ambush Protected Vehicles (MRAP)

Future Research Corp.
Transportation Coordinators' Automated Information for Movement System II (TC-AIMS II)

General Atomics, Aeronautical Systems Inc.
Extended Range Multipurpose (ERMP) Sky Warrior Unmanned Aircraft System (UAS)

General Dynamics
2.75" Family of Rockets
Abrams Upgrade
Advanced Field Artillery Tactical Data System (AFATDS)
Biometric Enterprise Core Capabilities (BECC)
Biometric Family of Capabilities for Full Spectrum Operations (BFCFSO)
Forward Area Air Defense Command and Control (FAAD C2)
Global Command and Control System–Army (GCCS–A)
Ground Soldier System (GSS)
Maneuver Control System (MCS)
Medical Communications for Combat Casualty Care (MC4)
Mounted Soldier
Prophet
Shadow Tactical Unmanned Aerial Vehicle (TUAV)
Small Caliber Ammunition
Stryker

General Dynamics Advanced Information Systems
Joint High Speed Vessel (JHSV)

General Dynamics Armament and Technical Products (GDATP) Division
2.75" Family of Rockets
Lightweight .50 cal Machine Gun
Joint Biological Point Detection System (JBPDS)
Small Arms–Crew Served Weapons

General Dynamics C4 Systems, Inc.
Air Warrior (AW)
Early Infantry Brigade Combat Team (E-IBCT) Capabilities
Common Hardware Systems (CHS)
Joint Tactical Radio System Airborne, Maritime/Fixed Station (JTRS AMF)
Joint Tactical Radio System Handheld, Manpack, and Small Form Fit (JTRS HMS)
Mortar Systems
Warfighter Information Network–Tactical (WIN–T) Increment 1
Warfighter Information Network–Tactical (WIN–T) Increment 2
Warfighter Information Network–Tactical (WIN–T) Increment 3

General Dynamics C4I Systems, Inc.
Mortar Systems

General Dynamics European Land Systems–Germany (GDELS–G)
Improved Ribbon Bridge (IRB)

General Dynamics Information Technology
Global Command and Control System–Army (GCCS–A)

General Dynamics Land Systems
Nuclear Biological Chemical Reconnaissance Vehicle (NBCRV)–Stryker

General Dynamics Land Systems–Canada
Mine Resistant Ambush Protected Vehicles (MRAP)

General Dynamics Ordnance and Tactical Systems
2.75" Family of Rockets
Excalibur (XM982)
Medium Caliber Ammunition
Small Caliber Ammunition
Tank Ammunition

General Dynamics Ordnance and Tactical Systems–Scranton Operations
Artillery Ammunition

General Dynamics SATCOM Tech
Warfighter Information Network–Tactical (WIN–T) Increment 1

General Electric (GE)
Black Hawk/UH-60

General Motors (GM)
High Mobility Multipurpose Wheeled Vehicle (HMMWV)

General Tactical Vehicle
Joint Light Tactical Vehicle (JLTV)

Gentex Corp.
Air Warrior (AW)

GEP
High Mobility Multipurpose Wheeled Vehicle (HMMWV)

Gibson and Barnes
Air Warrior (AW)

Global Defense Engineering
Force Provider (FP)

Group Home Foundation, Inc.
Joint Chem/Bio Coverall for Combat Vehicle Crewman (JC3)

GT Machine and Fabrication
Palletized Load System (PLS) and PLS Extended Service Program (ESP)

GTSI
Global Command and Control System–Army (GCCS–A)
Maneuver Control System (MCS)
Medical Communications for Combat Casualty Care (MC4)

Gulfstream
Fixed Wing

Gyrocam Systems LLC
Countermine

Hamilton Sundstrand
Black Hawk/UH-60

Harris Corp.
Defense Enterprise Wideband SATCOM Systems (DEWSS)
High Mobility Artillery Rocket System (HIMARS)
Joint Tactical Radio System Network Enterprise Domain (JTRS NED)
Multiple Launch Rocket System (MLRS) M270A1
Warfighter Information Network–Tactical (WIN–T) Increment 2
Warfighter Information Network–Tactical (WIN–T) Increment 3

Heckler and Koch Defense Inc.
Small Arms–Individual Weapons

HELLFIRE LLC
HELLFIRE Family of Missiles

Hewlett Packard
Common Hardware Systems (CHS)

Holland Hitch
Line Haul Tractor

Honeywell
Abrams Upgrade
Armored Knight
Early Infantry Brigade Combat Team (E-IBCT) Capabilities
Chinook/CH-47 Improved Cargo Helicopter (ICH)
Guided Multiple Launch Rocket System (GMLRS)
Kiowa Warrior

Howmet Castings
Lightweight 155mm Howitzer (LW155)

Hunter Man.
Force Provider (FP)

IBM
Distributed Learning System (DLS)

ICx™ Technologies Inc.
Joint Nuclear Biological Chemical Reconnaissance System (JNBCRS)

Idaho Technologies
Chemical Biological Medical Systems–Diagnostics

ILEX
Distributed Common Ground System (DCGS–Army)

iLumina Solutions
General Fund Enterprise Business Systems (GFEBS)

iRobot
Early Infantry Brigade Combat Team (E-IBCT) Capabilities

Institute for Defense Analysis
Aerial Common Sensor (ACS)

Intercoastal Electronics
Improved Target Acquisition System (ITAS)

Interstate Electronics
Stryker

ITT
Joint Tactical Radio System Network Enterprise Domain (JTRS NED)
Single Channel Ground and Airborne Radio System (SINCGARS)

ITT-CAS, Inc.
Counter-Rocket, Artillery and Mortar (C-RAM)
Forward Area Air Defense Command and Control (FAAD C2)

ITT Industries
Defense Enterprise Wideband SATCOM Systems (DEWSS)
Helmet Mounted Enhanced Vision Devices

JANUS Research
Secure Mobile Anti-Jam Reliable Tactical–Terminal (SMART–T)

JCB Inc.
High Mobility Engineer Excavator (HMEE)

JLG Industries, Inc.
All Terrain Lifter Army System (ATLAS)

Johns Hopkins University Applied Physics Laboratory
Defense Enterprise Wideband SATCOM Systems (DEWSS)
Medical Communications for Combat Casualty Care (MC4)

Kaegan Corp.
Close Combat Tactical Trainer (CCTT)

Kalmar RT Center LLC
Rough Terrain Container Handler (RTCH)

Kidde Dual Spectrum
Paladin/Field Artillery Ammunition Supply Vehicle (FAASV)

King Aerospace
Fixed Wing

Kipper Tool Company
Mobile Maintenance Equipment Systems (MMES)

Klune
Tube-Launched, Optically-Tracked, Wire-Guided (TOW) Missiles

Knight's Armaments Co.
Sniper Systems

Kongsberg Defence & Aerospace
Common Remotely Operated Weapon Station (CROWS)

L-3 Communications
Aviation Combined Arms Tactical Trainer (AVCATT)
Battle Command Sustainment Support Systems (BCS3)
Biometric Enterprise Core Capability (BECC)
Biometric Family of Capabilities for Full Spectrum Operations (BFCFSO)
Bradley Upgrade
Extended Range Multipurpose (ERMP) Sky Warrior Unmanned Aircraft System (UAS)
Force Protection Systems
Guardrail Common Sensor (GR/CS)
HELLFIRE Family of Missiles
Prophet
Raven Small Unmanned Aircraft System (SUAS)
Tank Ammunition
Warfighter Information Network–Tactical (WIN–T) Increment 2
Warfighter Information Network–Tactical (WIN–T) Increment 3

L-3 Communications Cincinnati Electronics
Lightweight Laser Designator Range Finder (LLDR)

L-3 Communications Electro-Optic Systems
Helmet Mounted Enhanced Vision Devices

L-3 Communications Integrated Systems, L.P.
Joint Cargo Aircraft

L-3 Communications Space & Navigation
High Mobility Artillery Rocket System (HIMARS)
Multiple Launch Rocket System (MLRS) M270A1

L-3 Communications Titan Group
Battle Command Sustainment Support System (BCS3)
Medical Communications for Combat Casualty Care (MC4)
Transportation Coordinators' Automated Information for Movement Systems II (TC-AIMS II)

L-3 Cyterra Corp.
Countermine

L-3 Global Communications Solutions, Inc.
Combat Service Support Communications (CSS Comms)

L-3 Interstate Electronics Corp
Precision Guidance Kit

L-3 Westwood
Tactical Electric Power (TEP)

L-3/IAC
Non Line of Sight–Launch System (NLOS–LS)

Lapeer Industries, Inc.
Armored Security Vehicle (ASV)

Letterkenny Army Depot
Force Provider (FP)
High Mobility Multipurpose Wheeled Vehicle (HMMWV)

Lex Products Corp.
Force Provider (FP)

Lincoln Labs
Secure Mobile Anti-Jam Reliable Tactical–Terminal (SMART–T)

Litton Advanced Systems
Airborne Reconnaissance Low (ARL)

LMI Consulting
Global Combat Support System–Army (GCSS–Army)

Lockheed Martin
Airborne Reconnaissance Low (ARL)
Battle Command Sustainment Support System (BCS3)
Distributed Learning System (DLS)
Global Command and Control System–Army (GCCS–A)
Guardrail Common Sensor (GR/CS)
Guided Multiple Launch Rocket System (GMLRS)
HELLFIRE Family of Missiles
High Mobility Artillery Rocket System (HIMARS)
Javelin
Joint Air-to-Ground Missile (JAGM)
Joint Light Tactical Vehicle (JLTV)
Joint Tactical Radio System Airborne, Maritime/Fixed Station (JTRS AMF)
Longbow Apache
Maneuver Control System (MCS)
Multiple Launch Rocket System (MLRS) M270A1
Non Line of Sight–Launch System (NLOS–LS)
One Semi-Automated Forces (OneSAF) Objective System
PATRIOT (PAC-3)

Lockheed Martin Information Systems
Joint Land Component Constructive Training Capability (JLCCTC)

Lockeed Martin Missiles & Fire Control
Early Infantry Brigade Combat Team (E-IBCT) Capabilities

Lockheed Martin Mission Systems
Warfighter Information Network–Tactical (WIN–T) Increment 2
Warfighter Information Network–Tactical (WIN–T) Increment 3

Lockheed Martin Simulation, Training and Support
Close Combat Tactical Trainer (CCTT)

Longbow LLC
HELLFIRE Family of Missiles

LTI DataComm, Inc.
Combat Service Support Communications (CSS Comms)

M7 Aerospace
Fixed Wing

Maine Military Authority
High Mobility Multipurpose Wheeled Vehicle (HMMWV)

Marsh Industrial
Force Provider (FP)

Martin Diesel
Secure Mobile Anti-Jam Reliable Tactical–Terminal (SMART–T)

Marvin Land Systems
Paladin/Field Artillery Ammunition Supply Vehicle (FAASV)

MaTech
Mortar Systems

MEADS International
Medium Extended Air Defense System (MEADS)

Medical Education Technologies
Medical Simulation Training Center (MSTC)

Meggitt Defense Systems
Close Combat Tactical Trainer (CCTT)

Meridian Medical Technologies
Chemical Biological Medical Systems–Therapeutics

Meritor
Family of Medium Tactical Vehicles (FMTV)
Line Haul Tractor

Michelin
Heavy Expanded Mobility Tactical Truck (HEMTT)/HEMTT Extended Service Program (ESP)
Palletized Load System (PLS) and PLS Extended Service Program (ESP)

MICOR Industries, Inc.
Common Remotely Operated Weapon Station (CROWS)

Mil-Mar Century, Inc.
Load Handling System Compatible Water Tank Rack (Hippo)

MITRE
Aerial Common Sensor (ACS)
Distributed Common Ground System (DCGS–Army)

Mittal
Stryker

Moog
HELLFIRE Family of Missiles
Tube-Launched, Optically-Tracked, Wire-Guided (TOW) Missiles

Mountain High Equipment & Supply Co.
Air Warrior (AW)

MPRI (An L-3 Company)
Distributed Learning System (DLS)

MPRI L-3 Communications
Global Combat Support System–Army (GCSS–Army)

Naval Air Warfare Center Aircraft Division
Weapons of Mass Destruction Elimination

Navistar Defense
Mine Resistant Ambush Protected Vehicles (MRAP)

NIITEK
Countermine

Northrop Grumman
Air/Missile Defense Planning and Control System (AMDPCS)
Battle Command Sustainment Support (BCS3)
Command Post Systems and Integration (CPS&I)
Defense Enterprise Wideband SATCOM Systems (DEWSS)
Distributed Common Ground System (DCGS–Army)
Global Combat Support System–Army (GCSS–Army)
Guardrail Common Sensor (GR/CS)
Integrated Air and Missile Defense (IAMD)
Integrated Family of Test Equipment (IFTE)
Joint Tactical Radio System Airborne, Maritime/Fixed Station (JTRS AMF)
Joint Tactical Radio System Ground Mobile Radios (JTRS GMR)
Joint Tactical Radio System Network Enterprise Domain (JTRS NED)
Longbow Apache
Movement Tracking System (MTS)
Paladin/Field Artillery Ammunition Supply Vehicle (FAASV)

Northrop Grumman Electronic Systems
Joint Tactical Ground Stations (JTAGS)

Northrop Grumman Guidance and Electronics Company Inc., Laser Systems
Lightweight Laser Designator Range Finder (LLDR)

Northrop Grumman Information Technology (NGIT)
Joint Warning and Reporting Network (JWARN)
One Semi-Automated Forces (OneSAF) Objective System

Northrop Grumman Integrated Systems
Countermine

Northrop Grumman Mission Systems
Counter-Rocket, Artillery and Mortar (C-RAM)
Joint Effects Model (JEM)

Northrop Grumman Space & Mission Systems Corp.
Force XXI Battle Command Brigade-and-Below (FBCB2)
Forward Area Air Defense Command and Control (FAAD C2)

Olin Corp.
Small Caliber Ammunition

Oppenheimer
Armored Knight

Oshkosh Truck Corp.
Dry Support Bridge (DSB)
Heavy Expanded Mobility Tactical Truck (HEMTT)/HEMTT Extended Service Program (ESP)
Improved Ribbon Bridge (IRB)
Palletized Load System (PLS) and PLS Extended Service Program (ESP)

Osiris Therapeutics
Chemical Biological Medical Systems–Therapeutics

Overwatch Systems
Early Infantry Brigade Combat Team (E-IBCT) Capabilities
Distributed Common Ground System (DCGS–Army)

Oxygen Generating Systems International
Air Warrior (AW)

Pennsylvania State University
Meteorological Measuring Set–Profiler (MMS–P)

PharmAthene
Chemical Biological Medical Systems–Prophylaxis

Pierce Manufacturing
Line Haul Tractor

Pine Bluff Arsenal
Screening Obscuration Device (SOD) –Visual Restricted (Vr)

PKMM
Forward Area Air Defense Command and Control (FAAD C2)

Power Manufacturing, Inc.
Mobile Maintenance Equipment Systems (MMES)

Precision Castparts Corp.
Lightweight 155mm Howitzer (LW155)

Radix
Aerial Common Sensor (ACS)

Rapiscan Systems
Non-Intrusive Inspection Systems (NII)

Raytheon
Advanced Field Artillery Tactical Data System (AFATDS)
Armored Knight
Bradley Upgrade
Early Infantry Brigade Combat Team (E-IBCT) Capabilities
Distributed Common Ground System (DCGS–Army)
Ground Soldier System (GSS)
Excalibur (XM982)
Improved Target Acquisition System (ITAS)
Integrated Air and Missile Defense (IAMD)
Javelin
Joint Air-to-Ground Missile (JAGM)
Joint Land Attack Cruise Missile Defense Elevated Netted Sensor (JLENS)
Joint Tactical Radio System Airborne, Maritime/Fixed Station (JTRS AMF)
Non Line of Sight–Launch System (NLOS–LS)
PATRIOT (PAC-3)
Secure Mobile Anti-Jam Reliable Tactical–Terminal (SMART–T)
Surface Launched Advanced Medium Range Air-to-Air Missile (SLAMRAAM)
Thermal Weapon Sight

Raytheon Missile Systems
Tube-Launched, Optically-Tracked, Wire-Guided (TOW) Missiles

Raytheon Technical Services, Inc.
Air Warrior (AW)

Red River Army Depot
High Mobility Multipurpose Wheeled Vehicle (HMMWV)

Remington
Sniper Systems

Robertson Aviation
Chinook/CH-47 Improved Cargo Helicopter (ICH)

Rock Island Arsenal
Mobile Maintenance Equipment Systems (MMES)

Rockwell Collins
Black Hawk/UH-60
Chinook/CH-47 Improved Cargo Helicopter (ICH)
Close Combat Tactical Trainer (CCTT)
Ground Soldier System (GSS)
Joint Tactical Radio System Ground Mobile Radios (JTRS GMR)
Joint Tactical Radio System Handheld, Manpack, and Small Form Fit (JTRS HMS)
Joint Tactical Radio System Network Enterprise Domain (JTRS NED)
NAVSTAR Global Positioning System (GPS)
Shadow Tactical Unmanned Aerial Vehicle (TUAV)

Rohde and Schwarz
Test Equipment Modernization (TEMOD)

Rolls Royce Corp.
Kiowa Warrior

Schutt Industries
Light Tactical Trailer (LTT)
Tactical Electric Power (TEP)

Science Applications International Corp. (SAIC)
Army Key Management System (AKMS)
Early Infantry Brigade Combat Team (E-IBCT) Capabilities
Calibration Sets Equipment (CALSETS)
Distributed Common Ground System (DCGS–Army)
Installation Protection Program (IPP) Family of Systems
Instrumentable–Multiple Integrated Laser Engagement System (I–MILES)
Non-Intrusive Inspection Systems (NII)
One Semi-Automated Forces (OneSAF) Objective System

Science and Engineering Services, Inc. (SESI)
Air Warrior (AW)

Integrated Family of Test Equipment (IFTE)
Joint Biological Standoff Detection System (JBSDS)

Secure Communications Systems, Inc.
Air Warrior (AW)

Segovia Global IP Services
Combat Service Support Communications (CSS Comms)

Sierra Nevada Corp.
Airborne Reconnaissance Low (ARL)
Army Key Management System (AKMS)
Shadow Tactical Unmanned Aerial Vehicle (TUAV)

Sikorsky
Black Hawk/UH-60

Sikorsky Aircraft
Light Utility Helicopter (LUH)

Silver Eagle Manufacturing Company (SEMCO)
Light Tactical Trailer (LTT)

Simulation, Training & Instrumentation (STRICOM)
Abrams Upgrade

Skillsoft
Distributed Learning System (DLS)

Smiths Detection, Inc.
Chemical Biological Protective Shelter (CBPS)
Joint Chemical Agent Detector (JCAD)

Meteorological Measuring Set–Profiler (MMS–P)

Snap-on Industrial
Mobile Maintenance Equipment Systems (MMES)

SNC Technologies
Artillery Ammunition
Small Caliber Ammunition

SNVC
General Fund Enterprise Business Systems (GFEBS)
Global Combat Support System–Army (GCSS–Army)

Southwest Research Institute
Chemical Biological Medical Systems–Therapeutics

Summa Technologies
Palletized Load System (PLS) and PLS Extended Service Program (ESP)

Sun MicroSystems
Common Hardware Systems (CHS)

Sypris
Army Key Management System (AKMS)

Systems Technologies (Systek), Inc.
Combat Service Support Communications (CSS Comms)

Tapestry Solutions
Battle Command Sustainment Support System (BCS3)
Joint Land Component Constructive Training Capability (JLCCTC)

Taylor-Wharton
Air Warrior (AW)

TCOM
Joint Land Attack Cruise Missile Defense Elevated Netted Sensor System (JLENS)

Tecom
Shadow Tactical Unmanned Aerial Vehicle (TUAV)

Teledyne
Secure Mobile Anti-Jam Reliable Tactical–Terminal (SMART–T)

Telephonics Corp.
Air Warrior (AW)

Telos Corp.
Combat Service Support Communications (CSS Comms)

Textron Defense Systems
Early Infantry Brigade Combat Team (E-IBCT) Capabilities
Spider

Textron Marine & Land Systems
Armored Knight
Armored Security Vehicle (ASV)

Thales
Tube-Launched, Optically-Tracked, Wire-Guided (TOW) Missiles

Thales Communications
Joint Tactical Radio System Handheld, Manpack, and Small Form Fit (JTRS HMS)

Thales Raytheon Systems
Sentinel

The Aegis Technology Group Inc.
One Semi-Automated Forces (OneSAF) Objective System

The Research Associates
Biometric Enterprise Core Capability (BECC)
Biometric Family of Capabilities for Full Spectrum Operations (BFCFSO)

Titan Corp.
Advanced Field Artillery Tactical Data System (AFATDS)

Tobyhanna Army Depot
Combat Service Support Communications (CSS Comms)
Forward Area Air Defense Command and Control (FAAD C2)

Tri-Tech USA Inc.
Force Provider (FP)

Triumph Systems Los Angeles
Lightweight 155mm Howitzer (LW155)

UAV Engines Limited
Shadow Tactical Unmanned Aerial Vehicle (TUAV)

Ultra, Inc.
Air/Missile Defense Planning and Control System (AMDPCS)

UNICOR
Single Channel Ground and Airborne Radio System (SINCGARS)

UNICOR Protective Materials Co.
Interceptor Body Armor

Universal Systems and Technology
Instrumentable–Multiple Integrated Laser Engagement System (I–MILES)

URS Corp.
Chemical Demilitarization

U.S. Army Information Systems Engineering Command
Defense Enterprise Wideband SATCOM Systems (DEWSS)

US Divers
Air Warrior (AW)

Vertigo Inc.
Force Provider (FP)

Vertu Corp.
Small Arms–Individual Weapons

ViaSat
Multifunctional Information Distribution System (MIDS)–Joint Tactical Radio System

Vickers
High Mobility Artillery Rocket System (HIMARS)

Viecore
Maneuver Control System (MCS)

Vision Technology Miltope Corp.
Integrated Family of Test Equipment (IFTE)

Watervliet Arsenal
Lightweight 155mm Howitzer (LW155)
Mortar Systems

WESCAM
Airborne Reconnaissance Low (ARL)

Westwind Technologies, Inc.
Air Warrior (AW)

Wexford Group International
Battle Command Sustainment Support (BCS3)

Williams Fairey Engineering Ltd.
Dry Support Bridge (DSB)

Wolf Coach, Inc., an L-3 Communications Company
Weapons of Mass Destruction Elimination

XMCO
Dry Support Bridge (DSB)
Heavy Loader
High Mobility Engineer Excavator (HMEE)

ZETA
Guardrail Common Sensor (GR/CS)

Contractors by State

Alabama
Anniston Army Depot
Austal USA
BAE Systems
Boeing
CAS, Inc.
CMI
DRS Technologies
Dynetics, Inc.
Dynetics, Millennium, Davidson (DMD)
Future Research Corp.
General Dynamics
General Dynamics Ordnance and Tactical Systems
ITT-CAS, Inc.
L-3 Communications Electro-Optic Systems
Lockheed Martin
MICOR Industries, Inc.
Northrop Grumman
Northrop Grumman Mission Systems
Raytheon
Science & Engineering Services, Inc. (SESI)
Science Applications International Corp. (SAIC)
Summa Technologies
Taylor-Wharton
URS Corp.
Vision Technology Miltope Corp.
Westwind Technologies, Inc.

Arizona
Alliant Techsystems, Inc.
BAE Systems
Boeing
General Dynamics
General Dynamics C4 Systems, Inc.
Honeywell
Intercoastal Electronics
L-3 Communications Electro-Optic Systems
Lockheed Martin
Raytheon
Raytheon Missile Systems
Robertson Aviation
U.S. Army Information Systems Engineering Command

Arkansas
Aerojet
Lockheed Martin
General Dynamics Armament and Technical Products (GDATP)
Pine Bluff Arsenal
URS Corp.

California
Aerojet
Aerovironment, Inc.
Agilent Technologies, Inc.
All American Racers, Inc.
ArgonST Radix
Armacel Armor
Armtec Defense
BAE Systems
BAE Systems Land & Armaments
Boeing
Ceradyne, Inc.
Cisco
Cubic Defense Systems
Ducommun AeroStructures
FLIR Systems, Inc.
General Atomics, Aeronautical Systems, Inc.
General Dynamics Ordinance and Tactical Systems
Gentex Corp.
Gibson and Barnes
Indigo Systems Corp.
Interstate Electronics
Kidde Dual Spectrum
L-3 Communications
L-3 Interstate Electronics Corp.
L-3/IAC
Marvin Land Systems
Northrop Grumman
Northrop Grumman Mission Systems
Northrop Grumman Space & Mission Systems Corp.
Radix
Rapiscan Systems
Raytheon
Science Applications International Corp. (SAIC)
Secure Communications Systems, Inc.
Sun MicroSystems
Tapestry Solutions
Tecom
Thales Raytheon Systems
Triumph Systems Los Angeles
Vertigo, Inc.
ViaSat
US Divers

Colorado
ITT Industries
Northrop Grumman Electronic Systems

Connecticut
BAE Systems
Colt's Manufacturing
DRS Fermont
Hamilton Sundstrand
Lex Products Corp.
Sikorsky
Sikorsky Aircraft

Delaware
ANP Technologies

Florida
CAE USA
Chenega
Computer Sciences Corp. (CSC)
DRS Optronics, Inc.
DRS Tactical Systems
DRS Technologies
DSE (Balimoy) Corp.
Elbit Systems
General Dynamics
General Dynamics Ordnance and Tactical Systems
Gyrocam Systems LLC
Harris Corp.
HELLFIRE LLC
Honeywell
Kaegan Corp.
Knight's Armaments Co.
L-3 Communications
L-3 CyTerra Corp.
Lockheed Martin
Lockheed Martin Information Systems
Lockheed Martin Simulation, Training and Support
Longbow LLC

MEADS International
Medical Education Technologies
Northrop Grumman
Northrop Grumman Guidance and Electronics Company, Inc., Laser Systems
Northrop Grumman Information Technology (NGIT)
Northrop Grumman Integrated Systems
Pierce Manufacturing
Raytheon
Science Applications International Corp. (SAIC)
Simulation, Training, and Instrumentation Command (STRICOM) (Orlando, FL)
Sypris
Thales Raytheon Systems
The Aegis Technology Group, Inc.
UNICOR Protective Materials Co.

Georgia
CSS
Data Path, Inc.
General Dynamics Information Technology
General Dynamics SATCOM Tech
Gulfstream
JCB, Inc.
Kipper Tool Company
Meggitt Defense Systems

Illinois
Caterpillar
Caterpillar Defense and Federal Products (OEM)
CDW-G
E.D. Etnyre and Co.
General Dynamics Ordnance and Tactical Systems
L-3 Communications
Navistar Defense
Northrop Grumman
Olin Corp.
Rock Island Arsenal
Snap-on Industrial

Indiana
Allison
Allison Transmissions
AM General (AMG)
ITT
Raytheon
Raytheon Technical Services, Inc.
Rolls Royce Corp.

Iowa
American Ordnance
Data Link Solutions
Rockwell Collins

Kansas
Detroit Diesel

Kentucky
DRS Technologies

Louisiana
Textron Marine & Land Systems

Maine
General Dynamics Armament and Technical Products (GDATP) Division
Group Home Foundation, Inc.
Maine Military Authority

Maryland
AAI Corp.
BAE Systems
Binary Group
Bruhn New-Tech
COMTECH
COMTECH Mobile Datacom
DynPort Vaccine
ECBC
ECS
FASCAN International
Global Defense Engineering
ICx™ Technologies, Inc.
iLumina Solutions
Johns Hopkins University Applied Physics Laboratory
Litton Advanced Systems
Lockheed Martin
Lockheed Martin Mission Systems
MaTech
Meridian Medical Technologies
Naval Air Warfare Center Aircraft Division
Northrop Grumman
Osiris Therapeutics
PharmAthene
Rohde and Schwarz
Science & Engineering Services, Inc. (SESI)
Sierra Nevada Corp
Smiths Detection, Inc.
TCOM
Thales Communications

Massachusetts
American Science & Engineering, Inc.
BAE Systems
General Dynamics
General Dynamics C4 Systems
General Electric
iRobot
ITT Industries
L-3 Communications
L-3 Communications Electro-Optic Systems
L-3 CyTerra Corp.
Lincoln Labs
Raytheon
Textron Defense Systems
Wolf Coach, Inc., an L-3 Communications Company

Michigan
AAR Mobility Systems
AM General
Avon Protection Systems
Detroit Diesel
General Dynamics
General Dynamics Land Systems
General Motors
General Tactical Vehicle
Holland Hitch
Howmet Castings
Lapeer Industries, Inc.
L-3 Communications
Marsh Industrial
Meritor
XMCO

Minnesota
Alliant Techsystems
Cummins Power Generation
General Dynamics C4 Systems, Inc.

Mississippi
American Eurocopter
BAE Systems
Engineering Research and Development Command
Thales Raytheon Systems
Vickers

Missouri
Alliant Techsystems
Boeing
DRS Sustainment Systems, Inc. (DRS-SSI)

Nevada
PKMM
Sierra Nevada Corp.

New Hampshire
BAE Systems
Skillsoft

New Jersey
Accenture
Airborne Systems North America
Airflyte Electronics Co.
AMT
BAE Systems
Booz Allen Hamilton
CACI
CECOM Software Engineering Center
Computer Sciences Corp. (CSC)
Dewey Electronics
DRS Technologies
Engineering Solutions & Products, Inc.
ILEX
ITT
JANUS Research
L-3 Communications
L-3 Communications Space & Navigation
Lockheed Martin
MITRE
Northrop Grumman
Systems Technologies (Systek), Inc.
Viecore

New Mexico
Aerojet General
EG&G
Hewlett Packard
Honeywell

New York
ADSI
Bren-Tronics
Carleton Technologies, Inc.
L-3 Global Communications Solutions, Inc.
Lockheed Martin
Oxygen Generating Systems International
Remington
Telephonics Corp.
The Research Associates
Watervliet Arsenal

North Carolina
General Dynamics Armament and Technical Products (GDATP)

Ohio
BAE Systems
Composix
Defiance
DRS Mobile Environmental Systems
General Dynamics
GEP
Hunter Manufacturing
L-3 Communications Cincinnati Electronics
Martin Diesel
Mil-Mar Century, Inc.

Oklahoma
BAE Systems
Engineering Professional Services
L-3 Westwood
Titan Corp.

Oregon
Daimler Truck, North America/Freightliner
Mountain High Equipment and Supply Co.
Precision Castparts Corp.
Silver Eagle Manufacturing Company (SEMCO)
URS Corp.

Pennsylvania
Action Manufacturing
BAE Systems
BAE Systems Land & Armaments, Ground Systems Division
Boeing
Fidelity Technologies Corp.
General Dynamics
General Dynamics Ordnance and Tactical Systems
General Dynamics Ordnance and Tactical Systems–Scranton Operations
JLG Industries, Inc.
Kongsberg Defence & Aerospace
L-3 Communications
Letterkenny Army Depot
Mittal
Oppenheimer
Pennsylvania State University
Tobyhanna Army Depot

South Carolina
Caterpillar
Charleston Marine Containers
Fabrique National Manufacturing, LLC
Force Protection Industries, Inc.
Michelin

Tennessee
American Ordnance
BAE/Holston
Barrett Firearms Manufacturing
Cummins Mid-South LLC
Power Manufacturing, Inc.
Teledyne

Texas
American Eurocopter
Apptricity Corp.
BAE Systems
BAE-TVS
Bell Helicopter, Textron
Casteel Manufacturing
Critical Solutions International, Inc.
Dell
DRS Technologies
DynCorp
Elbit Systems of America
Kalmar RT Center LLC
King Aerospace
L-3 Communications
L-3 Communications Electro-Optic Systems
L-3 Communications Integrated Systems, L.P.
Lockheed Martin
Lockheed Martin Missiles & Fire Control
M7 Aerospace
Oshkosh Truck Corp.
Overwatch Systems
Raytheon
Red River Army Depot
Southwest Research Institute
Thales Raytheon Systems
Ultra, Inc.

Utah
Idaho Technologies
Klune
L-3 Communications
Moog
Rockwell Collins
URS Corp.

Vermont
General Dynamics
Tri-Tech USA, Inc.

Virginia
Accenture
Aerial Machine and Tool, Inc.
Alliant Techsystems
ATK
Booz Allen Hamilton
CACI Technologies
CACI
Computer Sciences Corp. (CSC)
Deloitte LLP
DRS Technical Solutions
EADS North America
Fairfield
General Dynamics
General Dynamics Advanced Information Systems
General Dynamics Information Technology
GTSI
Heckler and Koch Defense, Inc.
IBM
Institute for Defense Analysis
ITT Industries
L-3 Communications
L-3 Communications (Titan Group)
LMI Consulting
Lockheed Martin
LTI DataComm, Inc.
MPRI (an L-3 Company)
MPRI (L-3 Communications Division)
NIITEK
Northrop Grumman
Rockwell Collins
Science Applications International Corp. (SAIC)
Segovia Global IP Services
SNVC
Telos Corp.
Universal Systems and Technology
Vertu Corp.
Wexford Group International
ZETA

Washington
Berg Companies, Inc.
Fluke Corp.

Washington, DC
UNICOR

West Virginia
ATK
Alliant Techsystems

Wisconsin
Alliant Techsystems
AMTEC Corp.
Oshkosh Truck Corp.
Schutt Industries

INTERNATIONAL CONTRACTORS

Canada
Bracco Diagnostics, Inc.
CDL Systems
General Dynamics Land Systems
GT Machine and Fabrication
SNC Technologies
WESCAM

England
Atlantic Inertial Units

Germany
General Dynamics European Land Systems-Germany

Italy
Alenia Aeronautica

Sweden
BAE Systems Bofors Defense

UK
BAE Systems
FBM Babcock Marine
Thales
UAV Engines Limited
Williams Fairey Engineering, Ltd.

Points of Contact

2.75" Family of Rockets
JAMS Project Office
ATTN: SFAE-MSL-JAMS
Redstone Arsenal, AL 35898-8000

Abrams Upgrade
ATTN: SFAE-GCS-CS-A
6501 E. 11 Mile Rd.
Warren, MI 48397-5000

AcqBusiness
Greggory Judge
P 703-797-8870
F 703-797-8989
greggory.judge@us.army.mil

Advanced Field Artillery Tactical Data System (AFATDS)
Product Director
Fire Support Command and Control
ATTN: SFAE-C3T-BC-FSC2
Building 2525
Fort Monmouth, NJ 07703-5404

Aerial Common Sensor (ACS)
PM Aerial Common Sensor
ATTN: SFAE-IEWS-ACS
Building 288
Fort Monmouth, NJ 07703

Air Warrior (AW)
PM Air Warrior
ATTN: SFAE-SDR-AW
Redstone Arsenal, AL 35898

Air/Missile Defense Planning and Control System (AMDPCS)
C-RAM Program Office
ATTN: SFAE-C3T-CR-AMD
Redstone Arsenal, AL 35898-5000

Airborne Reconnaissance Low (ARL)
PM Aerial Common Sensor
ATTN: SFAE-IEWS-ACS
Building 288
Fort Monmouth, NJ 07703

All Terrain Lifter Army System (ATLAS)
Product Manager
Combat Engineer/MHE
ATTN: SFAE-CSS-FP-C
Warren, MI 48397-5000

Armored Knight
PM HBCT
PM-Fire Support Platforms
ATTN: SFAE-GCS-HBCT-F
6501 East 11 Mile Rd.
Warren, MI 43897-5000

Armored Security Vehicle (ASV)
PD Armored Security Vehicle
SFAE-CSS-TV-A
6501 11 Mile Rd.
Warren, MI 48397-5000

Army Key Management System (AKMS)
PD, NETOPS-CF
ATTN: SFAE-C3T-WINT-NETOPS-CF
Fort Monmouth, NJ 07703

Artillery Ammunition
PM Combat Ammunition Systems
ATTN: SFAE-AMO-CAS
Picatinny Arsenal, NJ 07806

Aviation Combined Arms Tactical Trainer (AVCATT)
Project Manager
Combined Arms Tactical Trainers
12350 Research Parkway
Orlando, FL 32826-3276
407-384-3600

Battle Command Sustainment Support System (BCS3)
PM Battle Command Sustainment Support System (BCS3)
ATTN: SFAE-C3T-GC-BCS-3
10109 Gridley Rd.
Fort Belvoir, VA 22060

Biometric Enterprise Core Capability (BECC)
Project Manager
PM DoD Biometrics
ATTN: SFAE-PS-BI
Building 1445
Ft. Belvoir, VA 22060-5526

Biometric Family of Capabilities for Full Spectrum Operations (BFCFSO)
Project Manager
PM DoD Biometrics
ATTN: SFAE-PS-BI
Building 1445
Ft. Belvoir, VA 22060-5526

Black Hawk/UH-60
Utility Helicopter PP&C Branch Chief:
Mr. Rick Hubert
256-955-8771

Bradley Upgrade
6501 East 11 Mile Rd.
ATTN:SFAE-GCS-CS\
Warren, MI 43897-5000

Calibration Sets Equipment (CALSETS)
Product Director
Test, Measurement, and Diagnostic Equipment
Building 3651
Redstone Arsenal, AL 35898

Chemical Biological Medical Systems–Diagnosics
ATTN: JPM CBMS
64 Thomas Johnson Drive
Frederick, MD 21702

Chemical Biological Medical Systems–Prophylaxis
ATTN: JPM CBMS
64 Thomas Johnson Drive
Frederick, MD 21702

Chemical Biological Medical Systems–Therapeutics
ATTN: JPM CBMS
64 Thomas Johnson Drive
Frederick, MD 21702

Chemical Biological Protective Shelter (CBPS)
JPEO CBD
5203 Leesburg Pike
Skyline #2, Suite 1609
Falls Church, VA 22041

Chemical Demilitarization
Chemical Materials Agency (CMA)
ATTN: AMSCM-D
5183 Blackhawk Road
APG-EA, MD 21010-5424

Chinook/CH-47 Improved Cargo Helicopter (ICH)
PM Cargo Helicopters
ATTN: SFAE-AV-CH-ICH
Building 5678
Redstone Arsenal, AL 35898

Close Combat Tactical Trainer (CCTT)
Project Manager
Combined Arms Tactical Trainers
12350 Research Parkway
Orlando, FL 32826-3276

Combat Service Support Communications (CSS COMMS)
PM Defense Communications and Army Transmission Systems
6700 Springfield Center Dr.
Suite E
Springfield, VA 22150

Command Post Systems and Integration (CPS&I)
Project Manager Command Posts
ATTN: SFAE-C3T-CP
Building 456
Fort Monmouth, NJ 07703

Common Hardware Systems (CHS)
Product Director Common Hardware Systems (PD-CHS)
ATTN: SFAE-C3T-CP-CHS
Building 457
Fort Monmouth, NJ 07703

Common Remotely Operated Weapon Station (CROWS)
PM Soldier Weapons
(SFAE-SDR-SW)
PEO Soldier
Picatinny Arsenal, NJ 07806

Counter-Rocket, Artillery and Mortar (C-RAM)
C-RAM Program Office
ATTN: SFAE-C3T-CR
Redstone Arsenal, AL
35898-5000

Countermine
LTC Pete Lozis
PM Countermine & EOD
ATTN: SFAE-AMO-CCS-
Fort Belvoir, VA 22060-5811

Defense Enterprise Wideband SATCOM Systems (DEWSS)
PM Defense Communications and Army Transmission Systems
Building 209
Fort Monmouth, NJ 07703-5509

Distributed Common Ground System (DCGS-Army)
ATTN: SFAE-IEWS-DCGS-A
Building 550
Saltzman Ave.
Fort Monmouth, NJ 07703-5301

Distributed Learning System (DLS)
PM DLS, ATTN: SFAE-PS-DL, 11846 Rock Landing Dr., Suite B, Newport News, VA 23606

Dry Support Bridge (DSB)
PM Bridging Systems
SFAE-CSS-FP-H MS 401
6501 East 11 Mile Rd.
Warren, MI 43897-5000

Engagement Skills Trainer (EST) 2000
Project Manager
Combined Arms Tactical Trainers
12350 Research Parkway
Orlando, FL 32826-3276
407-384-3600

Excalibur (XM982)
PM Combat Ammo Systems
ATTN: SFAE-AMO-CAS-EX
Picatinny Arsenal, NJ 07806

Extended Range Multipurpose (ERMP) Unmanned Aircraft System (UAS)
PM Unmanned Aircraft Systems
ATTN: SFAE-AV-UAS
Redstone Arsenal, AL 35898

Family of Medium Tactical Vehicles (FMTV)
ATTN: SFAE-CSS
6501 East 11 Mile Rd.
Warren, MI 43897-5000

Fixed Wing
DA Systems Coordinator-Fixed Wing
ASA (ALT) Aviation-Intelligence & Electronic Warfare
ATTN: SAAL-SAI, Room 10006
2511 S. Jefferson Davis Highway
Arlington, VA 22202

Force Protection Systems
ATTN: SFAE-CBD-Guardian
5109 Leesburg Pike
Falls Church VA 22041

Force Provider (FP)
PM Force Sustainment Systems
LTC Rick Harger
(508) 233-5312
Rick.Harger@us.army.mil

Force XXI Battle Command Brigade-and-Below (FBCB2)
PM, FBCB2
Building 2525, Bay 1
Fort Monmouth, NJ 07703-5408

Forward Area Air Defense Command and Control (FAAD C2)
C-RAM Program Office
ATTN: SFAE-C3T-CR
Redstone Arsenal, AL 35898-5000

Future Tank Main Gun Ammunition
PM Maneuver Ammunition Systems
ATTN: SFAE-AMO-MAS
Picatinny Arsenal, NJ 07806

General Fund Enterprise Business Systems (GFEBS)
5911 Kingstowne Village Parkway
Suite 600
Alexandria, VA 22315

Global Combat Support System–Army (GCSS–Army)
Program Manager GCSS–Army
3811 Corporate Rd Suite C
Petersburg, VA 23805

Program Manager Army Enterprise
Systems Integration Program
9350 Hall Road
Fort Belvoir, VA 22060

Global Command and Control System–Army (GCCS-A)
Product Manager, Strategic Battle
Command
ATTN: SFAE-C3T-BC-SBC
Bldg 2525
Ft Monmouth, NJ 07703

Guardrail Common Sensor (GR/CS)
PM Aerial Common Sensor
ATTN: SFAE-IEWS-ACS
Building 288
Fort Monmouth, NJ 07703

Guided Multiple Launch Rocket System (GMLRS)
Precision Fires Rocket and Missile
Systems Project Office
ATTN: SFAE-MSL-PF-PGM/R
Building 5250
Redstone Arsenal, AL 35898

Heavy Expanded Mobility Tactical Truck (HEMTT)/HEMTT Extended Service Program (ESP)
PM Heavy Tactical Vehicles
ATTN: SFAE-CSS-TV-H
Mail Stop 429
6501 East Eleven Mile Road
Warren, MI 48397-5000

Heavy Loader
PM for Combat Engineer
Materiel Handling Equipment
6501 East 11 Mile Road,
Warren, MI 48397-5000

Hellfire Family of Missiles
JAMS Project Office
ATTN: SFAE-MSLS-JAMS
Redstone Arsenal, AL 35898

Helmet Mounted Enhanced Vision Devices
PM Soldier Sensors and Lasers
ATTN: SFAE-SDR-SSL
10170 Beach Road
Building 325
Fort Belvoir, VA 22060

High Mobility Artillery Rocket System (HIMARS)
Precision Fires Rocket and Missile
Systems Project Office
ATTN: SFAE-MSL-PF-FAL
Building 5250
Redstone Arsenal, AL 35898

High Mobility Multipurpose Wheeled Vehicle (HMMWV)
PM Light Tactical Vehicles
ATTN: SFAE-CSS-TV-L
6501 11 Mile Rd. MS 245
Warren, MI 43897

Improved Ribbon Bridge (IRB)
PM Bridging Systems
(SFAE-CSS-FP-H) MS 401
6501 East 11 Mile Rd.
Warren, MI 43897

Improved Target Acquisition System (ITAS)
PM Close Combat Weapon Systems
Project Office
ATTN: SFAE-MSL-CWS-J
Redstone Arsenal, AL 35898

Installation Protection Program (IPP) Family of Systems
ATTN: SFAE-CBD-Guardian
5109 Leesburg Pike
Falls Church, VA 22041

Instrumentable–Multiple Integrated Laser Engagement System (I-MILES)
Project Manager
Training Devices
12350 Research Parkway
Orlando, FL 32826
407-384-5200

Integrated Air and Missile Defense (IAMD)
PEO Missiles and Space
ATTN: SFAE-MSLS-IAMD
Huntsville, AL 35807

Integrated Family of Test Equipment (IFTE)
Product Director
Test, Measurement, and Diagnostic
Equipment
Building 3651
Redstone Arsenal, AL 35898

Interceptor Body Armor
ATTN: SFAE-SDR-EQ
10170 Beach Road
Building 325
Fort Belvoir, VA 22060

Javelin
PM Close Combat Weapon Systems
Project Office
ATTN: SFAE-MSL-CWS-J
Redstone Arsenal, AL 35898

Joint Air-to-Ground Missile (JAGM)
Joint Air to Ground Missile Product
Office
ATTN: SFAE-MSL-JAMS-M
5250 Martin Rd.
Redstone Arsenal, AL 35898

Joint Biological Point Detection System (JBPDS)
ATTN: SFAE-CBD-NBC-D
5183 Blackhawk Rd.
APG, MD 21010

Joint Biological Standoff Detection System (JBSDS)
ATTN: SFAE-CBD-NBC-D
5183 Blackhawk Rd.
APG, MD 21010

Joint Cargo Aircraft (JCA)
DA Systems Coordinator-JCA
ASA (ALT) Aviation-Intelligence & Electronic Warfare
ATTN: SAAL-SAI, Room 10006
2511 S. Jefferson Davis Highway
Arlington, VA 22202

Joint Chem/Bio Coverall for Combat Vehicle Crewman (JC3)
JPEO CBD
5203 Leesburg Pike
Skyline #2, Suite 1609
Falls Church, VA 22041

Joint Chemical Agent Detector (JCAD)
ATTN: SFAE-CBD-NBC-D
5183 Blackhawk Rd.
APG, MD 21010

Joint Chemical Biological Radiological Agent Water Monitor (JCBRAWM)
ATTN: SFAE-CBD-NBC-R
5183 Blackhawk Rd.
APG, MD 21010

Joint Effects Model (JEM)
JPEO CBD
5203 Leesburg Pike
Skyline #2, Suite 1609
Falls Church, VA 22041

Joint High Speed Vessel (JHSV)
Product Director
Army Watercraft Systems
ATTN: SFAE-CSS-FP-W
Warren, MI 48397-5000

Joint Land Attack Cruise Missile Defense Elevated Netted Sensor System (JLENS)
PEO Missiles and Space
ATTN: SFAE-MSLS-CMDS-JLN
P.O. Box 1500
Huntsville, AL 35807

Joint Land Component Constructive Training Capability (JLCCTC)
Project Manager
Constructive Simulation
12350 Research Parkway
Orlando, FL 32826
407-384-3650

Joint Light Tactical Vehicle (JLTV)
PM Joint Light Tactical Vehicle (JLTV)
ATTN: SFAE-CSS-JC-JL/MS 640
Bldg 326/3rd Floor
29865 Mitchell St.
Harrison Twp., MI 48045-4941

Joint Nuclear Biological Chemical Reconnaissance System (JNBCRS)
ATTN: SFAE-CBD-NBC-R
5183 Blackhawk Rd.
APG, MD 21010

Joint Precision Airdrop System (JPADS)
PM Force Sustainment Systems, LTC
Daryl P. Harger
508-223-5312
Rick.Harger@us.army.mil

Joint Service General Purpose Mask (JSGPM)
JPEO CBD
5203 Leesburg Pike
Skyline #2, Suite 1609
Falls Church, VA 22041

Joint Service Personnel/Skin Decontamination System (JSPDS)
JPEO CBD
5203 Leesburg Pike
Skyline #2, Suite 1609
Falls Church, VA 22041

Joint Service Transportable Decontamination System (JSTDS)–Small Scale (SS)
JPEO CBD
5203 Leesburg Pike
Skyline #2, Suite 1609
Falls Church, VA 22041

Joint Tactical Ground Stations (JTAGS)
PEO Missiles and Space
Lower Tier Project Office
ATTN: SFAE-MSLS-LT
P.O. Box 1500
Huntsville, AL 35807

Joint Tactical Radio System Airborne and Maritime/Fixed Station (JTRS AMF)
Joint Program Executive Office (JPEO)
Joint Tactical Radio System (JTRS)
33000 Nixie Way Bldg. 50
Suite 339
San Diego, CA 92147

Joint Tactical Radio System Ground Mobile Radios (JTRS GMR)
Joint Program Executive Office (JPEO)
Joint Tactical Radio System (JTRS)
33000 Nixie Way Bldg. 50
Suite 339
San Diego, CA 92147

Joint Tactical Radio System (JTRS) Handheld, Manpack, and Small Form Fit (HMS)
Joint Program Executive Office(JPEO)
Joint Tactical Radio System (JTRS)
33000 Nixie Way Bldg. 50
Suite 339
San Diego, CA 92147

Joint Tactical Radio System Multifunctional Information Distribution System (MIDS)
Joint Program Executive Office (JPEO)
Joint Tactical Radio System (JTRS)
33000 Nixie Way Bldg. 50
Suite 339
San Diego, CA 92147

Joint Tactical Radio Systems (JTRS)–NED
Joint Program Executive Office (JPEO)
Joint Tactical Radio System (JTRS)
33000 Nixie Way Bldg. 50
Suite 339
San Diego CA 92147

Joint Warning and Reporting Network (JWARN)
JPEO CBD
5203 Leesburg Pike
Skyline #2, Suite 1609
Falls Church, VA 22041

Kiowa Warrior
Product Manager
ATTN: SFAE-AV-ASH-KW
5681 Wood Road
Redstone Arsenal, AL 35898

Light Tactical Trailer (LTT)
PM Light Tactical Vehicles
ATTN: SFAE-CSS-TV-L
6501 11 Mile Rd. MS 245
Warren, MI 43897

Light Utility Helicopter (LUH)
LTC James B. Brashear
LUH PM
(256) 842-8000
james.b.brashear@us.army.mil

Light Weight 155mm Howitzer (LW155)
ATTN: SFAE-GCS-JLW
Picatinny Arsenal, NJ 07806

Lightweight .50 cal Machine Gun
PM Soldier Weapons
ATTN: SFAE-SDR-SW
PEO Soldier
Picatinny Arsenal, NJ 07806

Lightweight Laser Designator Range Finder (LLDR)
PM Soldier Sensors and Lasers
ATTN: SFAE-SDR-SEQ-SSL
10170 Beach Rd.
Building 325
Fort Belvoir, VA 22060

Line Haul Tractor
PM Heavy Tactical Vehicles
ATTN: SFAE-CSS-TV-H
Mail Stop 429
6501 East Eleven Mile Road
Warren, MI 48397-5000

Load Handling System Compatible Water Tank Rack (Hippo)
PM Petroleum and Water Systems
6501 East 11 Mile Rd.
Mail Stop 111
Warren, MI 43897

Longbow Apache
PM Apache
Building 5681
Redstone Arsenal, AL 35898

Maneuver Control System (MCS)
PdM TBC
ATTN: SFAE-C3T-BC-TBC
Fort Monmouth, NJ 07703

Medical Communications for Combat Casualty Care (MC4)
PM Medical Communications for Combat Casualty Care (MC4)
524 Palacky St.
Fort Detrick, MD 21702

Medical Simulation Training Center (MSTC)
Project Manager
Combined Arms Tactical Trainers
12350 Research Parkway
Orlando, FL 32826-3276
407-384-3600

Medium Caliber Ammunition
PM Maneuver Ammunition Systems
ATTN: SFAE-AMO-MAS
Picatinny Arsenal, NJ 07806

Medium Extended Air Defense System (MEADS)
PATRIOT/MEADS Combined Aggregate Program (CAP)
PEO Missiles and Space
ATTN: SFAE-MSLS-LT-CAP
P.O. Box 1500
Huntsville, AL 35807

Meteorological Measuring Set–Profiler (MMS-P)
Product Director for Target Identification & Meteorological Sensors
ATTN: SFAE-IEWS-NS-TIMS
Avenue of Memories (563)
Ft Monmouth, NJ 07703

Mine Protection Vehicle Family (MPVF)
LTC Charles Dease
6501 East 11 Mile Rd
ATTN:SFAE-CSS-FP-AMS
Warren, MI 43897-5000

Mobile Maintenance Equipment Systems (MMES)
PM-SKOT
ATTN: SFAE-CSS-JC-SK
Building 104, 1st Floor
Rock Island, IL 61299-7630

Modular Fuel System (MFS)
PM Petroleum and Water Systems
501 East 11 Mile Rd.
Mail Stop 111
Warren, MI 48397

Mortar Systems
PM Combat Ammunition Systems
ATTN: SFAE-AMO-CAS-MS
Picatinny Arsenal, NJ 07806

Movement Tracking System (MTS)
PM Logistics Information Systems
800 Lee Ave., Bldg. 5100
Fort Lee, VA 23801

Multiple Launch Rocket System (MLRS) M270A1
Precision Fires Rocket and Missile Systems Project Office
ATTN: SFAE-MSL-PF-FAL
Building 5250
Redstone Arsenal, AL 35898

NAVSTAR Global Positioning System (GPS)
PM GPS
328 Hopkins Road
Building 246
Aberdeen Proving Ground, MD 21005

Non-Intrusive Inspection (NII) Systems
ATTN: SFAE-CBD-Guardian
5109 Leesburg Pike
Falls Church VA 22041

Non Line of Sight–Launch System (NLOS–LS)
NLOS–LS Project Office
ATTN: SFAE-MSLS-NL
Building 112, Room 304
Redstone Arsenal, AL 35898

Nuclear Biological Chemical Reconnaissance Vehicle (NBCRV)- Stryker
ATTN: SFAE-CBD-NBC-R
5183 Blackhawk Rd.
APG, MD 21010

One Semi-Automated Forces (OneSAF)
Project Manager
Constructive Simulation
12350 Research Parkway
Orlando, FL 32826
407-384-3650

One Tactical Engagement Simulation System (OneTESS)
Project Manager
Training Devices
12350 Research Parkway
Orlando, FL 32826
407-384-5200

Paladin/ Field Artillery Ammunition Supply Vehicle (FAASV)
Project Manager HBCT
ATTN: SFAE-GCS-HBCT
6501 East 11 Mile Rd.
Warren, MI 48397

PATRIOT (PAC-3)
PEO Missiles and Space
Lower Tier Project Office
ATTN: SFAE-MSLS-LT
P.O. Box 1500
Huntsville, AL 35807

Precision Guidance Kit
PM Combat Ammunition Systems
ATTN: SFAE-AMO-CAS
Picatinny Arsenal, NJ 07806

Prophet
PM SW
ATTN: SFAE-IEWS&S-G
Building 288
Sherrill Ave.
Monmouth, NJ 07703

Raven Small Unmanned Aircraft System (SUAS)
Product Manager
Small Unmanned Aircraft Systems
ATTN: SFAE-AV-UAS-SU

Rough Terrain Container Handler (RTCH)
Product Manager
Combat Engineer/MHE
ATTN: SFAE-CSS-FP-C
Warren, MI 48397-5000

Screening Obscuration Device (SOD) - Visual Restricted (Vr)
ATTN: SFAE-CBD-NBC-R
5183 Blackhawk Rd.
APG, MD 21010

Secure Mobile Anti-Jam Reliable Tactical – Terminal (SMART-T)
PM WIN-T
ATTN: SFAE-C3T-WIN-MST
Fort Monmouth, NJ 07703

Sentinel
PEO Space and Missile Defense
ATTN: SFAE-MSLS
Redstone Arsenal, AL 35898

Single Channel Ground and Airborne Radio System (SINCGARS)
PM Command Posts
Building 456
Fort Monmouth, NJ 07703

Small Arms–Crew Served Weapons
PM Soldier Weapons
ATTN: SFAE-SDR-SW
PEO Soldier
Picatinny Arsenal, NJ 07806

Small Arms–Individual Weapons
PM Soldier Weapons
(SFAE-SDR-SW)
PEO Soldier
Picatinny Arsenal, NJ 07806

Small Caliber Ammunition
Project Manager Maneuver Ammunition Systems
ATTN: SFAE-AMO-MAS
Picatinny Arsenal, NJ 07806

Sniper Systems
PM Soldier Weapons
(SFAE-SDR-SW)
PEO Soldier
Picatinny Arsenal, NJ 07806

Spider
COL Raymond H. Nulk
PM Close Combat Systems
ATTN: SFAE-AMO-CCS
Picatinny Arsenal, NJ 07806

Stryker
ATTN: SFAE-GCS-BCT MS 325
6501 East 11 Mile Rd.
Warren, MI 48397

Surface Launched Advanced Medium Range Air-to-Air Missile (SLAMRAAM)
PEO Missile and Space
ATTN: SFAE-MSLA-CMDS
Redstone Arsenal, AL 35898

Shadow Tactical Unmanned Aerial Vehicle (TUAV)
Product Manager
Unmanned Aircraft Systems
ATTN: SFAE-AV-UAS

Tactical Electric Power (TEP)
5850 Delafield Road
Fort Belvoir, VA 22060-5809

Tank Ammunition
PM Maneuver Ammunition Systems
ATTN: SFAE-AMO-MAS
Picatinny Arsenal, NJ 07806

Test Equipment Modernization (TEMOD)
Product Director Test, Measurement, and Diagnostic Equipment
Building 3651
Redstone Arsenal, AL 35898

Thermal Weapon Sight
PM Soldier Sensors and Lasers
ATTN: SFAE-SDR-SSL
10170 Beach Road
Building 325
Fort Belvoir, VA 22060

Transportation Coordinators' Automated Information for Movement System II (TC-AIMS II)
PM TIS
200 Stovall St., Suite 9S23
Alexandria, VA 22314

Tube-Launched, Optically-Tracked, Wire-Guided (TOW) Missiles
PM Close Combat Weapon Systems
Project Office
ATTN: SFAE-MSL-CWS-J
Redstone Arsenal, AL 35898

Unit Water Pod System (Camel)
PM Petroleum and Water Systems
ATTN: LTC Michael Receniello
6501 East 11 Mile Rd.
Mail Stop 111
Warren, MI 43897

Warfighter Information Network–Tactical (WIN-T) Increment 1
Project Manager, WIN-Tactical
ATTN: SFAE-C3T-WIN
Building 918
Murphy Dr.
Fort Monmouth, NJ 07703

Warfighter Information Network–Tactical (WIN-T) Increment 2
Project Manager, WIN-Tactical
ATTN: SFAE-C3T-WIN
Building 918
Murphy Dr.
Fort Monmouth, NJ 07703

Warfighter Information Network–Tactical (WIN-T) Increment 3
Project Manager, WIN-Tactical
ATTN: SFAE-C3T-WIN
Building 918
Murphy Dr.
Fort Monmouth, NJ 07703

Weapons of Mass Destruction Elimination
ATTN: SFAE-CBD-Guardian
5109 Leesburg Pike
Falls Church VA 22041